D1197842

PERGAMON INTERNATIONAL LIBRARY
of Science, Technology, Engineering and Social Studies
The 1000-volume original paperback library in aid of education, industrial training and the enjoyment of leisure
Publisher: Robert Maxwell, M.C.

INTRODUCTION TO HEALTH PHYSICS

Other Titles of Related Interest

REPORTS AND RECOMMENDATIONS OF THE INTERNATIONAL COMMISSION ON RADIOLOGICAL PROTECTION

Publication	
No 2	Permissible Dose for Internal Radiation
No 5	The Handling and Disposal of Radioactive Materials in Hospitals and Medical Research Establishments
No 7	Principles of Environmental Monitoring Related to the Handling of Radioactive Materials
No 8	The Evaluation of Risks from Radiation
No 9	Recommendations of the ICRP (Adopted 17th September 1965)
No 10	Evaluation of Radiation Doses to Body Tissues from Internal Contamination due to Occupational Exposure
No 10a	The Assessment of Internal Contamination Resulting from Recurrent or Prolonged Uptakes
No 11	A Review of the Radiosensitivity of the Tissues in Bone
No 12	General Principles of Monitoring for Radiation Protection of Workers
No 13	Radiation Protection in Schools

Publication	
No 14	Radiosensitivity and Spatial Distribution of Dose
No 15	Protection Against Ionizing Radiation from External Sources
No 21	Supplement to Publication No 15
No 16	Protection of the Patient in X-Ray Diagnosis
No 17	Protection of the Patient in Radionuclide Investigation
No 18	The RBE for High-LET Radiations with Respect to Mutagenesis
No 19	The Metabolism of Compounds of Plutonium and other Actinides
No 20	Alkaline Earth Metabolism in Adult Man
No 22	Implications of Commission Recommendations that Doses be Kept as Low as Readily Achievable
No 23	Reference Man: Anatomical, Physiological and Metabolic Characteristics

Introduction to Health Physics

BY

HERMAN CEMBER
Northwestern University

PERGAMON PRESS

OXFORD · NEW YORK · TORONTO

SYDNEY · PARIS · FRANKFURT

U.K.	Pergamon Press Ltd., Headington Hill Hall, Oxford OX3 0BW, England
U.S.A.	Pergamon Press Inc., Maxwell House, Fairview Park, Elmsford, New York 10523, U.S.A.
CANADA	Pergamon of Canada, P.O. Box 9600, Don Mills M3C 2T9, Ontario, Canada
AUSTRALIA	Pergamon Press (Aust.) Pty. Ltd., 19a Boundary Street, Rushcutters Bay, N.S.W. 2011, Australia
FRANCE	Pergamon Press SARL, 24 rue des Ecoles, 75240 Paris, Cedex 05, France
WEST GERMANY	Pergamon Press GmbH, 6242, Kronberg-Taunus, Pferdstrasse 1, Frankfurt-am-Main, West Germany

First edition 1969

Reprinted 1976

Library of Congress Catolog Card No. 68-8528

Printed in Great Britain by Biddles Ltd., Guildford, Surrey

ISBN 0 08 012821 1

This book is respectfully dedicated to the memory of

DR. ELDA E. ANDERSON

and

DR. THOMAS PARRAN

CONTENTS

CHAPTER 1

INTRODUCTION

HEALTH physics, or radiological health, as it is frequently called, is that area of environmental health engineering that deals with the protection of the individual and population groups against the harmful effects of ionizing radiation. The health physicist is responsible for the safety aspects in the design of processes, equipment, and facilities utilizing radiation sources, so that radiation exposure to personnel will be minimized, and will at all times be within acceptable limits; and he must keep personnel and the environment under constant surveillance in order to ascertain that his designs are indeed effective. If control measures are found to be ineffective, or if they break down, he must be able to evaluate the degree of hazard, and to make recommendations regarding remedial action.

The scientific and engineering aspects of health physics are concerned mainly with: (1) the physical measurements of different types of radiation and radioactive materials, (2) the establishment of quantitative relationships between radiation exposure and biological damage, (3) the movement of radioactivity through the environment, and (4) the design of radiologically safe equipment, processes, and environments. Clearly, health physics is a professional field that cuts across the basic physical, life, and earth sciences, as well as such applied areas as toxicology, industrial hygiene, medicine, public health, and engineering. The professional health physicist, therefore, in order to perform effectively, must be competent in the wide spectrum of disciplines that bridge the fields between industrial operations and technology on one hand, and modern health science on the other hand. He must have an appreciation of the complex interrelationships between man and the physical, chemical, biological, and even social components of the environment; as well as a quantitative understanding of group phenomena. In addition to these general prerequisites, he must be technically competent in the subject matter unique to his speciality.

The main purpose of this book is to lay the groundwork for attaining technical competency in health physics. Because of the nature of the subject matter and the topics covered, however, it is hoped that the book will be a useful source of information to workers in environmental health as well as to those who will use radiation as a tool. For the latter group, it is also hoped that this book will impart an appreciation for radiation safety as well as an understanding of the philosophy of environmental health.

CHAPTER 2

REVIEW OF PHYSICAL PRINCIPLES

Mechanics

Units and Dimensions

Health Physics is a science, and hence is a systematic organization of knowledge about the interaction between radiation and organic and inorganic matter. Quite clearly, the organization must be quantitative as well as qualitative, since control of radiation hazards implies a knowledge of the dose response relationship between radiation and the biological effects of radiation.

Quantitative relationships are based on measurements, which in reality are comparisons of the attribute under investigation to a standard. A measurement includes two components: a number and a unit. When measuring the height of a person, for example, the result is given as 70 inches if the British system of units is used, or as 177.8 centimeters if the metric system is used. The units "inches" in the first case and "centimeters" in the second tell us what the criterion for comparison is, and the number tells us how many of these units are included in the quantity being measured. Although 70 inches means exactly the same thing as 177.8 centimeters, it is clear that, without an understanding of the units, the information contained in the number above would be meaningless. The British system of units is used chiefly in engineering, while the metric system is widely used in science.

Three attributes are considered basic in the physical sciences: length, mass, and time. In the British system of units, these attributes are measured in feet, pounds, and seconds respectively, while the metric system is divided into two subsystems—the mks, in which the three qualities are specified in meters, kilograms, and seconds—and the cgs, in which the centimeter, gram, and second are used to designate length, mass, and time. In health physics, the cgs system is used most frequently.

All other units, such as force, energy, power, etc., are derived from these three basic units for mass, length, and time (M, L, T). All the derived units, therefore, can be expressed in dimensions of M, L and T. For example, the unit of force in the cgs system is the dyne, and is defined as follows:

> One dyne is that unbalanced push or pull which will accelerate a mass of 1 gram at a rate of 1 centimeter per second per second.

Expressed mathematically:

$$\text{force} = \text{mass} \times \text{acceleration},$$

$$f = ma, \tag{2.1}$$

and the units are:

$$\text{dynes} = \text{grams} \times \frac{\text{cm}}{\text{sec}}/\text{sec}.$$

Since dimensions may be treated in exactly the same way as numbers, the dimension for acceleration is written as cm/sec^2. The dimensions for force, therefore, in the cgs system are

$$\text{dynes} = \frac{\text{g cm}}{\text{sec}^2}.$$

Work and Energy

Energy is defined as the ability to do work. Since all work requires the expenditure of energy, the two terms are expressed in the same units, and consequently have the same dimensions. Work is done, or energy expended, when a force f is exerted through some distance r.

$$W = fr. \tag{2.2}$$

In the cgs system, the erg is the unit of work and energy, and is defined as follows:

> One erg of work is done when a force of 1 dyne is exerted through a distance of 1 centimeter.

Since work is defined as the product of a force and a distance, the dimensions for work and energy are:

$$\text{ergs} = \text{dynes} \times \text{cm}$$

$$= \frac{\text{g cm}}{\text{sec}^2} \times \text{cm} = \frac{\text{g cm}^2}{\text{sec}^2}.$$

Although the erg is the basic unit of energy in the cgs system, it is not too practical for many measurements in the field of health physics because it is an extremely large unit in terms of the energies encountered in the microscopic world of the atom. For many purposes, a more practical unit, called the electron volt (abbreviated eV), is used. The electron volt is a *unit of energy*, and is defined as follows:

$$1 \text{ electron volt} = 1.6 \times 10^{-12} \text{ ergs}.$$

When work is done on a body, the energy expended in doing the work is added to the energy of the body. For example, if a mass is lifted from one elevation to another, the energy that was expended during the performance of the work is converted to potential energy. On the other hand, when work is done to accelerate a body, the energy that was expended appears as kinetic energy in the moving body. In the case where work was done in lifting a body, the mass possesses more potential energy at the higher elevation than it did before it was lifted. Work was done, in this case, against the force of gravity, and the total increase in potential energy of the mass is equal to its weight, which is merely the force with which the mass is attracted to the earth, multiplied by the height through which the mass was raised. *Potential energy* is defined as energy that a body possesses by virtue of its position in a force field. *Kinetic energy* is defined as energy possessed by a moving body as result of its motion. For bodies of mass m, moving with a velocity v (for the case where v is less than about 3×10^9 cm/sec), the kinetic energy is given by

$$E_k = \tfrac{1}{2} mv^2. \tag{2.3}$$

Relativity Effects

According to the system of classical mechanics that was developed by Newton and the other great thinkers of the Renaissance period, mass is an immutable property of matter; it could be changed in size, shape, or state, but it could neither be created nor destroyed. Although this law of conservation of mass seems to be true for the world which we can perceive with our senses, it is in fact only a special case for conditions of large masses and slow speeds. In the sub-microscopic world of the atom, where masses are measured in units of 9×10^{-28} g, where distances are measured in units of 10^{-8} cm, and where velocities are measured in terms of the velocity of light, classical mechanics is not applicable. Einstein, in his special theory of relativity, postulated that the velocity of light in a vacuum, 3×10^{10} cm/sec, is an upper limit of speed that no material body can ever attain; the velocity of light may be asymptotically approached by any mass, but can never be reached. Furthermore, according to Einstein, the mass of a moving body is not constant, as was previously thought, but rather is a function of the velocity with which the body is moving. As the velocity increases, the mass increases, and when the velocity of the body approaches the velocity of light, the mass increases very rapidly. The mass, m, of a moving object whose velocity is v is related to its rest mass, m_0, by the equation

$$m = \frac{m_0}{\sqrt{\left(1 - \frac{v^2}{c^2}\right)}}, \tag{2.4}$$

where c is the velocity of light.

Example 2.1

Compute the mass of an electron moving at 10% and 99% of the speed of light. The rest mass of an electron is 9.11×10^{-28} g. At $v = 0.1\, c$,

$$m = \frac{9.11 \times 10^{-28}\text{ g}}{\sqrt{\left(1 - \frac{(0.1\, c)^2}{c^2}\right)}} = 9.16 \times 10^{-28}\text{ g},$$

and at $v = 0.99\, c$

$$m = \frac{9.11 \times 10^{-28}\text{ g}}{\sqrt{\left(1 - \frac{(0.99\, c)^2}{c^2}\right)}} = 64.6 \times 10^{-28}\text{ g}.$$

This illustration shows that whereas an electron suffers a mass increase of only ½% when it is moving at 10% of the speed of light, its mass increases about seven-fold when the velocity is increased to 99% of the velocity of light.

Kinetic energy of a moving body can be thought of as the income from work put into the body, or energy input, in order to bring the body up to its final velocity.

Expressed mathematically, we have

$$W = E_k = fr = \tfrac{1}{2}\, mv^2. \tag{2.5}$$

The expression for kinetic energy in equations (2.3) and (2.5) is a special case, however, since the mass is assumed to remain constant during the time that the body is undergoing acceleration from its initial to its final velocity. If the final velocity is sufficiently high to produce observable relativistic effects (this is usually taken as $v = 0.1\, c = 3 \times 10^9$ cm/sec), then equations (2.3) and (2.5) are no longer valid.

As the body gains velocity under the influence of an unbalanced force, its mass continuously increases until it attains the value given by equation (2.4). This particular value for the mass is thus applicable only to one point during the time that the body was undergoing acceleration, that is, only after the body has completed its acceleration. The magnitude of the unbalanced force, therefore, must be continuously increased during the accelerating process to compensate for the increasing inertia of the body due to its continuously increasing mass. Equations (2.2) and (2.5) assume the force to be constant and therefore are not applicable to cases where relativistic effects must be considered. One way in which to overcome this difficulty is to divide the total distance r into many smaller distances, $\Delta r_1, \Delta r_2, \ldots, \Delta r_n$, as shown in Fig. 2.1, and then multiply each of these small distances by the average force exerted while traversing the small distance, and then summing the products. This process may be written as

$$W = f_1\Delta r_1 + f_2\Delta r_2 + \ldots f_n\Delta r_n, \tag{2.6A}$$

and abbreviated as

$$W = \sum_{n=1}^{n} f_n\Delta r_n . \tag{2.6B}$$

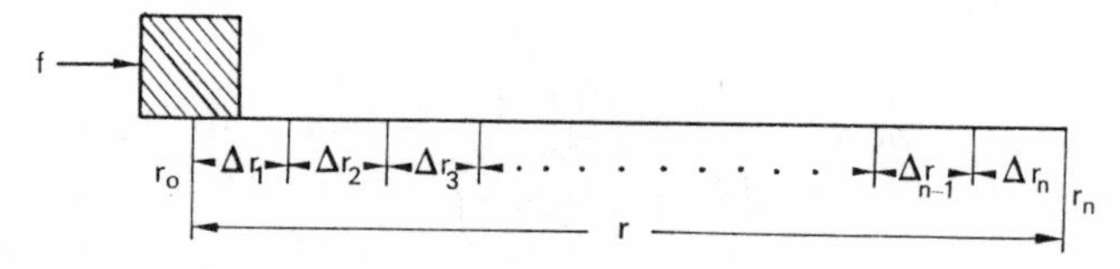

FIG. 2.1. Diagram illustrating that the total work done in accelerating a body is $W = \sum_{n=1}^{n} f_n \Delta r_n$.

As r is successively divided into smaller and smaller lengths, the calculation, using equation (2.6), of the work done becomes more accurate. A limiting value for W may be obtained by letting each small distance, Δr_n, in equation (2.6) approach zero, that is, by considering such small increments of distance that the force remains approximately constant during the specified interval. In the notation of the calculus, such an infinitesimally small quantity is called a differential, and is specified by prefixing the symbol for the quantity with the letter "d". Thus, if r represents distance, dr represents an infinitesimally small distance, and the differential of work done, which is the product of the force and the infinitesimally small distance, is

$$\mathrm{d}W = f\,\mathrm{d}r. \tag{2.7}$$

The total energy expended in going from the point r_0 to point r_n, then, is merely the sum of all the products of the force and the infinitesimally small distances through which it acted. This sum is indicated by the mathematical notation

$$W = \int_{r_0}^{r_n} f\,\mathrm{d}r. \tag{2.8}$$

The ratio of two differentials is called a derivative, and the process in which a derivative is obtained is called differentiating. Since acceleration is defined as the rate of change of velocity with respect to time,

$$a = \frac{v_2 - v_1}{t_2 - t_1} = \frac{\Delta v}{\Delta t}, \tag{2.9}$$

where v_1 and v_2 are the respective velocities at times t_1 and t_2 then equation (2.1) may be written as

$$f = m \frac{\Delta v}{\Delta t}, \tag{2.10}$$

and, by letting Δt approach zero, thereby obtaining the instantaneous rate of change of velocity, or the derivative of velocity with respect to time, we have, using the differential notation,

$$f = m \frac{dv}{dt}, \tag{2.11}$$

which is the expression of Newton's second law of motion for the case where the mass remains constant. Newton's second law states that the rate of change of momentum of an accelerating body is proportional to the unbalanced force acting on the body. For the general case, where mass is not constant, Newton's second law is therefore written as

$$f = \frac{d\,(mv)}{dt}. \tag{2.12}$$

Substitution of the value of f from equation (2.12) into equation (2.8) gives

$$W = \int_0^r \frac{d\,(mv)}{dt}\, dr. \tag{2.13}$$

Since $v = dr/dt$, equation (2.13) can be written as

$$W = \int_0^r \frac{d(mv)}{dt} \cdot v\, dt = \int_0^{mv} v\, d\,(mv), \tag{2.14}$$

and substituting $m = m_0/\sqrt{(1-v^2/c^2)}$, we have

$$W = \int_0^v v\, d\left(\frac{m_0\, v}{\sqrt{1 - v^2/c^2}}\right). \tag{2.15}$$

Differentiating the term in the parenthesis gives

$$W = m_0 \int_0^v \left[\frac{v}{(1 - v^2/c^2)^{1/2}} + \frac{v^3/c^2}{(1 - v^2/c^2)^{3/2}}\right] dv. \tag{2.16}$$

Now, multiply the numerator and denominator of the first term in equation (2.16) by $1-v^2/c^2$ to give

$$W = m_0 \int_0^v \left[\frac{v - v^3/c^2}{(1 - v^2/c^2)^{3/2}} + \frac{v^3/c^2}{(1 - v^2/c^2)}\right] dv \tag{2.17}$$

$$= m_0 \int_0^v \frac{v}{(1 - v^2/c^2)^{3/2}} dv = m_0 \int v\,(1 - v^2/c^2)^{-3/2}\, dv. \tag{2.18}$$

The integrand in equation (2.18) is almost in the form

$$\int_a^b u^n\, du = \left.\frac{u^{n+1}}{n+1}\right|_a^b, \tag{2.19}$$

where $u^n = \left(1 - \frac{v^2}{c^2}\right)^{-3/2}$, and $du = -\frac{2v}{c^2}\, dv.$

To convert equation (2.18) into the form for integration given by equation (2.19), it is necessary only to complete du. This is done by multiplying the integrand by $-2/c^2$ and the entire expression by $-c^2/2$ in order to keep the total value of equation (2.18) unchanged. The solution of equation (2.18), which gives the kinetic energy of a body that was accelerated from zero velocity to a velocity v, is

$$E_k = W = m_0c^2\left(\frac{1}{\sqrt{(1 - v^2/c^2)}} - 1\right) = m_0c^2\left(\frac{1}{\sqrt{(1 - \beta^2)}} - 1\right), \tag{2.20}$$

where $\beta = v/c$.

Equation (2.20) is the exact expression for kinetic energy, and must be used whenever the moving body experiences observable relativistic effects.

Example 2.2

(a) What is the kinetic energy of the electron in Example 2.1 that travels at 99% of the velocity of light?

$$E_k = m_0c^2\left(\frac{1}{\sqrt{(1 - \beta^2)}} - 1\right)$$

$$= 9.11 \times 10^{-28}\ \text{g}\left(3 \times 10^{10}\ \frac{\text{cm}}{\text{sec}}\right)^2\left(\frac{1}{\sqrt{(1 - (0.99)^2)}} - 1\right)$$

$$= 4.97 \times 10^{-6}\ \text{ergs}.$$

(b) How much additional energy is required to increase the velocity of this electron to 99.9% of the velocity of light, an increase in velocity of only 0.91%?

$$E_k = 9.11 \times 10^{-28}\ \text{g} \left(3 \times 10^{10}\ \frac{\text{cm}}{\text{sec}}\right)^2 \left(\frac{1}{\sqrt{(1-(0.999)^2)}} - 1\right)$$

$$= 17.55 \times 10^{-6}\ \text{ergs}.$$

The additional work necessary to increase the kinetic energy of the electron from 99% to 99.9% of the velocity of light is

$$W = (17.55 - 4.97)\ 10^{-6}\ \text{ergs}$$

$$= 12.58 \times 10^{-6}\ \text{ergs}.$$

(c) What is the mass of the electron whose β is 0.999?

$$m = \frac{m_0}{\sqrt{(1-\beta^2)}}$$

$$= \frac{9.11 \times 10^{-28}\ \text{g}}{\sqrt{(1-(0.999)^2)}} = 204 \times 10^{-28}\ \text{g}.$$

The relativistic expression for kinetic energy given by equation (2.20) is rigorously true for particles moving at all velocities, while the non-relativistic expression for kinetic energy, equation (2.3), is applicable only to cases where the velocity of the moving particle is much less than the velocity of light. It can easily be shown that the relativistic expression reduces to the non-relativistic expression for low velocities by expanding the expression $1/\sqrt{(1-\beta^2)}$ in equation (2.20) according to the binomial theorem, and then dropping higher terms that become insignificant when $v \ll c$. According to the binomial theorem,

$$(a+b)^n = a^n + na^{n-1}b + \frac{n\,(n-1)\,a^{n-2}b^2}{2\,!} + \ldots. \tag{2.21}$$

The expansion of $1/\sqrt{(1-\beta^2)}$, or $(1-\beta^2)^{-\frac{1}{2}}$ according to equation (2.21) is accomplished by letting $a = 1$, $b = -\beta^2$, and $n = -\frac{1}{2}$.

$$(1-\beta^2)^{-1/2} = 1 + \tfrac{1}{2}\,\beta^2 + \tfrac{3}{8}\,\beta^4 + \ldots. \tag{2.22}$$

Since $\beta = v/c$, then, if $v \ll c$, terms from β^4 and higher will be insignificantly small, and may therefore be dropped. When this is done, and the first two terms from equation (2.22) are substituted into equation (2.20), we have

$$E_k = m_0c^2 \left(1 + \tfrac{1}{2}\,\frac{v^2}{c^2} - 1\right)$$

$$= \tfrac{1}{2}\,m_0v^2, \tag{2.3}$$

which is the non-relativistic case. Equation (2.3) is applicable when $v \ll c$.

In Example 2.1 it was shown that, at a very high velocity ($\beta = 0.99$), a kinetic energy increase of 153% resulted in a velocity increase of the moving body by only 0.91%. In non-relativistic cases, the increase in velocity is directly proportional to the square root of the work done on the moving body or, in other words, to the kinetic energy of the body. In the relativistic case, the velocity increase due to additional energy is smaller than in the non-relativistic case because the additional energy serves to increase the mass of the moving body rather than its velocity. This equivalence of mass and energy is one of the most important consequences of Einstein's theory of relativity. According to Einstein, the relationship between mass and energy is

$$E = mc^2, \tag{2.23}$$

where E is the total energy of a piece of matter whose mass is m, and c is the velocity of light in a vacuum. The theory of relativity tells us that all matter contains potential energy by virtue of its mass. It is this energy source which is tapped to obtain nuclear energy. The main virtue of this energy source is the vast amount of energy that can be derived from conversion into its energy equivalent of small amounts of nuclear fuel.

Example 2.3

(a) How much energy can be obtained from one gram of nuclear fuel?

$$\begin{aligned} E &= mc^2 \\ &= 1 \text{ g} \times \left(3 \times 10^{10} \frac{\text{cm}}{\text{sec}}\right)^2 \\ &= 9 \times 10^{20} \text{ ergs.} \end{aligned}$$

Since there are 2.78×10^{-14} kilowatt hours per erg, 1 g of nuclear fuel yields

$$\begin{aligned} E &= 9 \times 10^{20} \text{ ergs} \times 2.78 \times 10^{-14} \frac{\text{kWh}}{\text{erg}}, \\ &= 2.5 \times 10^7 \text{ kilowatt hours.} \end{aligned}$$

(b) How much coal, whose heat content is 13,000 Btu per pound, must be burned to liberate the same amount of energy as one gram of nuclear fuel?

$$1 \text{ Btu} = 2.93 \times 10^{-4} \text{ kWh.}$$

∴ Amount of coal required is:

$$1.3 \times 10^4 \frac{\text{Btu}}{\text{lb}} \times 2.93 \times 10^{-4} \frac{\text{kWh}}{\text{Btu}} \times 2 \times 10^3 \frac{\text{lb}}{\text{ton}} \times C \text{ tons}$$

$$= 2.5 \times 10^7 \text{ kWh.}$$

$$\therefore \quad C = \frac{2.5 \times 10^7}{1.3 \times 10^4 \times 2.93 \times 10^{-4} \times 2 \times 10^3}$$

$$= 3280 \text{ tons.}$$

The loss in mass accompanying ordinary energy transformations is not detectable because of the very large amount of energy released per unit mass, and the consequent small change in mass for ordinary reactions. In the case of coal, for example, the above example shows a loss in mass of 1 g per 3280 tons. The fractional mass loss is

$$f = \frac{\Delta m}{m} = \frac{1 \text{ g}}{3.28 \times 10^3 \text{ tons} \times 2 \times 10^3 \text{ lb/ton} \times 4.54 \times 10^2 \text{ g/lb}}$$

$$= 3.3 \times 10^{-10}.$$

Such a small fractional loss in mass is not detectable by any of our weighing techniques.

Electricity

Electrical Charge: The Statcoulomb

All matter is electrical in nature, and consists of extremely small charged particles called protons and electrons. The mass of the proton is 1.6723×10^{-24} g and the mass of the electron is 9.1085×10^{-28} g. These two particles have charges of exactly the same magnitude, but are qualitively different. A proton is said to have a positive charge and an electron a negative charge. Under normal conditions, matter is electrically neutral because the positive and negative charges are homogeneously (on a macroscopic scale) dispersed in equal numbers in a manner that results in no net charge. However, it is possible, by suitable treatment, to induce either net positive or negative charges on bodies. Combing the hair, for example, with a hard rubber comb transfers electrons to the comb from the hair, leaving a net negative charge on the comb.

Charged bodies exert forces on each other by virtue of their electrical fields. Bodies with like charges repel each other, while those with unlike charges attract each other. In the case of point charges, the magnitude of these electrical forces is proportional to the product of the charges and inversely proportional to the square of the distance between the charged bodies. This relationship was described by Coulomb, and is known as Coulomb's law. Expressed algebraically, it is

$$f = \frac{q_1 q_2}{r^2}. \tag{2.24}$$

This relationship is used to define the magnitude of the statcoulomb.

One statcoulomb is that quantity of electrical charge which, when placed at a distance of 1 cm from a similar charge, experiences a repulsive force of 1 dyne.

The statcoulomb is often referred to as the electrostatic unit, or e.s.u., of charge. In this book, the symbol for the statcoulomb will be sC in order to prevent confusion with the electrostatic unit of potential, which is also frequently referred to as an e.s.u. (of potential). Since the cgs system is used, the force is expressed in dynes and the distance in centimeters. The dimensions of the statcoulomb are:

$$\text{dyne} = \frac{\text{sC}^2}{\text{cm}^2},$$

$$\text{sC}^2 = \text{dyne cm}^2$$

$$= \frac{\text{g cm}}{\text{sec}^2} \cdot \text{cm}^2$$

$$\text{sC} = \frac{\text{cm}^{\frac{3}{2}}}{\text{sec}} \cdot \text{g}^{\frac{1}{2}}.$$

The smallest known quantity of charge is that on a proton and an electron, $\pm 4.8 \times 10^{-10}$ sC.

Example 2.4

Compare the electrical and gravitational forces of attraction between a proton and an electron that are separated by a distance of 5×10^{-9} cm.

$$f = \frac{q_1 q_2}{r^2}$$

$$= \frac{4.8 \times 10^{-10} \text{ sC} \times 4.8 \times 10^{-10} \text{ sC}}{25 \times 10^{-18} \text{ cm}^2}$$

$$= 9.2 \times 10^{-3} \text{ dynes.}$$

The gravitational force between two bodies follows the same mathematical formulation as Coulomb's law for electrical forces. In the case of gravitational forces, the force is always attractive. The gravitational force is given by

$$F = \frac{Gm_1 m_2}{r^2}. \quad (2.25)$$

G is a universal constant that is equal to 6.67×10^{-8} dyne cm^2/g^2, and must be used because the unit of force, the dyne, was originally defined using "inertial" mass, according to Newton's second law of motion given by equation (2.1). The mass in equation (2.25), is commonly called "gravitational" mass. Despite the two different designations, however, it should be emphasized that inertial mass and gravitational mass are equivalent. It should also be pointed out that F in equation (2.25) gives the weight of an object of mass m_1 when m_2 represents the mass of the earth and r is the distance from the object to the center of the earth. Weight is merely a measure of the gravitational attractive force between an object and the earth, and therefore varies from point to point on the surface of the earth, according to the distance of the point from the earth's center. On the surface of another planet, the weight of the same object would be different from that on earth because of the different size and mass of that planet and its consequent different attractive force. In outer space, if the object is not under the gravitational influence of any heavenly body, it must be weightless. Mass, on the other hand, is a measure of the amount of matter and its numerical value is therefore independent of the point in the universe where it is measured.

The gravitational force between the electron and the proton is

$$F = \frac{6.67 \times 10^{-8} \text{ dyne cm}^2/\text{g}^2 \times 9.11 \times 10^{-28} \text{ g} \times 1.67 \times 10^{-24} \text{ g}}{(5 \times 10^{-9} \text{ cm})^2}$$

$$= 4.05 \times 10^{-42} \text{ dynes.}$$

It is immediately apparent that in the interaction between charged particles, gravitational forces are extremely small in comparison to the electrical forces acting between the particles, and may be completely neglected in most instances.

Electrical Potential: The Statvolt

If one charge is held rigidly, and another charge is placed in the electrical field of the first charge, it will have a certain amount of potential energy relative to any other point within the electric field. In the case of electrical potential energy, the reference point is taken at an infinite distance from the charge that sets up the electric field; that is, at a point far enough away from the charge so that its effect is negligible. As a consequence of the fact that these charges do not interact electrically, a value of zero is arbitrarily assigned to the potential energy in the system of charges; the charge at an infinite distance from the one that sets up the electric field has no electrical potential energy. If the two charges are of the same sign, then, to bring them closer together requires work, or the expenditure of energy, in order to overcome the repulsive force between the two charges. Since work was done in bringing the two charges together, the potential energy in the system of charges is now greater than it was initially. On the other hand, if the two charges are of opposite sign, then a decrease in distance between them occurs spontaneously because of the attractive forces, and work is done by the system. The potential energy of the system consequently decreases; that is, the potential energy of the freely moving charge, with respect to the rigidly held charge, decreases.

This is exactly analogous to the case of a freely falling mass whose potential energy decreases as it approaches the surface of the earth. In the case of the mass in the earth's gravitational field, however, the reference point for potential energy of the mass is arbitrarily set on the surface of the earth. This means that the mass has no potential energy when it is lying right on the earth's surface. All numerical values for potential energy of the mass, therefore, are positive numbers. In the case of electrical potential energy, however, as a consequence of the arbitrary convention that the point of the zero numerical value is at an infinite distance from the charge that sets up the electric field, the numerical values for the potential energy of a charge, due to attractive electrical forces, must be negative.

The quantitative aspects of electrical potential energy may be investigated with the aid of Fig. 2.2, which shows a charge $+Q$ that sets up an electric field which extends uniformly in all directions. Another charge, $+q$, is used to explore the electric field set up by Q. When the exploring charge is at point a, at a distance r_a cm from Q, it has an amount of potential energy that depends on the magnitudes of Q, q, and r_a If the charge q is now to be moved

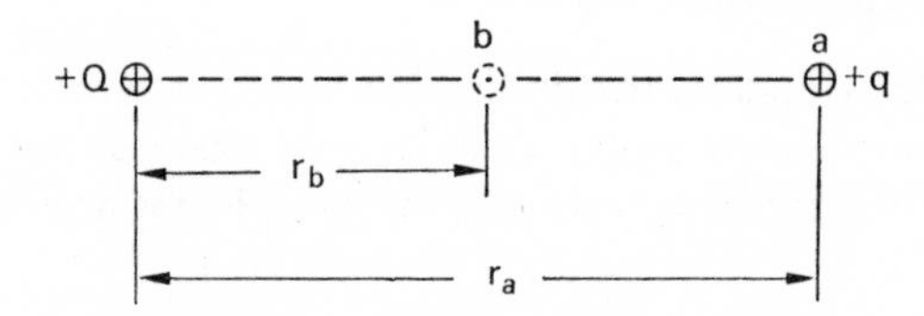

FIG. 2.2. Diagram illustrating work done in moving a charge between two points of different potential in an electric field.

to point b, which is closer to Q, then, because of the repulsive force between the two charges, work is done in moving the charge from point a to point b. The amount of work that is done in moving charge q from point a to point b may be calculated by multiplying the force exerted on the charge q by the distance through which it was moved, in accordance with equation (2.2). From equation (2.24), however, it is seen that the force is not constant, but varies inversely with the square of the distance between the charges. The magnitude of the force, therefore, increases rapidly as the charge q approaches Q, and increasingly greater amounts of work are done when the exploring charge q is moved a unit distance. The movement of the exploring charge may be accomplished by a series of infinitesimally small movements, during each of which an infinitesimally small amount of work is done. The total energy expenditure, or increase in potential energy of the exploring charge, is then merely equal to the sum of all the infinitesimal increments of work. This infinitesimal energy increment is given by

$$dW = -f\,dr \qquad (2.7)$$

(the minus sign is used here because an increase in potential energy results from a decrease in distance between the charges) and, if the value for f from equation (2.24) is substituted into equation (2.7), we have

$$dW = -\frac{Qq}{r^2}\,dr, \tag{2.26}$$

$$W = -Qq\int_{r_a}^{r_b}\frac{dr}{r^2}. \tag{2.27}$$

Integration of equation (2.27) gives

$$W = Qq\left(\frac{1}{r_b} - \frac{1}{r_a}\right). \tag{2.28}$$

If the distances a and b are measured in centimeters, and if the charges are given in statcoulombs, then the energy W is given in ergs.

Example 2.5

If Q is $+2$ statcoulombs, q is $+1$ statcoulomb, and r_a and r_b are 2 and 1 cm respectively, then the work done is, from equation (2.28),

$$\begin{aligned} W &= 2\text{ sC} \times 1\text{ sC}\,(1/1\text{ cm} - \tfrac{1}{2}\text{ cm}) \\ &= (2\text{ sC}^2)\,(\tfrac{1}{2}\text{ cm}^{-1}) \\ &= 1\text{ erg}. \end{aligned}$$

In this example, 1 erg of work was expended in moving one statcoulomb of charge from a to b. The potential difference between points a and b is therefore said to be one statvolt, and point b is the point of higher potential.

> One statvolt of potential difference exists between any two points in an electric field if one erg of work is expended in moving a unit charge between the two points.

Expressed more concisely, the definition of a statvolt is

$$1\text{ statvolt} = \frac{1\text{ erg}}{1\text{ statcoulomb}}.$$

The electrical potential at any point due to an electric field from a point charge Q is defined as the potential energy that a unit positive exploring charge, $+q$, would have if it were brought from a point at an infinite distance from Q to the point in question. The electrical potential at point b in Fig. 2.2 can be computed from equation (2.28) by setting point a equal to infinity.

The potential at point b, V_b, which is defined as the potential energy per unit positive charge at b, is, therefore,

$$V_b = \frac{W}{q} = \frac{Q}{r_b}. \tag{2.29}$$

Example 2.6

(a) What is the potential at a distance of 5×10^{-9} cm from a proton?

$$V = \frac{Q}{r} = \frac{4.8 \times 10^{-10} \text{ sC}}{5 \times 10^{-9} \text{ cm}}$$

$$= 0.096 \text{ statvolts.}$$

(b) What is the potential energy of another proton at this point?

According to equation (2.29), the potential energy of the proton is equal to the product of the potential at the point in the electric field and its charge.

$$\therefore \quad W = Vq = 9.6 \times 10^{-2} \text{ sV} \times 4.8 \times 10^{-10} \text{ sC}$$

$$= 4.6 \times 10^{-11} \text{ ergs.}$$

(c) What is the potential energy of an electron at a distance of 5×10^{-9} cm from a proton?

The potential energy of an electron at a distance of 10^{-8} cm from a proton is, according to equation (2.29),

$$W = v \times q$$

$$= 9.6 \times 10^{-2} \text{ sV } (-\ 4.8 \times 10^{-10} \text{ sC})$$

$$= -\ 4.6 \times 10^{-11} \text{ ergs.}$$

In the system of electrical units commonly employed in everday usage, the unit of potential is the practical volt, or simply the volt. The practical volt is exactly analogous to the statvolt; it differs only in magnitude. The practical volt is much smaller than the statvolt, one practical volt is equal to 1/300 sV, or

$$1 \text{ statvolt} = 300 \text{ practical volts.}$$

If two electrodes are connected to the terminals of a source of voltage, as shown in Fig. 2.3, then a charged particle anywhere in the electric field between

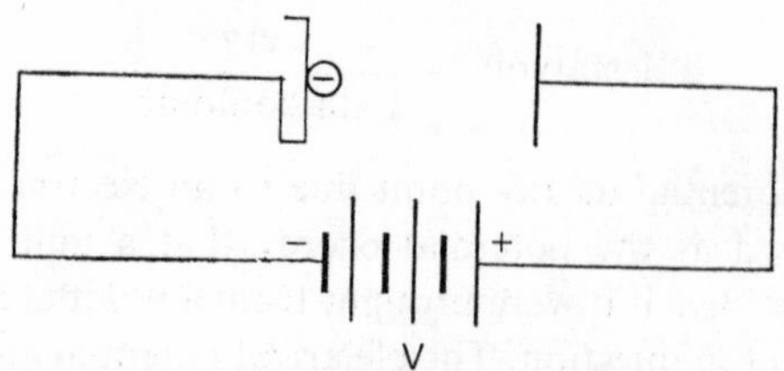

FIG. 2.3. Diagram showing the potential energy in an electric field.

the two plates will have an amount of potential energy given by equation (2.29).

$$W = Vq,$$

where V is the electrical potential at the point occupied by the charged particle. If, for example, the cathode in Fig. 2.3 is one volt negative with respect to the anode and the charged particle is an electron on the surface of the cathode, then the potential energy of the electron with respect to the anode is

$$W = Vq$$

$$= \frac{-1\ \text{V}\,(-4.8 \times 10^{-10}\ \text{sC})}{300\ \text{V/sV}}$$

$$= 1.6 \times 10^{-12}\ \text{ergs.}$$

This amount of energy, 1.6×10^{-12} ergs, is called an *electron volt*, and is abbreviated eV. Since the magnitude of the electron volt is convenient when dealing with the energetics of atomic and nuclear mechanics, this quantity of energy is taken as a unit and, consequently, is frequently used in health physics. Multiples of the electron volt are the keV, 10^3 eV; the MeV, 10^6; and the BeV, 10^9 eV.

Example 2.7

How many electron volts of energy correspond to the mass of a resting electron?

$$E = mc^2$$

$$= 9.11 \times 10^{-28}\ \text{g}\,(3 \times 10^{10}\ \text{cm/sec})^2$$

$$= 81.99 \times 10^{-8}\ \text{ergs.}$$

Since there are 1.6×10^{-12} ergs/eV,

$$E = \frac{81.99 \times 10^{-8}\ \text{ergs}}{1.6 \times 10^{-12}\ \text{ergs/eV}}$$

$$= 0.51 \times 10^6\ \text{eV.}$$

It should be emphasized that, although the numerical value for the electron volt was calculated by computing the potential energy of an electron at a potential of 1 volt, the electron volt is not a unit of electrons or volts; it is a unit of energy, and may be interchanged (after numerical correction) with any other unit of energy.

Example 2.8

How many electron volts (eV) of heat must be added to change 1 liter of water, whose temperature is 50°C, to completely dry steam?

The specific heat of water is 1 calorie per gram, and the heat of vaporization of water is 539 calories per gram.

$$\therefore \text{ Heat added} = 1000 \text{ g} \left(1 \frac{\text{cal}}{\text{g deg}} (100 - 50) \text{ deg} + 539 \frac{\text{cal}}{\text{g}}\right)$$
$$= 589{,}000 \text{ calories.}$$

Since there are $4.186 \times 10^7 \dfrac{\text{ergs}}{\text{cal}}$ and $1.6 \times 10^{-12} \dfrac{\text{ergs}}{\text{eV}}$, we have:

$$\text{heat added} = \frac{5.89 \times 10^5 \text{ cal} \times 4.186 \times 10^7 \text{ ergs/cal}}{1.6 \times 10^{-12} \text{ ergs/eV}}$$
$$= 1.54 \times 10^{25} \text{ eV.}$$

The answer to Example 2.8 is an astronomically large number (but not very much energy on the scale of ordinary physical and chemical reactions) and shows why the electron volt is a useful energy unit only for reactions in the atomic world.

Example 2.9

An alpha particle, whose charge is $+(2 \times 4.8 \times 10^{-10})$ sC and whose mass is 6.601×10^{-24} g, is accelerated across a potential difference of 100,000 V. What is its kinetic energy, and how fast is it moving?

The potential energy of the alpha particle at the moment it begins to undergo acceleration is, from equation (2.29),

$$W = Vq$$
$$= \frac{10^5 \text{ V}}{300 \text{ V/sV}} \times 2 \times 4.8 \times 10^{-10} \text{ sC}$$
$$= 3.2 \times 10^{-7} \text{ ergs,}$$

or, in terms of electron volts

$$W = \frac{3.2 \times 10^{-7} \text{ ergs}}{1.6 \times 10^{-12} \text{ ergs/eV}}$$
$$= 200{,}000 \text{ eV.}$$

Since the potential energy of the alpha particle is all converted into kinetic energy after the alpha particle falls through the 100,000-volt potential difference, the kinetic energy must then also be 200,000 eV.

The velocity of the alpha particle may be computed by equating its potential and kinetic energies.

$$\frac{Vq}{300} = \tfrac{1}{2} mv^2, \qquad (2.30)$$

and solving for v:

$$v = \sqrt{\left(\frac{2\ \mathrm{Vq}}{300\ \mathrm{m}}\right)}$$

$$= \sqrt{\left(\frac{2 \times 105\ \mathrm{V} \times 9.6 \times 10^{-10}\ \mathrm{sC}}{300\ \mathrm{V/sV} \times 6.601 \times 10^{-24}\ \mathrm{g}}\right)}$$

$$= 3.12 \times 10^{8}\ \frac{\mathrm{cm}}{\mathrm{sec}}.$$

Electric Field

The term "electric field" was used in the preceding sections of the chapter without explicitly defining the term. Implicit in the use of the term, however, was the connotation by the context that an electric field is any region where electric forces act. An electric field is not merely a descriptive term; to define an electric field requires a number to specify the magnitude of the electric forces that act in the electric field and a direction in which these forces act. The strength of an electric field is called the electric field intensity, and may be

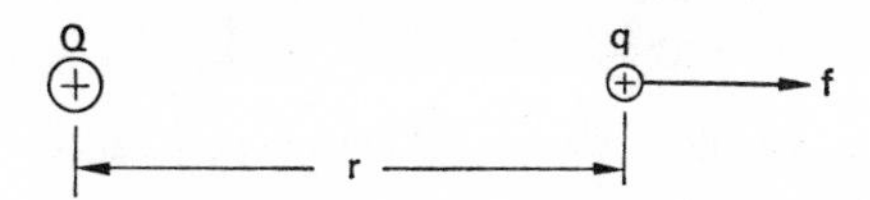

FIG. 2.4. The force on an exploring charge q in the electric field of charge Q.

defined in terms of the force (magnitude and direction) that acts on a unit-exploring charge which is placed into the electric field. Consider an isolated charge, $+Q$, that sets up an electric field, and an exploring charge, $+q$, that is used to investigate the electric field, as shown in Fig. 2.4. The exploring charge will experience a force in the direction shown and of a magnitude given by equation (2.24).

$$f = \frac{Qq}{r^2}. \tag{2.24}$$

The force per unit charge at the point r cm from charge Q is the electric field intensity at that point, and is given by the equation

$$\mathscr{E} = \frac{f}{q}\,\frac{\mathrm{dynes}}{\mathrm{sC}} = \frac{Q}{r^2}\,\frac{\mathrm{sC}}{\mathrm{cm}^2}. \tag{2.31}$$

According to equation (2.31) electric field intensity is expressed in units of dynes per statcoulomb (which is equivalent to statcoulombs per square centimeter). It should be emphasized that $\mathscr{E}$ is a vector quantity, that is, that it has direction as well as magnitude.

Example 2.10

(a) What is the electric field intensity at point P due to the two charges, +6 and +3 sC, shown in Fig. 2.5? The electric field intensity at point P due to the +6 sC charge is

$$\mathscr{E}_1 = \frac{Q_1}{r_1^2} = \frac{6 \text{ sC}}{4 \text{ cm}^2} = 1.5 \frac{\text{dynes}}{\text{sC}},$$

and acts in the direction shown in Fig. 2.5. (The magnitude of the field intensity is usually shown graphically by a vector whose length is propor-

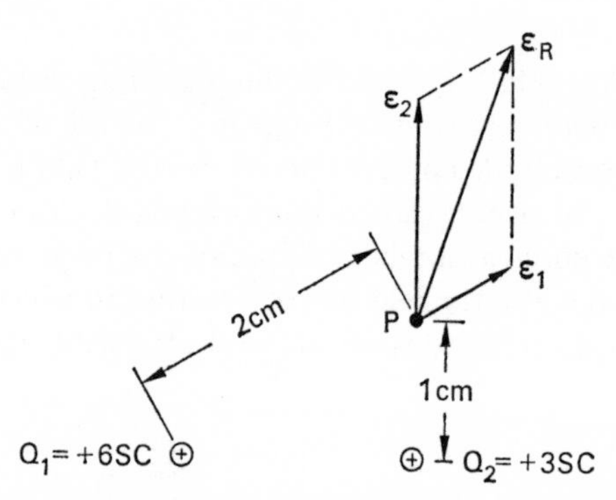

FIG. 2.5a. Resultant electric field from two positive charges.

tional to the field intensity. In Fig. 2.5 the scale is 1 cm = 1 dyne/sc. $\mathscr{E}_1$ therefore is drawn 1.5 cm long.) $\mathscr{E}_2$, the electric field intensity at P due to the +3 statcoulomb charge, alone is

$$\mathscr{E}_2 = \frac{Q_2}{r_2^2} = \frac{3 \text{ sC}}{1 \text{ cm}^2} = \frac{3 \text{ dynes}}{\text{sC}},$$

and acts along the line $Q_2 P$, as shown in the illustration. The resultant electrical intensity at point P is the vector sum of $\mathscr{E}_1$, and $\mathscr{E}_2$. If these two vectors are accurately drawn, both in magnitude and direction, then the resultant may be obtained graphically by completing the parallelogram of forces and drawing the diagonal $\mathscr{E}_R$. The length of the diagonal is proportional to the magnitude of the resultant electric field intensity and its direction shows the direction of the electric field at point P. In this case, since 1 dyne/sC is equal to 1 cm, the resultant electric field intensity is found to be about 4 dynes/sC, and it acts in a direction 30° clockwise from the vertical. The value of $\mathscr{E}_R$ may also be determined analytically from the law of cosines

$$a^2 = b^2 + c^2 - 2bc \cos A, \tag{2.32}$$

where b and c are two adjacent sides of a triangle, A is the included angle, and a is the side opposite angle A. In this case, b is 3, c is 1.5, angle A is

120°, and a is the resultant whose magnitude is to be found, the electric field intensity $\mathscr{E}_R$. The direction of the resultant can be computed from the geometrical arrangement of the vectors. From equation (2.32) the resultant is found to be:

$$\mathscr{E}_R^2 = (3)^2 + (1.5)^2 - 2\,(3)\,(1.5) \cos 120°,$$

$$\mathscr{E}_R = 3.97 \frac{\text{dynes}}{\text{sC}}.$$

(b) What is the magnitude and direction of $\mathscr{E}$ if the 3-statcoulomb charge is negative and the 6-statcoulomb charge is positive?

In this case, the magnitudes of $\mathscr{E}_1$ and $\mathscr{E}_2$ would be exactly the same as in part (a) of the problem; the direction of $\mathscr{E}_1$ would also remain unchanged,

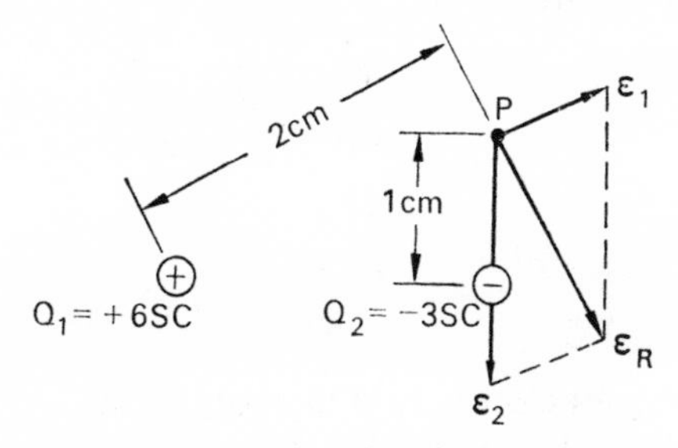

FIG. 2.5b. Resultant electric field from two opposite charges.

but the direction of $\mathscr{E}_2$ would be toward the 3-statcoulomb charge, as shown in Figure 2.5b. From the geometrical arrangement, it is seen that the resultant intensity acts in a direction 120° clockwise from the vertical. The magnitude of $\mathscr{E}_R$ is, from equation (2.31),

$$\mathscr{E}_R^2 = (3)^2 + (1.5)^2 - 2\,(3)\,(1.5) \cos 60°,$$

$$\mathscr{E}_R = 2.6 \frac{\text{dynes}}{\text{sC}}.$$

Point charges result in non-uniform electric fields. A uniform electric field may be produced by applying a potential difference across two large parallel

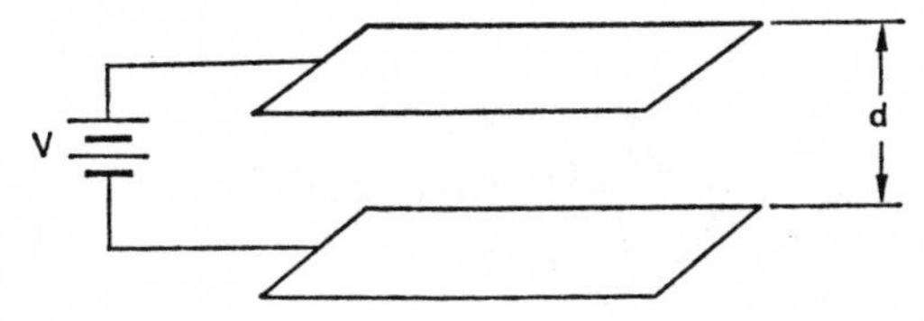

FIG. 2.6. Conditions for producing a relatively uniform electric field. The field will be quite uniform throughout the region between the two plates, but will be distorted at the edges of the plates.

plates made of electrical conductors separated by an insulator, as shown in Fig. 2.6. The electric intensity throughout the region between the two plates is $\mathscr{E}$ dynes per statcoulomb. The force acting on any charge of statcoulombs within this field, therefore, is

$$f = \mathscr{E}q \text{ dynes.} \tag{2.33}$$

If the charge q happens to be positive, then, to move it across the distance d, from the negative to the positive plates, against the electrical force in the uniform field requires the expenditure of energy given by the equation

$$W = fd = \mathscr{E}qd. \tag{2.34}$$

However, since potential difference is defined as work per unit charge, equation (2.34) may be written as

$$V = \frac{W}{q} = \mathscr{E}d, \tag{2.35}$$

or

$$\mathscr{E} = \frac{V}{d}\,\frac{\text{V}}{\text{cm}}. \tag{2.36}$$

Equation (2.36) expresses electric field intensity in the units most commonly used for this purpose: volts per centimeter.

A non-uniform electric field that is of interest to the health physicist (in instrument design) is that due to a potential difference applied across two coaxial conductors, as shown in Fig. 2.7. If the radius of the inner conductor is a centimeters, that of the outer conductor b cm, then the electric intensity at any point between the two conductors, r cm from the center, is given by

$$\mathscr{E} = \frac{1}{r}\,\frac{V}{\ln b/a}\,\frac{\text{V}}{\text{cm}}, \tag{2.37}$$

where V is the potential difference between the two conductors.

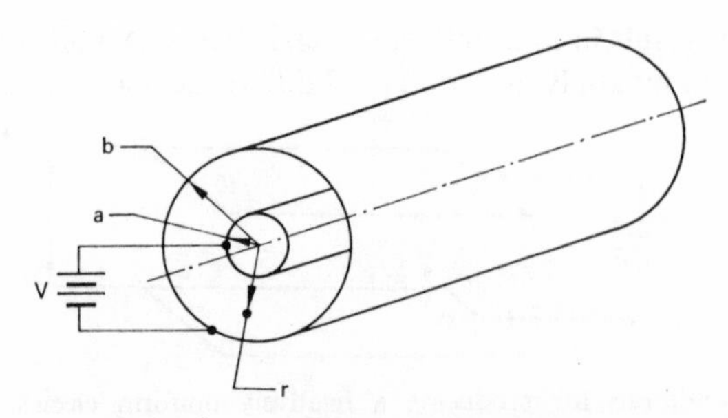

FIG. 2.7. Conditions for the non-uniform electric field between two coaxial conductors given by equation (2.37).

Example 2.11

A Geiger counter is constructed of a wire anode whose diameter is 0.1 mm and a cathode, coaxial with the anode, whose diameter is 2 cm. If the voltage across the tube is 1000 V, what is the electric field intensity (a) at a distance of 0.03 mm from the surface of the anode, and (b) at a point midway between the center of the tube and the cathode?

$$\text{(a)} \quad \mathscr{E} = \frac{1}{r}\frac{V}{\ln b/a}$$

at $r = \frac{1}{2}(0.01) + 0.003 = 0.008$ cm, we have

$$\mathscr{E} = \frac{1}{0.008 \text{ cm}} \times \frac{1000 \text{ V}}{\ln 1/0.005}$$

$$= 23{,}600 \frac{\text{V}}{\text{cm}}.$$

(b) At $r = 1$ cm,

$$\mathscr{E} = \frac{1}{1 \text{ cm}} \times \frac{1000 \text{ V}}{\ln 1/0.005}$$

$$= 189 \frac{\text{V}}{\text{cm}}.$$

It should be noted that in the case of coaxial geometry, extremely intense electric fields may be obtained with relatively small potential differences. Such large fields require mainly a large ratio of outer to inner electrode radii.

Energy Transfer

In a quantitative sense, the biological effects of radiation depends on the amount of energy absorbed by living matter from a radiation field and by the spatial distribution, in tissue, of the absorbed energy. In order to comprehend the physics of tissue irradiation, therefore, some pertinent mechanisms of energy transfer must be understood.

Elastic Collision

An elastic collision is defined as a collision between two bodies in which kinetic energy and momentum are conserved; that is, the sum of the kinetic energy of the two bodies before the collision is equal to the sum after collision, and the sum of the momenta before and after the collision is the same. In an elastic collision, the total kinetic energy is redistributed between the colliding bodies; one body gains energy at the expense of the other. A simple case is illustrated in the example below.

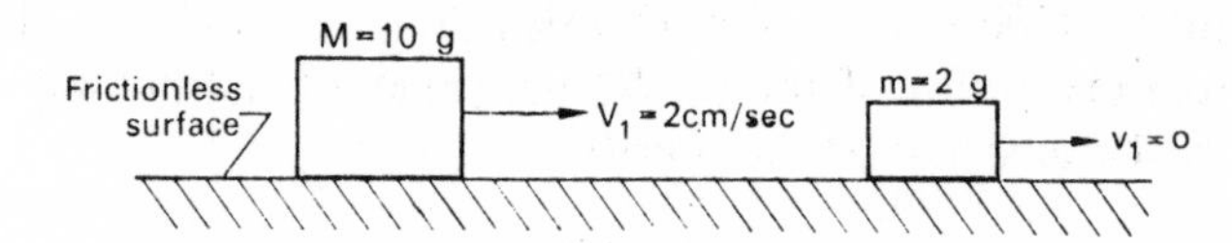

FIG. 2.8. Elastic collision between blocks M and m, in which the sum of both kinetic energy and momenta of the two blocks before and after the collision are the same.

Example 2.12

A block made of perfectly elastic material, whose mass is 10 g, slides on a frictionless surface with a velocity of 2 cm/sec, and strikes a stationary elastic block whose mass is 2 g. How much energy was transferred from the large block to the small block during the collision?

If V_1, v_1, V_2, and v_2 are the respective velocities of the large and small blocks before and after the collision, then, according to the laws of conservation of energy and momentum, we have

$$\tfrac{1}{2} MV_1^2 + \tfrac{1}{2} mv_1^2 = \tfrac{1}{2} MV_2^2 + \tfrac{1}{2} mv_2^2 \tag{2.38}$$

$$MV_1 + mv_1 = MV_2 + mv_2. \tag{2.39}$$

Since $v_1 = 0$, equations (2.38) and (2.39) may be solved simultaneously to give $V_2 = 1\tfrac{1}{3}$ cm/sec and $v_2 = 3\tfrac{1}{3}$ cm/sec.

The kinetic energy transferred during the collision is

$$\tfrac{1}{2} MV_1^2 - \tfrac{1}{2} MV_2^2 = \tfrac{1}{2} \times 10 \left(4 - \frac{16}{9}\right) = 11\tfrac{1}{9} \text{ ergs,}$$

and this, of course, is the energy gained by the smaller block:

$$\tfrac{1}{2} mv^2 = \tfrac{1}{2} \times 2 \times \frac{100}{9} = 11\tfrac{1}{9} \text{ ergs.}$$

Note that the magnitude of the force exerted by the larger block on the smaller block during the collision was not considered in the solution of Example 2.12. The reason for not explicitly considering the force in the solution can be seen from equation (2.10), which may be written as

$$f \Delta t = m \Delta v. \tag{2.10A}$$

According to equation (2.10A), the force necessary to change the momentum of a block is dependent on the time during which it acts. The parameter of importance in this case is the product of the force and the time. This parameter is called the impulse; equation (2.10A) may be written in words as

$$\text{impulse} = \text{change of momentum}.$$

The length of time during which the force acts depends on the relative velocity of the system of moving masses and on the nature of the mass. Generally, the more "give" in the colliding blocks, the greater will be the time of application of the force and the smaller, consequently, will be the magnitude of the force. For this reason, for example, a baseball player who catches a ball moves his hand back at the moment of impact, thereby increasing the time during which the stopping force acts and decreasing the shock to his hand. For this same reason, a jumper flexes his knees as his feet strike the ground, thereby increasing the time that his body comes to rest and decreasing the force on his body. For example, a man who jumps down a distance of 1 meter is moving with a velocity of 443 cm/sec at the instant that he strikes the floor. If he weighs 70 kg, and if he lands rigidly flat footed and is brought to a complete stop in 0.01 sec, then the stopping force, from equation (2.10A), is 3.1×10^9 dynes, or 6980 lb. If, however, he lands on his toes, then lowers his heels, and flexes his knees as he strikes, thereby increasing his actual stopping time to 0.5 sec, the average stopping force is only about 140 lb.

In the case of the two blocks in Example 2.12, if the time of contact is 0.01 sec, then the average force of the collision during this time interval is

$$f = \frac{10 \text{ g} \times 0.67 \text{ cm/sec}}{0.01 \text{ sec}} = 670 \text{ dynes}.$$

The instantaneous forces acting on the two blocks vary from zero at the instant of impact to a maximum value at some time during the collision, then to zero again as the second block leaves the first one. This may be graphically shown in Fig. 2.9, a curve of force vs. time during the collision. The average force during the collision is the area under the curve divided by the time that the two blocks are in contact.

In the case of a collision between two masses, such as that described above, one block exerts a force on the other only while the two blocks are in "contact". During "contact" the two blocks seem to be physically touching each

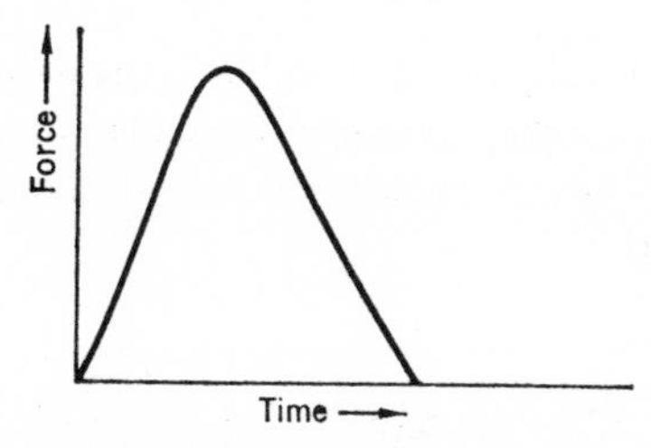

FIG. 2.9. The variation in time of the force between the colliding bodies.

other. Actually, however, the two blocks are merely very close together, too close in fact to perceive any space between them. Under this condition the two blocks repel each other by very short-range forces that are thought to be electrical in nature. (These forces will be discussed again in Chapter 3.) This concept of a "collision" without actual contact between the colliding masses may be easily demonstrated with the aid of magnets. If magnets are affixed to the two blocks in Example 2.12, as shown in Fig. 2.10, then the

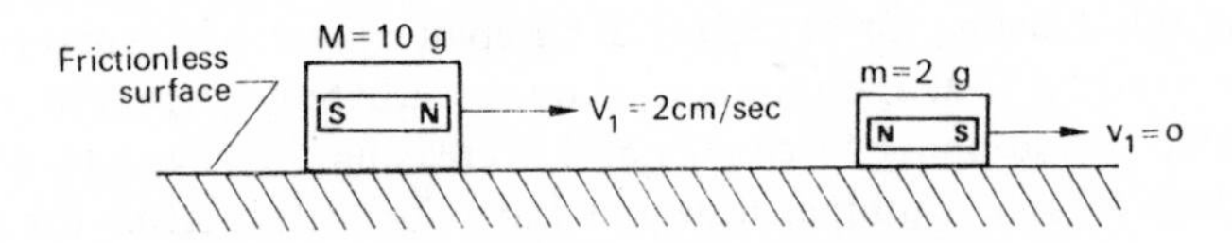

FIG. 2.10. "Collision" between two magnetic fields.

magnetic force, which acts over relatively long distances, will repel the two blocks, and block *B* will move. If the total mass of each block, including the magnet, remains the same as in example 2.12, then the calculations and results of Example 2.12 are applicable. The only difference between the physical "collision" and magnetic "collision" is that the magnitude of the force in the former case is greater than in the latter instance, but the time during which the forces are effective is greater in the case of the magnetic "collision". In both instances, the product of average force and time is exactly the same.

Inelastic Collision

If the conditions in Fig. 2.8 are modified by fastening block *B* to the floor with a rubber band, then, in order to break the rubber band and cause the block to slide freely, block *A* must transfer at least sufficient energy to break the rubber band. Any additional energy transferred would then appear as kinetic energy of block *B*. If the energy necessary to break the rubber band is called the binding energy of block *B*, then the kinetic energy of block *B* after it is struck by block *A* is equal to the difference between the energy lost by *A* and the binding energy of *B*. Algebraically, this may be written as

$$E_b = E_a - \phi, \tag{2.40}$$

where E_a is the energy lost by block *A* and ϕ is the binding energy of block *B*. In a collision of this type, where energy is expended to free one of the colliding bodies, kinetic energy is not conserved, and the collision is therefore not elastic, i.e. it is inelastic.

Example 2.13

A stationary block *B*, whose mass is 2 g, is held by a rubber band whose elastic constant is 10 dynes/mm and whose ultimate strength is 40 dynes.

Another block A, whose mass is 10 g, is moving with a velocity of 2 cm/sec on a frictionless surface. If block A strikes block B, with what velocity will block B move after the collision?

From Example 2.12 it is seen that the energy lost by block A in this collision is $11\frac{1}{9}$ ergs. The energy expended in breaking the rubber band may be calculated from the product of the force needed to break the rubber band and the distance that the rubber band stretches before breaking. In the case of a spring, rubber band, or any other substance that is elastically deformed, the deforming force is opposed by a restoring force whose magnitude is proportional to the deformation, that is,

$$f = kr, \tag{2.41}$$

where f is the force needed to deform the elastic body a distance r, and k is the "spring constant", or the force per unit deformation. Since equation (2.41) shows that the force is not constant, but rather is proportional to the deformation of the rubber band, the work done in stretching the rubber band must be computed by application of the calculus. The infinitesimal work done in stretching the rubber band through a distance dr is

$$\mathrm{d}W = f\,\mathrm{d}r, \tag{2.7}$$

and the total work done in stretching the rubber band from 0 to r is given by equation (2.8)

$$W = \int_0^r f\,\mathrm{d}r. \tag{2.8}$$

Substituting equation (2.41) for f, we have

$$W = \int_0^r kr\,\mathrm{d}r, \tag{2.42}$$

and solving equation (2.45) shows the work done in stretching the rubber band to be

$$W = \frac{kr^2}{2}. \tag{2.43}$$

Since in this example k is equal to 10 dynes/mm, the ultimate strength of the rubber band, 40 dynes, is reached when the rubber is extended 4 mm. With these numerical values, equation (2.43) may be solved.

$$W = \frac{10 \text{ dynes/mm} \times 10 \text{ mm/cm } (0.4 \text{ cm})^2}{2}$$

$$= 8 \text{ dyne cm}$$

$$= 8 \text{ ergs.}$$

Of the $11\frac{1}{9}$ ergs lost by block A in its collision with block B, 8 ergs are dissipated in breaking the rubber band that holds block B (the binding energy). The kinetic energy of block B, therefore, is, from equation (2.40),

$$E_b = 11\tfrac{1}{9} - 8 = 3\tfrac{1}{9} \text{ ergs.}$$

If block A would have had less than 8 ergs of kinetic energy, then the rubber band would not have been broken; the restoring force in the rubber band would have pulled block B back, and would have caused it to oscillate about its equilibrium position. (For this oscillation to actually occur, block A would have to be withdrawn immediately after the collision, otherwise block B, on its rebound, would transfer its energy back to block A, and send it back with the same velocity that it had before the first collision. The net effect of the two collisions, then, would have been only the reversal of the direction in which block A traveled.)

Waves

Energy may be transmitted by disturbing a "medium", permitting the disturbance to travel through the medium, and then collecting the energy with a suitable receiver. For example, if work is done in raising a stone, and the stone is dropped into water, the potential energy of the stone before being dropped is converted into kinetic energy, which is then transferred to the water when the stone strikes. The energy gained by the water disturbs the water and causes it to move up and down. This disturbance spreads out from the point of the initial disturbance at a velocity characteristic of the medium (in this case the water). The energy can be "received" at a remote distance from the point of the initial disturbance by a bob that floats on the water. The wave, in passing by the bob, will cause the bob to move up and down, thereby imparting energy to it. It should be noted here that the *water* moves only in a vertical direction; while the *disturbance* moves in the horizontal direction.

Displacement of water upwards from the undisturbed surface produces a crest, while downward displacement results in a trough. The amplitude of a wave is a measure of the vertical displacement, and the distance between corresponding points on adjacent disturbances is called the wavelength. (The wavelength is usually represented by the Greek letter lambda, λ.) The

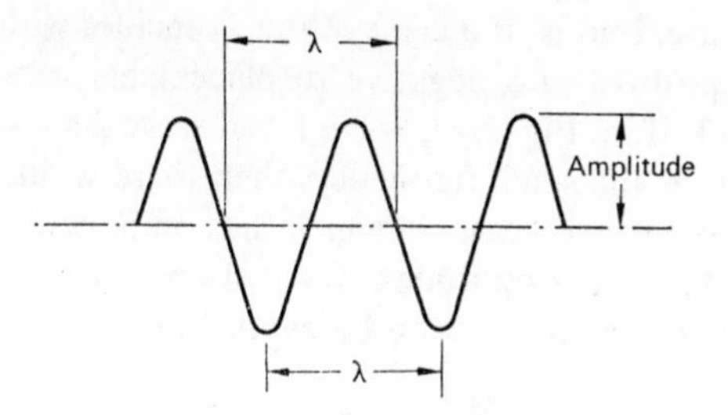

FIG. 2.11. Graphical representation of a wave.

number of disturbances per second at any point in the medium is called the frequency. The velocity with which a wave (disturbance) travels is equal to the product of the wavelength and the frequency,

$$v = f\lambda. \tag{2.44}$$

Example 2.14

Sound waves, which are disturbances in the air, travel through air at a velocity of 344 meters per second. Middle C has a frequency of 264 c/sec. Calculate the wavelength of this note.

$$\lambda = \frac{v}{f}$$

$$= \frac{344 \text{ m/sec}}{264 \text{ sec}^{-1}}$$

$$= 1.3 \text{ m.}$$

If more than one disturbance passes through a medium at the same time, then, where the respective waves meet, the total displacement of the medium is equal to the algebraic sum of the two waves. For example, if two rocks are dropped into a pond, then, if the crests of the two waves should coincide as the waves pass each other, the resulting crest is equal to the height of the two separate crests, and the trough is as deep as the sum of the two individual troughs, as shown in Fig. 2.12. If, on the other hand, the two waves are

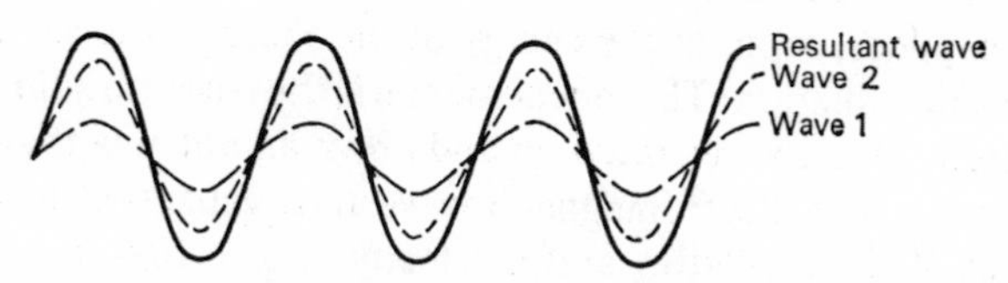

FIG. 2.12. The addition of two waves of equal frequency, and in phase.

exactly out of phase, that is, if a crest of one coincides with the trough of the other, then the positive and negative displacements cancel each other, as shown in Fig. 2.13. If, in Fig. 2.13, wave 1 and wave 2 are of exactly the same amplitude as well as the same frequency, then there would be no net disturbance. For the more general case, in which the component waves are of different frequencies, different amplitudes, and only partly out of phase, complex wave forms may be formed, as seen below in Fig. 2.14.

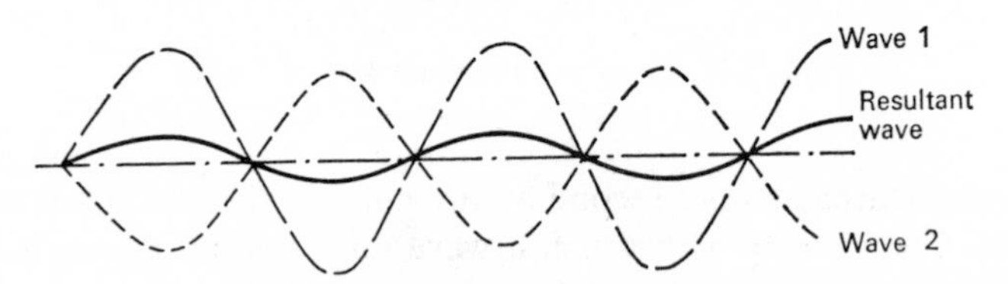

FIG. 2.13. The addition of two waves of equal frequency, but different amplitude, and 180° out of phase.

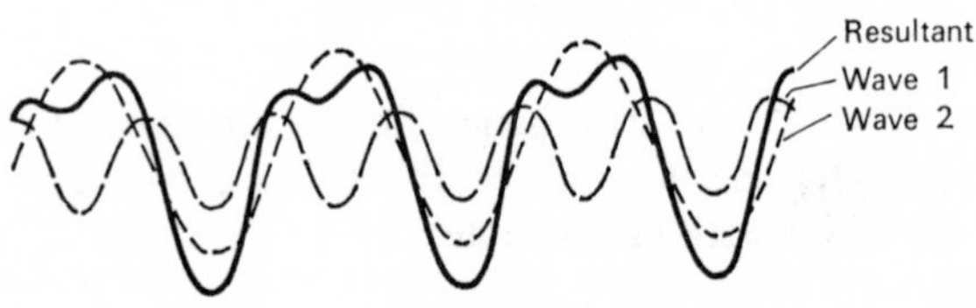

FIG. 2.14. Complex wave formed by the algebraic addition of two different pure waves.

Electromagnetic Waves

In 1820 Oersted, a Danish physicist, discovered that electricity and magnetism were intimately related; that magnetic fields always are generated by electric currents (that is, with moving charges). Faraday, a Scottish experimental physicist, found in 1831 that electricity could be generated from a magnetic field. In 1864 Maxwell, a Scottish theoretical physicist, published a general theory that related the experimental findings of Oersted and Faraday. His theory states that a changing electric field is always associated with a changing magnetic field, and that a changing magnetic field is always associated with an electric field. Together, these electric and magnetic fields constitute an electromagnetic wave.

An electromagnetic wave may be initiated by accelerating a charged particle. When this happens, some of the energy of the charged particle is radiated as electromagnetic radiation. This phenomenon is the basis of radio transmission, in which electrons are accelerated up and down an antenna that is connected to an oscillator. The electromagnetic wave thus generated has a frequency equal to that of the oscillator, and a velocity of 3×10^{10} cm/sec. The waves consist of oscillating electric and magnetic fields that are perpendicular to each other, and are mutually perpendicular to the direction of propagation of the wave, Figure 2.15. The energy carried by the waves depends on the strength of the associated electric and magnetic fields.

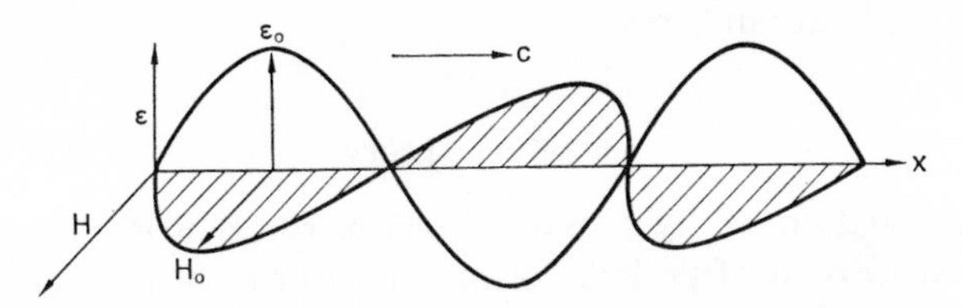

FIG. 2.15. Schematic representation of an electromagnetic wave. The electric intensity E and the magnetic intensity H are at right angles to each other, and the two are mutually perpendicular to the direction of propagation of the wave. The velocity of propagation is C, the electric intensity is $E = E_0 \sin 2\pi/\lambda\ (x\text{-}ct)$, and the magnetic intensity is $H = H_0 \sin 2\pi/\lambda\ (x\text{-}ct)$.

Radio waves, microwaves (radar), infrared radiation, visible light, ultraviolet light and X-rays, are all electromagnetic radiations. They are qualitatively alike, and differ only in wavelength to form a continuous electromagnetic spectrum.

All these radiations are transmitted through space and through the atmosphere (which is, for this purpose, almost empty space) at a velocity very close to 3×10^{10} cm/sec. Because of the reciprocal relationship between frequency and wavelength given by equation (2.47), therefore, specification of either the frequency or the wavelength of an electromagnetic wave is equivalent to specifying both. Wavelengths may range from 5×10^{8} cm for 60 c/sec electric waves through visible light (green light has a wavelength about 5×10^{-5} cm and a frequency of 6×10^{14} c/sec to short wavelength X and gamma radiation (whose wavelengths are on the order of 10^{-8} cm and less). There is no sharp cut-off in wavelength at either end of the spectrum, neither is there a sharp dividing line between the various portions of the electromagnetic spectrum. Each portion blends into the next, and the lines of demarcation shown below in Fig. 2.16 are arbitrarily placed to show the approximate wavelength span of the regions of the electromagnetic spectrum.

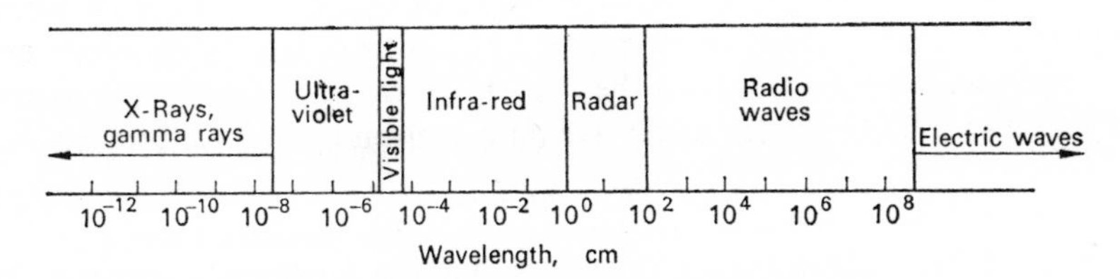

FIG. 2.16. The electromagnetic spectrum.

The health physicist is interested in that portion of the electromagnetic spectrum that is commonly called X-rays or gamma-rays. X-rays and gamma-rays are exactly the same radiations. They differ only in their manner of origin. These radiations are frequently called "ionizing radiations" for reasons which will be discussed later in Chapter 5.

Quantum Theory

The representation of light and other electromagnetic radiations as a continuous succession of periodic disturbances or as a "wave train" in an electromagnetic field is a satisfactory model which can be used to explain many physical phenomena, and which can serve as a basis for the design of apparatus for transmitting and receiving electromagnetic energy. According to this wave theory, or the classical theory as it is often called, the energy transmitted by a wave is proportional to the square of the amplitude of the wave. This model, despite its wide usefulness, nevertheless fails to predict certain phenomena in the field of modern physics. Accordingly, therefore, a new model for electromagnetic radiation was postulated, one which could be used to "explain" certain phenomena which are not amenable to explanation by the wave theory. It should be emphasized that to hypothesize a new theory is not synonymous with abandonment of the old theory. Models or theories are useful in so far as they describe observed phenomena and permit prediction of consequences of certain acts. Philosophically, most scientists subscribe to the school of thought known as "logical positivism". According to this philosophy, there is no way to discover or verify an absolute truth. Science is not concerned with absolute truth or with reality—it is concerned with giving the simplest possible unified description of as many experimental findings as possible. According to this philosophy of science, the acceptance of several different theories on the nature of electromagnetic radiations (or for the nature of matter, electricity, etc.) is perfectly acceptable, provided each theory is capable of describing experimental facts that the others cannot describe.

The more recent theory of the nature of electromagnetic radiation is called the quantum theory. According to the quantum theory, electromagnetic radiation consists of "corpuscles" or "particles" of energy which travel at a velocity of 3×10^{10} cm/sec. Each particle, or "quantum" as it is called, contains a discrete quantity of electromagnetic energy. The energy contained in a quantum is proportional to the "color" of the radiation, or to its frequency when it is considered as a wave, and is given by the relationship

$$E = hf. \tag{2.45}$$

The symbol h is called Planck's constant, and is a fundamental constant of nature whose magnitude in the cgs system is $h = 6.614 \times 10^{-27}$ erg sec. E, in equation (2.48), is given in ergs and f, the frequency, is in cycles per second. Using the universal constant $c = 3 \times 10^{10}$ cm/sec to represent the velocity in space of all electromagnetic radiations, equation (2.44) may be written as

$$f\lambda = c, \tag{2.46}$$

and, by substituting the value of f from equation (2.46) into equation (2.45), we have

$$E = h\frac{c}{\lambda}. \tag{2.47}$$

A quantum of electromagnetic energy is also called a photon. Equations (2.48) and (2.50) show that a photon is completely described when either its energy, or its frequency, or its wavelength is given.

Example 2.15

(a) Radio station KDKA in Pittsburgh, Pennsylvania, broadcasts on a carrier frequency of 1020 kilocycles per second.

(1) What is the wavelength of the carrier frequency?

$$\lambda f = c,$$

$$\lambda = \frac{c}{f} = \frac{3 \times 10^{10} \text{ cm/sec}}{1.02 \times 10^3 \text{ kc/sec} \times 10^3 \text{ cycle/kc}}$$

$$= 2.94 \times 10^4 \frac{\text{cm}}{\text{cycle}}.$$

(2) What is the energy of a KDKA photon, in ergs and in electron volts?

$$E = hf$$

$$= 6.614 \times 10^{-27} \text{ erg/sec} \times 1.02 \times 10^6 \text{ sec}^{-1}$$

$$= 6.75 \times 10^{-21} \text{ ergs},$$

or

$$E = \frac{6.614 \times 10^{-21} \text{ ergs}}{1.6 \times 10^{-12} \text{ ergs/e}V}$$

$$= 4.14 \times 10^{-9} \text{ e}V.$$

(b) What is the energy, in electron volts, of an X-ray photon whose wavelength is 1×10^{-8} cm?

$$E = \frac{hc}{\lambda}$$

$$= \frac{6.614 \times 10^{-27} \text{ erg sec} \times 3 \times 10^{10} \text{ cm/sec}}{10^{-8} \text{ cm} \times 1.6 \times 10^{-12} \text{ ergs/eV}}$$

$$= 1.24 \times 10^{5} \text{ eV}.$$

The wavelengths of X-rays and gamma-rays are very short; on the order of 10^{-8} cm or less. Because of this, and in order to avoid writing the factor 10^{-8} repeatedly, another unit, called the angstrom unit, is used in X-ray work and in health physics. The angstrom unit, which is symbolized by Å, is a unit of length which is equal to 1×10^{-8} cm.

It may seem strange that, having found the wave theory of electromagnetic radiation inadequate to explain certain physical phenomena, part of the wave model should be incorporated into the quantum model of electromagnetic radiation. This dualism, however, seems to be inherent in the "explanations" of atomic and nuclear physics. Mass and energy, particle and wave in the case of electromagnetic energy, and, as will be shown below in subsequent paragraphs, wave and particle in the case of sub-atomic particulates, all seem to be part of a dualism in nature; either aspect of this dualism can be demonstrated in the laboratory by appropriate experiments.

In the case of the photon, some degree of correspondence with the classical picture of electromagnetic radiation can be demonstrated by a simple thought experiment. It is conceivable that, given a very large number of waves that differ from each other in frequency and amplitude, a wave packet, or quantum, could result from reinforcement of the waves over a very limited region, and complete interference ahead and behind the region of reinforcement. Fig. 2.17 is an attempt to portray graphically such a phenomenon.

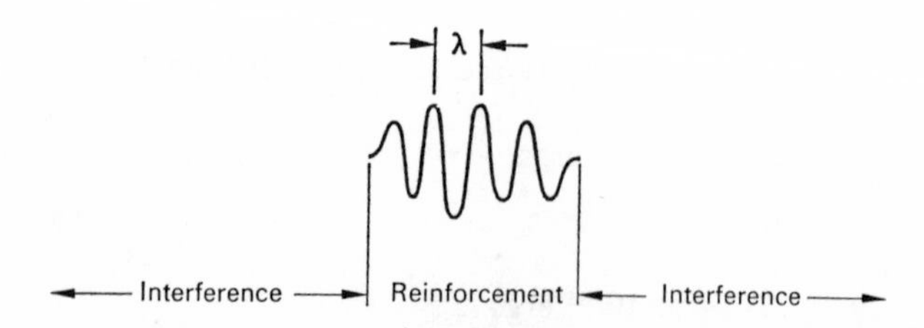

Fig. 2.17. Possible combination of electromagnetic waves to produce a wave packet, a quantum of electromagnetic energy called a *photon*. The energy content of the photon is $E = hc/\lambda$.

The model of a photon shown in Fig. 2.17 combines wave properties and particle properties. Furthermore, this model suggests that a photon may be considered a moving particle that is guided in its path by the waves which combine to produce the particle. The "mass" of a photon may be found by equating its energy with the relativistic energy of a moving particle:

$$E = hf = mc^2 \tag{2.48A}$$

$$\therefore \quad m = \frac{hf}{c^2}. \tag{2.48B}$$

The "momentum" p, of the photon, therefore, is

$$p = mc = \frac{hf}{c}, \tag{2.49}$$

and, if the value of f from equation (2.46) is substituted into equation (2.46), we have

$$p = \frac{h}{\lambda}. \tag{2.50}$$

The duality of nature was further emphasized when two experimenters in the Bell Telephone Laboratories, Davisson and Germer, found a beam of electrons to behave like a wave. When they bombarded a nickel crystal with a fine beam of electrons whose kinetic energy was 54 eV, they found the electrons to be reflected only in certain directions, rather than isotropically as expected. Only at angles of 50 and 0 (backscatter) were scattered electrons detected. Such a behaviour was unexplainable if the bombarding electrons were considered to be particulate in nature. By assuming them to be waves, however, the observed distribution of scattered electrons could easily be explained. The answer simply was that the electron "waves" underwent destructive interference at all angles except those at which they were observed; there the electron waves reinforced each other.

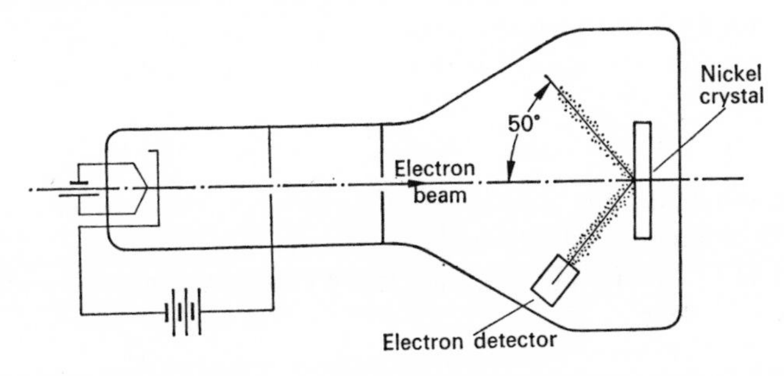

FIG. 2.18. Experiment of Davisson and Germer suggesting the wave nature of electrons.

Matter Waves

In 1924 the French physicist, Louis de Broglie, suggested that not only electrons, but all moving particles were associated with wave properties. The length of these waves, according to de Broglie, was inversely proportional to the momentum of the moving particle and, furthermore, that the constant of proportionality was Planck's constant, h. The length of these matter waves is given by equation (2.50). Since momentum p is equal to mv, equation (2.50) may be rewritten as

$$\lambda = \frac{h}{mv}. \tag{2.51}$$

Here, too, as in the case of the photon, we have in the same equation properties characteristic of particles and properties characteristic of waves. The mass of the moving particle, m, in equation (2.51), represents a particle concept, while λ, the wavelength of the "matter" wave associated with the moving particle, is, quite clearly, a wave concept.

The fact that moving particles possess wave properties is the basis of the electron microscope. In any kind of a microscope, whether used with beams of light waves or beams of de Broglie "matter" waves, the resolving power (the ability to separate two points that are close together, or the ability to see the edges of a very small object sharply and distinctly) is an inverse function of the wavelength of the probing beam; a shorter wavelength permits better resolution than a longer wavelength. For this reason, optical microscopes are usually illuminated with blue light, since blue is near the short wavelength end of the visible spectrum. Under optimum conditions, the limit of resolution of an optical microscope, using blue light whose wavelength is 4000 Å, is on the order of 0.1 micron. Since high velocity electrons are associated with very much shorter wavelengths than blue light, an electron microscope, which uses a beam of electrons instead of a beam of light, has a much greater resolving power than even the best optical microscope. Since useful magnification is limited by resolving power, the increased resolution possible with an electron microscope permits much greater useful magnification than could be obtained with the optical microscope.

Example 2.16

What is the de Broglie wavelength of an electron that is accelerated across a potential difference of 100,000 V?

According to equation (2.20), the kinetic energy of a particle moving with a velocity $v = \beta c$ is

$$E_k = m_0 c^2 \left(\frac{1}{\sqrt{(1 - \beta^2)}} - 1 \right),$$

from which it follows that

$$\frac{m_0c^2}{E_k + m_0c^2} = \sqrt{(1 - \beta^2)}. \tag{2.52}$$

In Example 2.8 it was shown that m_0c^2 for an electron is 5.1×10^5 eV. We therefore have, after substituting the appropriate numerical values into equation (2.52),

$$\frac{5.1 \times 10^5 \text{ eV}}{1 \times 10^5 \text{ eV} + 5.1 \times 10^5 \text{ eV}} = \sqrt{(1 - \beta 2)},$$

$$0.837 = \sqrt{(1 - \beta^2)},$$

$$(0.837)^2 = 1 - \beta^2,$$

$$\therefore \quad \beta = \sqrt{\{1 - (0.837)^2\}}$$

$$= 0.55.$$

The momentum of the electron is

$$p = mv = \frac{m_0}{\sqrt{(1 - \beta^2)}} \beta C$$

$$= \frac{9.11 \times 10^{-28} \text{ g} \times 0.55 \times 3 \times 10^{10} \text{ cm/sec}}{0.837}$$

$$= 1.795 \times 10^{-17} \text{ g cm/sec},$$

and the de Broglie wavelength, consequently, is

$$\lambda = \frac{h}{p}$$

$$= \frac{6.614 \times 10^{-27} \text{ erg sec}}{1.795 \times 10^{-17} \text{ g cm/sec}}$$

$$= 3.69 \times 10^{-10} \text{ cm}$$

$$= 0.0369 \text{ Å}.$$

The wave-particle dualism may seem especially abstract when it is extended to include particles of matter whose existence may be confirmed by our experience, by our senses, and by our intuition. At first it may seem that the wave properties of particles are purely mathematical figments of the imagination whose only purpose is to quantitatively describe experimental phenomena that are not otherwise amenable to theoretical analysis. In this regard, of course, the wave properties of matter serve a useful purpose. However, it is possible to give a physical interpretation of matter waves. According to the physical theory of waves, the intensity of a wave is proportional to the square of the amplitude of the wave. In 1926 Max Born applied this concept to the wave properties of matter. In the case of a beam of electrons, the square of the amplitude of the electron waves was postulated to be proportional to the intensity of the beam, or to the number of electrons per square centimeter per second incident on a plane perpendicular to the direction of the beam. If this beam strikes a crystal, as in the case shown in Fig. 2.18, then the reflected electron waves either reinforce or interfere with each other, and cause the observed interference pattern. The waves are reinforced most strongly at certain points, and at other points are exactly out of phase, thereby cancelling each other out. Where reinforcement occurs, a maximum electron density is observed, whereas interference results in a decrease in electron density. The exact distribution of the interference pattern is determined by the crystalline structure of the scatterer, and therefore is uniquely representative of the scatterer. This method of "finger-printing" substances is the basis of electron diffraction methods used by physical chemists to identify unknowns.

For a beam of electrons, this relationship between wave and particle properties seems reasonable. In the case of a single electron, however, the electron-wave must be interpreted differently. If instead of bombarding the crystal in Fig. 2.18 with a beam of electrons, the electrons were fired at the crystal one at a time and each separate scattered electron was detected, it would be found that the single electrons would be scattered through the same angles as the beam of electrons. However, the exact coordinates of any particular scattered electron would not be known until it is "seen" by the electron detector. After firing the same total number of single electrons as those in a beam, and plotting the positions of each scattered electron, exactly the same "interference" pattern would be observed as in the case of a beam of electron waves. This experiment shows that, although the behaviour of a single electron cannot be precisely predicted, the behaviour of a group of electrons can be predicted. From this it may be inferred that the square of the amplitude at any point of the curve of position versus electron intensity gives the probability of any single electron being scattered through that point.

Uncertainty Principle

It should be noted that the implications of Born's probability interpretation of the wave properties of matter are truly revolutionary. According to classical physics, if the mass and velocity of a particle, as well as the external forces acting on the particle, are known, then, in principle at least, all of its future actions could be precisely predicted. According to the wave-mechanical model, however, precise predictions are not possible—probability replaces certainty. Heisenberg, in 1927, developed these ideas still further, and postulated his uncertainty theory, in which he said that it is impossible, in principle, to know both the exact location and momentum of a moving particle at any point in time. Either one of these two quantities could be determined to any desired degree of accuracy. The accuracy of the other quantity, however, decreases as precision of the first quantity increases. The product of the two uncertainties Heinsenberg showed to be proportional to Planck's constant h, and to be given by the relationship

$$\Delta x \times \Delta p = \frac{h}{2\pi}. \tag{2.53}$$

It should be emphasized that the uncertainty expressed by Heisenberg is not due to faulty measuring tools or technics, or to experimental errors; it is a fundamental limitation of nature which is due to the fact that any measurement must disturb the object being measured. Precise knowledge about anything can therefore never be attained. In many instances, this inherent uncertainty can be understood intuitively. When a student is being tested, for example, it is common knowledge that he may become tense or suffer some other physio-psychological stress which may cast some doubt on the accuracy of the test results. In any case, the testor cannot be absolutely certain that the psychologic strain of the examination did not influence the actions of the student and thereby influence the results of the examination. The uncertainty expressed by equation (2.53) can be illustrated by an example in which we try to locate a particle by looking at it. To "see" a particle means that light is reflected from the particle into the eye. When the quantum of light strikes the particle and is reflected toward the eye, some energy is transferred from the quantum to the particle, thereby changing both the position and momentum of the particle. The reflected photon tells the observer where the particle *was*, not where it *is*.

Problems

1. Two blocks, of mass 100 g and 200 g, approach each other along a frictionless surface. at velocities 40 and 100 cm/sec respectively. If the blocks collide, and remain together, calculate their joint velocity after the collision.

2. A bullet whose mass is 50 g travels at a velocity of 500 m/sec. It strikes a rigidily fixed wooden block, and penetrates a distance of 20 cm before coming to a stop.

(a) What is the deceleration of the bullet?
(b) What was the decelerating force?
(c) What was the initial momentum of the bullet?
(d) What was the impulse of the collision?

3. Compute the mass of the earth, assuming it to be a sphere of 25,000 miles circumference, if at its surface it attracts a mass of 1 g with a force of 980 dynes.

4. An automobile weighing 2000 kg, and going at a speed of 60 km/hr, collides with a truck weighing 5 metric tons that was moving at right angles to the direction of the auto, at a speed of $41\frac{1}{2}$ km/hr. If the two vehicles become joined in the collision, what is the magnitude and direction of their velocity after the collision?

5. A small electrically charged sphere of mass 0.1 g hangs by a thread 100 cm long between two parallel vertical plates spaced 6 cm apart. If 100 volts are across the plates, and if the charge on the sphere is 10^{-9} coulombs, what angle does the thread make with the vertical direction?

6. A capacitor has a capacitance of 10 μF. How much charge must be removed to cause a decrease of 20 volts across the capacitor?

7. A small charged particle whose mass is 0.01 g remains stationary in space when it is placed in an upward directed electric field of 10 V/cm. What is the charge on the particle?

8. A 1-micron diameter droplet of oil, whose specific gravity is 0.9, is introduced into an electric field between two large parallel plates, separated by 5 mm, across which is placed a potential difference V volts. If the oil droplet carries a net charge of 100 electrons, how many volts must be across the plates if the droplet is to remain suspended in the space between the plates?

9. A diode vacuum tube consists of a cathode and an anode spaced 5 mm apart. If 300 volts are placed across the electrodes,

(a) What is the velocity of an electron midway between the electrodes, and at the instant of striking the plate, if the electrons are emitted by the cathode with zero velocity?

(b) If the plate current is 20 mA, what is the average force exerted on the anode?

10. Calculate the ratios v/c and m/m_0 for a 1-MeV electron and for a 1-MeV proton.

11. Assuming an uncertainty in the momentum of an electron equal to one half its momentum, calculate the uncertainty in position of a 1-MeV electron.

12. If light quanta have mass, they should be attracted by the earth's gravity. To test this hypothesis a parallel beam of light is directed horizontally at a receiver 10 miles away. How far would the photons have fallen during their flight, to the receiver, if quanta have mass?

13. The maximum wavelength of U.V. light observing the photoelectric effect in tungsten is 2730 Å. What will be the kinetic energy of photoelectrons produced by U.V. radiation of 1500 Å?

14. Calculate the uncertainty in position of an electron that was accelerated across a potential difference of 100,000 $\pm$ 100 volts.

15. (a) What voltage is required to accelerate a proton from zero velocity to a velocity corresponding to a de Broglie wavelength of 0.01 Å?

(b) What would be the kinetic energy of an electron with this wavelength?

(c) What is the energy of an X-ray photon whose wavelength is 0.01 Å?

16. A current of 25 mA flows through 25-gauge wire, 0.0179 in. (17.9 mils) in diameter. If there are 5×10^{22} free electrons per cm^3 in copper, calculate the average speed with which electrons flow in the wire.

17. An electron starts at rest on the negative plate of a parallel capacitor, and is accelerated by a potential of 1000 volts across a gap of 1 cm.

(a) With what velocity does the electron strike the positive plate?

(b) How long does it take the electron to travel the 1 cm distance?

18. A cylindrical capacitor is made of two coaxial conductors—the outer one has a diameter of 20.2 mm and the diameter of the inner one is 0.2 mm. The inner conductor is 1000 volts positive with respect to the outer conductor. Repeat parts (a) and (b) of problem 17, and compare the results to those of problem 17.

Suggested References

1. SEARS, F. W. and ZEMANSKY, M. W.: *University Physics*, Addison-Wesley, Cambridge-1955.
2. BORN, M.: *Atomic Physics*, Hafner, New York, 5th edition.
3. FEYNMAN, R. P., LEIGHTON, R. B. and SANDS, M.: *The Feynman Lectures on Physics*, Vols. I, II, and III, Addison-Wesley, Reading, 1965.
4. ROGERS, E. M.: *Physics for the Inquiring Mind*, D. Van Nostrand, Princeton, 1960.
5. FERRENCE, M., LEMON, H. B. and STEPHENSON, R. J.: *Analytical Experimental Physics*, University of Chicago Press, Chicago, 1956.

CHAPTER 3

ATOMIC AND NUCLEAR STRUCTURE

Atomic Structure

Matter, as we ordinarily know it, is electrically neutral. Yet the fact that matter can be easily electrified—by walking with rubber-soled shoes on a carpet, by sliding across a plastic auto-seat cover when the atmospheric humidity is low, and by numerous other commonplace means—testifies to the fact that matter is electrical in nature. The manner in which the positive and negative electrical charges were held together was a matter of concern to the physicists of the early twentieth century.

Rutherford's Nuclear Atom

The British physicist Rutherford had postulated, in 1911, that the positive charge in an atom was concentrated in a central massive point called the nucleus, and that the negative electrons were situated at some remote points, about one angstrom unit distant from the nucleus. In one of the all-time classical experiments of physics, two of Rutherford's students, Geiger and Marsden, in 1913 tested the validity of this hypothesis by bombarding an extremely thin (6×10^{-5} cm) gold foil with highly energetic, massive, positively charged projectiles called alpha particles. These projectiles, whose kinetic energy was 7.68 MeV, were emitted from the radioactive substance polonium. If Rutherford's idea had merit, then it was expected that most of the alpha particles would pass straight through the thin gold foil. Some of the alpha particles, however, those that would pass by a gold nucleus sufficiently close to permit a strong interaction between the electric field of the alpha particle and of the positive point charge in the gold nucleus, would be deflected as a result of the repulsive force between the alpha particle and the gold nucleus. An angular scan with an alpha particle detector about the point where the beam of alpha particles traversed the gold foil, as shown in Fig. 3.1, permitted Geiger and Marsden to measure the alpha particle intensity at various scattering angles. The experimental results verified Rutherford's hypothesis. Although most of the alpha particles passed undeflected through the gold foil, a continuous distribution of scattered alpha particles was observed as the alpha particle detector traversed a scattering angle from 0° to 150°. Similar results were obtained with other scatterers. The observed angular distributions of the scattered alpha particles agreed with those predicted by Rutherford's

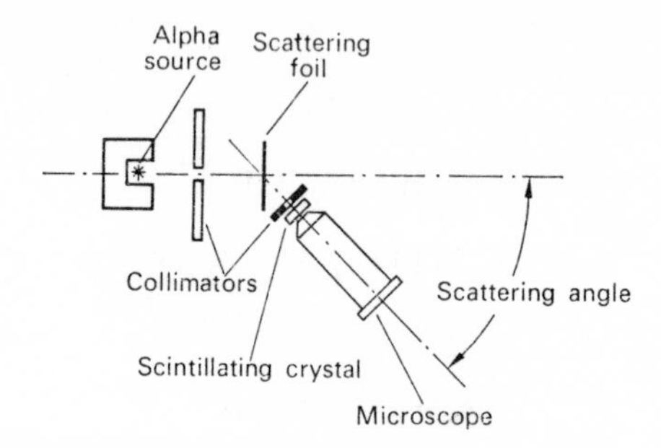

FIG. 3.1. Diagram showing principle of Rutherford's alpha-particle-scattering experiment. The alpha source, its collimator, and the scattering foil are fixed; the alpha particle detector, consisting of a collimator, a ZnS scintillating crystal, and a microscope rotates around the point where the alpha beam strikes the scattering foil.

theory, thereby providing experimental evidence for the nuclear atom. Matter was found to consist mainly of open space. A lattice of atoms, consisting of positively charged nuclei about 5×10^{-13} cm in diameter, and separated by distances on the order of 10^{-8} cm, was inferred from the scattering data. Detailed analyses of many experimental data later showed the nucleus to have a radius of

$$r = 1.2 \times 10^{-13} A^{1/3} \text{ cm}, \tag{3.1}$$

where A is the atomic mass number. The number of unit charges in the nucleus (1 unit charge is 4.8×10^{-10} statcoulombs) was found to be approximately equal to the atomic number of the atom, and to about one-half the atomic weight. Later work in Rutherford's laboratory by Moseley and by Chadwick in 1920 showed the number of positive charges in the nucleus to be exactly equal to the atomic number. These data implied that the proton, which carries one charge unit, is a fundamental building block of nature. The outer periphery of the atom, at a distance of about 5×10^{-9} cm from the nucleus, was thought to be formed by electrons, equal in number to the proton within the nucleus, distributed around the nucleus. However, no satisfactory theory to explain this structure of the atom was postulated by Rutherford. Any acceptable theory must answer two questions: first, how are the electrons held in place outside the nucleus despite the attractive electrostatic forces, and second, what holds the positive charges in the nucleus together in the face of the repulsive electrostatic forces?

Bohr's Atomic Model

A simple solar system-like model, with the negative electrons revolving about the positively charged nucleus, seemed inviting. According to such a model, the attractive force between the electrons and the nucleus could be balanced by the centrifugal force due to the circular motion of the electrons. Classical electromagnetic theory, however, predicted that such an atom is unstable. The electrons revolving in their orbits undergo continuous radial acceleration. Since classical theory predicts that charged particles radiate electromagnetic energy whenever they experience a change in velocity (either in speed or in direction), it follows that the orbital electrons should eventually spiral into the nucleus as they lose their kinetic energy by radiation. (The loss of kinetic energy by this mechanism is called *bremsstrahlung*, and is very important in health physics. This point will be taken up in more detail in later chapters.) The objection to the solar system-like atomic model, based on the argument of energy loss due to radial acceleration, was overcome in 1913 by the Danish physicist Niels Bohr simply by denying the validity of classical electromagnetic theory in the case of motion of orbital electrons.

Although this was a radical step, it was by no means without precedent. The German physicist, Max Planck, had already shown that a complete description of black body radiation could not be given with classical theory. To do

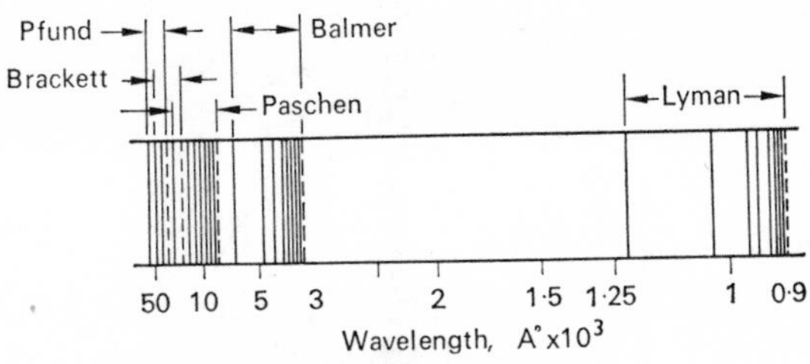

FIG. 3.2. Hydrogen spectrum.

this, he postulated a quantum theory of radiation, in which electromagnetic radiations are assumed to be particles whose energy depends only on the frequency of the radiation. Bohr adopted Planck's quantum theory, and used it to develop an atomic model that was consistent with the known atomic phenomena. The main source of experimental data from which Bohr inferred his model was atomic spectra. Each element, when excited by the addition of energy, radiates only certain colors that are unique to it. (This is the basis of "neon" signs. Neon, sealed in a glass tube, emits red light as a consequence of electrical excitation of the gas. Mercury vapor is used in the same way to produce blue light.) Because of the discrete nature of these colours, atomic spectra are called "sharp-line" spectra to distinguish them from white light or black body radiation, which has a continuous spectrum. Hydrogen, for example, emits electromagnetic radiation of several distinct frequencies when it is excited, as shown in Fig. 3.2. Some of these radiations are in the ultraviolet region, some are in the visible light region, and some are in the infrared region. The spectrum of hydrogen consists of several well-defined series of lines whose wavelengths were described empirically by physicists of the late nineteenth century by the equation

$$\frac{1}{\lambda} = R\left(\frac{1}{n_1^2} - \frac{1}{n_2^2}\right), \tag{3.2}$$

where R is a constant, named after Rydberg, whose numerical value is 109,700 cm^{-1}, n_1 is any whole number equal to or greater than 1, and n_2 is a whole number equal to or greater than $n_1 + 1$. The Lyman series, which lies in the ultraviolet region, is the series in which $n_1 = 1$, and $n_2 = 2, 3, 4, \ldots$. The longest wavelength in this series, obtained by setting n_2 equal to 2 in equation (3.2), is 1215 Å. Succeeding lines, when n_2 is 3 and 4, are 1026 and 972 Å respectively. The shortest line, called the series limit, is obtained by solving equation (3.2) with n_2 equal to infinity; in this case, the wavelength of the most energetic photon is 911 Å.

Bohr's atomic model, Fig. 3.3, is based on two fundamental postulates:

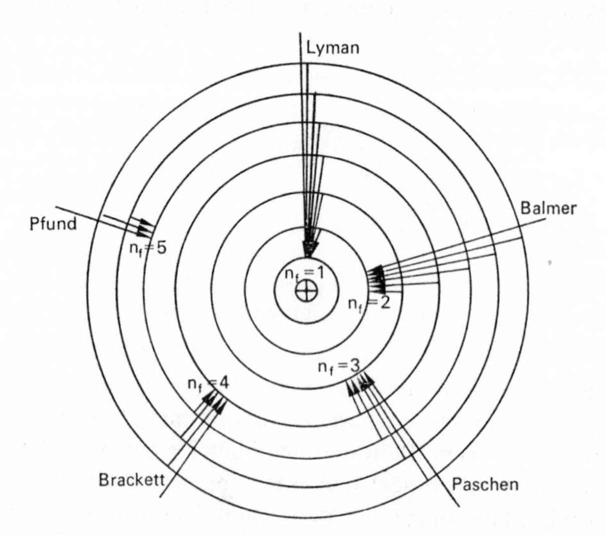

FIG. 3.3. Bohr's model of the hydrogen atom, showing the origin of the various series of lines seen in the hydrogen spectrum.

1. The orbital electrons can revolve around the nucleus only in certain fixed radii, called stationary states. These radii are such that the angular momentum of the revolving electrons must be integral multiples of $h/2\pi$,

$$mvr = \frac{nh}{2\pi}, \tag{3.3}$$

where m is the mass of the electron, v is its linear velocity, r is the radius of revolution, h is Planck's constant, and n is any integer 1, 2, . . .

2. A photon is emitted only when an electron falls from one orbit to another orbit of lower energy. The energy of the photon is equal to the difference between the energy levels of the electrons in the two orbits.

$$hf = E_2 - E_1, \tag{3.4A}$$

$$f = \frac{E_2}{h} - \frac{E_1}{h}, \tag{3.4B}$$

where f is the frequency of the emitted photon, E_2 and E_1 are the high- and low-energy orbits respectively.

When the electron revolves around the nucleus, the electrostatic force of attraction between the electron and the nucleus is balanced by the centrifugal force due to the revolution of the electron:

$$\frac{Ze . e}{r^2} = \frac{mv^2}{r}. \tag{3.5}$$

Z is the atomic number of the atom and e is the electronic and protonic charge; Ze, therefore, is the charge on the nucleus. Substituting the value for v from equation (3.3) into equation (3.5), and solving for r we have,

$$r = \frac{n^2 h^2}{4\pi^2 m e^2 Z}. \tag{3.6}$$

Equation (3.6) gives the radii of the electronic orbits that will satisfy the condition for the stationary states when whole numbers are substituted for n. The normal condition of the atom, or the ground state, is that state for which n is 1. In the ground state the atom is in its lowest possible energy state and therefore in its most stable condition. Transitions from the ground state to higher energy orbits are possible through the absorption of sufficient energy to raise the electron to a larger radius.

The energy in any orbit may be calculated by considering the kinetic energy of the electron due to its motion around the nucleus and the potential energy due to its position in the electric field of the nucleus. Since the kinetic energy of the electron is equal to $\frac{1}{2} mv^2$ (the electron does not revolve rapidly enough to consider relativistic effects), then, from equation (3.5), we have

$$E_k = \frac{1}{2} mv^2 = \frac{Ze^2}{2r}. \tag{3.7}$$

The potential energy is, from equation (2.29),

$$E_p = \frac{Ze}{r}(-e) = -\frac{Ze^2}{r}. \tag{3.8}$$

The total energy of the electron is

$$E = E_k + E_p$$

$$= \frac{Ze^2}{2r} - \frac{Ze^2}{r} = -\frac{Ze^2}{2r}. \tag{3.9}$$

The total energy given by equation (3.9) is negative simply as a result of the convention discussed in Chapter 2. By definition, the point of zero potential energy was set at an infinite distance from the nucleus. Since the force between the nucleus and the electron is attractive, it follows that at any point closer than infinity, the potential energy must be less than that at infinity, and therefore must have a negative numerical value.

The total energy in any permissible orbit is found by substituting the value for the radius, equation (3.6), into equation (3.9):

$$E = \frac{-2\pi^2 mZ^2e^4}{h^2} \times \frac{1}{n^2}. \tag{3.10}$$

Equation (3.10) may now be substituted into equation (3.4B) to get the frequency of the "light" that is radiated from an atom when an electron falls from an excited state into one of lower energy. Letting n_f and n_i represent respectively the orbit numbers for the lower and higher levels, we have for the frequency of the emitted radiation:

$$f = \frac{2\pi^2 mZ^2e^4}{h^3}\left(\frac{1}{n_f^2} - \frac{1}{n_i^2}\right), \tag{3.11}$$

and the reciprocal of the wavelength of this radiation is

$$\frac{1}{\lambda} = \frac{2\pi^2 mZ^2e^4}{ch^3}\left(\frac{1}{n_f^2} - \frac{1}{n_i^2}\right). \tag{3.12}$$

When numerical values are substituted into the first term of equation (3.12), the numerical value for Rydberg's constant is obtained.

Excitation and Ionization

The Bohr equation may be illustrated for the case of hydrogen by substituting $Z = 1$ into the equations. The radius of the ground state is found to be, from equation (3.6), 0.526×10^{-8} cm. The wavelength of the light emitted when the electron falls from the first excited state, $n_i = 2$, to the ground state, $n_f = 1$, may be calculated by substituting these values into equation (3.12), and is found to be 1215 Å. The energy of this photon is

$$E = \frac{hc}{\lambda} = \frac{6.6 \times 10^{-27} \text{ erg sec} \times 3 \times 10^{10} \text{ cm/sec}}{1.215 \times 10^{-5} \text{ cm} \times 1.6 \times 10^{-12} \text{ ergs/eV}} = 10.2 \text{ eV}.$$

This same amount of energy, 10.2 eV, is necessary to excite hydrogen to the first excited state. Precisely this amount of energy, no more and no less, may be used.

When a sufficient amount of energy is imparted to raise it to an infinitely great orbit, that is, to remove the electron from the electrical field of the nucleus, the atom is said to be *ionized*, and the negative electron together with the remaining positively charged atom, is called an *ion pair*. This process is called *ionization*. Ionization or excitation may occur when either a photon or a charged particle, such as an electron, a proton, or an alpha particle, collides with an orbital electron. This mechanism is of great importance in health physics because it is the avenue through which energy is transferred from radiation to matter. When living matter is irradiated, the primary event in the sequence of events leading to biological damage is either excitation or ionization. The *ionization potential* of an element is the amount of energy necessary to ionize the least tightly bound electron in an atom of that element. To remove a second electron requires considerably more energy than removal of the first electron. For most elements, the first ionization potential is on the order of several electron volts. In the case of hydrogen, the ionization potential may be calculated from equation (3.11) by setting n_l equal to infinity:

$$I = hf = \frac{2\pi^2 m Z^2 e^4}{1.6 \times 10^{-12} \text{ erg/eV}. h^2} \left(\frac{1}{1} - \frac{1}{\infty}\right)$$

$$= \frac{2\pi^2 \times 9.11 \times 10^{-28}\text{g} \times 1 \times (4.8 \times 10^{-10} \text{ SC})^4}{1.6 \times 10^{-12} \text{ erg/eV } (6.625 \times 10^{-27} \text{ erg sec})^2}$$

$$= 13.6 \text{ eV}.$$

A collision in which a rapidly moving particle transfers much more than 13.6 electron volts to the orbital electron of hydrogen results in the ionization of the hydrogen. The excess energy above 13.6 eV that is transferred in this collision appears as kinetic energy of the electron and of the resulting positive ion which recoils under the impact of the collision, in accordance with the requirements of the conservation of momentum. Such inelastic collisions occur only if the incident particle is sufficiently energetic, about 100 eV or greater, to meet this requirement. In those instances where the energy of the particle is insufficient to meet this requirement, an elastic collision with the atom as a whole occurs.

When a photon whose energy is great enough to ionize an atom collides with a tightly bound orbital electron, the photon disappears and the electron is ejected from the atom with a kinetic energy equal to the difference between the energy of the photon and the ionization potential. This mechanism is called the *photoelectric effect*, and is described by the equation

$$E_{pe} = hf \cdot \phi, \tag{3.13}$$

where E_{pe} is the kinetic energy of the photoelectron (the ejected electron), hf is the photon energy, and ϕ is the ionization potential (commonly called the work function). Einstein won the Nobel prize in 1921 for his work on the theoretical aspects of the photoelectric effect.

Example 3.1

An ultraviolet photon whose wavelength is 2000 Å strikes the outer orbital electron of sodium; the ionization potential of the atom is 5.41 eV. What is the kinetic energy of the photoelectron?

The energy of the photon is

$$\frac{hf \text{ ergs}}{1.6 \times 10^{-12} \text{ ergs/eV}} = \frac{hc}{1.6 \times 10^{-12}\,\lambda}$$

$$= \frac{6.625 \times 10^{-27} \text{ erg sec} \times 3 \times 10^{10} \text{ cm/sec}}{1.6 \times 10^{-12} \text{ ergs/eV} \times 2 \times 10^{-5} \text{ cm}}$$

$$= 6.2 \text{ eV}.$$

From equation (3.13), the kinetic energy of the photoelectron is found to be

$$E_{pe} = 6.20 - 5.41 = 0.79 \text{ eV}.$$

Modifications of the Bohr Atom

The atomic model proposed by Bohr "explains" certain atomic phenomena for hydrogen and for hydrogen-like atoms, such as singly ionized helium, He^{+} and doubly ionized lithium, Li^{++}. Calculation of spectra for other atoms according to the Bohr model is complicated by the screening effect of the other electrons, which effectively reduces the electrical field of the nucleus, and by electrical interactions among the electrons. The simple Bohr theory described above is inadequate even for the hydrogen atom. Examination of the spectral lines of hydrogen with spectroscopes of very high resolving power shows the lines to have a fine structure. The spectral lines are in reality each made of several lines very close together. This observation implies the existence of sub-levels of energy within the principal energy levels, and that these sub-levels were very close together. These sub-levels can be explained by assuming the orbits to be elliptical instead of circular, with the nucleus at one of the foci, and that ellipses of different eccentricities have slightly different energy levels. For any given principal energy level, the major axes of these ellipses are the same, eccentricity varies only by changes in the length of the minor axes. The eccentricity of these ellipses is restricted by quantum conditions. The angular momentum of an electron rotating in an elliptical orbit is an integral multiple of $h/2\pi$, as in the case of the circular Bohr orbit. However, the numerical value for this multiple is not the same as that for the circular orbit. In the case of the circular orbit, this multiple, or quantum number, is called the principal quantum number and is given the symbol n. For the elliptical orbit, the multiple is called the azimuthal quantum number, usually symbolized by the letter l, and may be any integral number between 0 and $n-1$ inclusive.

Elliptical orbits alone are insufficient to account for observed spectral lines. To explain the very fine structure, it was necessary to postulate that each orbital electron spins about its own axis in the same manner as the earth spins about its axis as it revolves around the sun. The angular momentum due to this spin also is quantized, and can only have a value equal to one-half a unit of angular momentum:

$$s = \pm \frac{1}{2}\frac{h}{2\pi}. \tag{3.14}$$

The orbital electron can spin in only one of two directions with respect to the direction of its revolution about the nucleus; either in the same direction or in the opposite direction. This accounts for the plus and minus signs in equation (3.14). The total angular momentum of the electron is therefore equal to the vector sum of the orbital and spin angular momenta.

One more fact must be considered before the description of the atom is complete. A closed loop through which an electric current flows has a magnetic moment that is proportional to the product of the current and the area of the loop. The electron revolving in its orbit is equivalent to a closed current-carrying loop, and therefore has a magnetic moment. The fact that the electron spins results in an additional magnetic moment, which may be either positive or negative, depending on the direction of spin relative to the direction of the orbital motion. The total magnetic moment is therefore equal to the sum of the orbital and spin magnetic moments. If the atoms of any substance are placed in a strong magnetic field, the orbital electrons, because of their magnetic moments, will orient themselves in definite directions relative to the applied magnetic field. These directions are such that the component of the vector representing the orbital angular momentum, l, that is parallel to the magnetic field must have an integral value of angular momentum. This integral number, which is given the symbol m, is called the magnetic quantum number; it can have numerical values ranging from l, l–1, l–2, . . . , 0 . . ., $-(l\text{–}2)$, $-(l\text{–}1)$, $-l$.

To completely describe an atom, it is necessary to specify four quantum numbers, which have the values given below, for each of the orbital electrons.

Symbol	Name	Value
n	principal quantum number	1, 2, . . .
l	azimuthal quantum number	0 to $n-1$
m	magnetic quantum number	$-l$ to 0 to $+l$
s	spin quantum number	$-\frac{1}{2}$, $+\frac{1}{2}$

By using Bohr's atomic model, and by assigning all possible numerical values to these four quantum numbers according to certain rules, it is possible to construct the periodic table of the elements.

Periodic Table of the Elements

The periodic table of the elements may be constructed with Bohr's atomic model by application of the *Pauli Exclusion Principle.* This principle states that no two electrons in any atom may have the same set of four quantum numbers. Hydrogen, the first element, has a nuclear charge of +1, and therefore has only one electron. Since the principal quantum number of this electron must be 1, l and m must be 0, and the spin quantum number s may be either plus or minus $\frac{1}{2}$. If now we go to the second element, helium, we must have two orbital electrons, since helium has a nuclear charge of +2. The first electron in the helium atom may have the same set of quantum numbers as the electron in the hydrogen atom. The second electron, however, must differ. This difference can be only in the spin, since we may have two different spins for the set of quantum numbers $n = 1$, $l = 0$, and $m = 0$. This second electron exhausts all the possibilities for $n = 1$. If now a third electron is added when we go to atomic number 3, lithium, it must have the principal quantum number 2. In this principal energy level, the orbit may be either circular or elliptical, that is, the azimuthal quantum number l may be either 0 or 1. In the case of $l = 0$, the magnetic quantum number m can only be equal to 0; when $l = 1$, m may be either -1, 0, or $+1$. Each of these quantum states may contain two electrons, one each with spins of plus and minus $\frac{1}{2}$. Eight different electrons, each with its own unique set of quantum numbers, are therefore possible in the second principal energy level. These eight different possibilities are utilized by the elements Li, Be, B, C, N, O, F, and Ne, atomic-numbered elements 3–10 inclusive. The additional electron for sodium, atomic number 11, must have the principal quantum number $n = 3$. By assigning all the possible combinations of the four quantum numbers to the electrons in the third principal energy level, it is found that eighteen electrons are possible. These energy levels are not filled successively as were those in the K and L shells. (The principal quantum levels corresponding to $n = 1, 2, 3, 4, 5, 6$, and 7 are called the K, L, M, N, O, P, and Q shells, respectively.) No outermost electron shell contains more than eight electrons. After the M shell contains eight electrons, as in the case of argon, the next element in the periodic table, potassium, atomic number 20, starts another principal energy level with one electron in the N shell. Subsequent elements then may add electrons either in the M or in the N shells, until the M shell contains its full complement of eighteen electrons. No electrons appear in the O shell until the N shell has eight electrons. The maximum number of electrons that may exist in any principal energy level is given by the product $2n^2$, where

n is the principal quantum number. Thus, the O shell may have a maximum of $2 \times 5^2 = 50$ electrons.

The fact that no outermost electron shell contains more than eight electrons is responsible for the periodicity of the chemical properties of many elements, and is the physical basis for the periodic table. Since chemical reactions involve the outer electrons, it is not surprising that atoms with similar outer electronic structures should have similar chemical properties. For example, Li, Na, K, Rb, and Cs behave chemically similarly because each of these elements has only one electron in its outermost orbit. The inert gases, He, Ne, A, Kr, Xe, and Rn have similar electronic structures too; they all have eight electrons in their outermost shells, and have all the inside shells filled to their theoretical maxima. Because all their electron shells are filled, these elements cannot undergo any chemical reactions. The elements are thought to have the electronic configurations given in Table 3.1.

Examination of Table 3.1 reveals certain interesting points. The first twenty elements successively add electrons to their outermost shells. The next eight elements, scandium to nickel, have four shells, but add successive electrons to the third shell until it is filled with the maximum number of 18. These elements are called *transition elements*. The same thing happens with elements 39 to 46 inclusive. Electrons are added to the fourth shell until they number 18, then the fifth shell increases until it contains eight electrons. In element number 55, cesium, the sixth principal electron orbit, the P shell, starts to fill. Instead of continuing, however, the N level starts to fill. Beginning with cerium and continuing through lutetium, electrons are successively added to the fourth electron shell, while the two outermost shells remain about the same. This group of elements is usually called the *rare earths*, and sometimes called the *lanthanides* because they begin immediately after lanthanum. The rare earths differ from the transition elements in the depth of the electronic orbit which is filling. While the transition elements fill the second outer orbit, the rare earths fill the third electron shell, which is deeper in the atom. Since, in the case of the rare earths, the two outermost electron shells are alike, it is extremely difficult to separate them by chemical means. They are of importance to the health physicist because they include a great number of the fission products. The concern of the health physicist with the rare earths is aggravated by the fact that the analytical chemistry of the rare earths is very difficult, and also by the relative dearth of knowledge regarding their metabolic pathways and toxicological properties. Despite their name, the rare earths are not rare; they are found to be widely distributed in nature, albeit in small concentrations. Another group of rare earths is found. In the elements starting with thorium, and continuing to lawrencium, the O shell fills while the P and Q shells remain about the same. These rare earths are frequently called the *actinide* elements. They are of importance to the health

TABLE 3.1. ELECTRONIC STRUCTURE OF THE ELEMENTS

Period	Element	Atomic no.	K Shell 1	L Shell 2	M Shell 3	N Shell 4	O Shell 5	P Shell 6	Q Shell 7
1	H	1	1						
	He	2	2						
2	Li	3	2	1					
	Be	4	2	2					
	B	5	2	3					
	C	6	2	4					
	N	7	2	5					
	O	8	2	6					
	F	9	2	7					
3	Ne	10	2	8					
	Na	11	2	8	1				
	Mg	12	2	8	2				
	Al	13	2	8	3				
	Si	14	2	8	4				
	P	15	2	8	5				
	S	16	2	8	6				
	Cl	17	2	8	7				
4	A	18	2	8	8				
	K	19	2	8	8	1			
	Ca	20	2	8	8	2			
	Sc	21	2	8	9	2			
	Ti	22	2	8	10	2			
	V	23	2	8	11	2			
	Cr	24	2	8	13	1			
	Mn	25	2	8	13	2			
	Fe	26	2	8	14	2			
	Co	27	2	8	15	2			
	Ni	28	2	8	16	2			
	Cu	29	2	8	18	1			
	Zn	30	2	8	18	2			
	Ga	31	2	8	18	3			
	Ge	32	2	8	18	4			
	As	33	2	8	18	5			
	Se	34	2	8	18	6			
	Br	35	2	8	18	7			
5	Kr	36	2	8	18	8			
	Rb	37	2	8	18	8	1		
	Sr	38	2	8	18	8	2		
	Y	39	2	8	18	9	2		
	Zr	40	2	8	18	10	2		
	Nb	41	2	8	18	12	1		
	Mo	42	2	8	18	13	1		
	Tc	43	2	8	18	14	1		
	Ru	44	2	8	18	15	1		

TABLE 3.1 (*cont.*)

Period	Element	Atomic no.	K Shell 1	L Shell 2	M Shell 3	N Shell 4	O Shell 5	P Shell 6	Q Shell 7
	Rh	45	2	8	18	16	1		
	Pd	46	2	8	18	18	0		
	Ag	47	2	8	18	18	1		
	Cd	48	2	8	18	18	2		
	In	49	2	8	18	18	3		
	Sn	50	2	8	18	18	4		
	Sb	51	2	8	18	18	5		
	Te	52	2	8	18	18	6		
	I	53	2	8	18	18	7		
6	Xe	54	2	8	18	18	8		
	Cs	55	2	8	18	18	8	1	
	Ba	56	2	8	18	18	8	2	
	La	57	2	8	18	18	9	2	
	Ce	58	2	8	18	19	9	2	
	Pr	59	2	8	18	20	9	2	
	Nd	60	2	8	18	22	8	2	
	Pm	61	2	8	18	23	8	2	
	Sm	62	2	8	18	24	8	2	
	Eu	63	2	8	18	25	8	2	
	Gd	64	2	8	18	25	9	2	
	Tb	65	2	8	18	26	9	2	
	Dy	66	2	8	18	28	8	2	
	Ho	67	2	8	18	29	8	2	
	Er	68	2	8	18	30	8	2	
	Tm	69	2	8	18	31	8	2	
	Yb	70	2	8	18	32	8	2	
	Lu	71	2	8	18	32	9	2	
	Hf	72	2	8	18	32	10	2	
	Ta	73	2	8	18	32	11	2	
	W	74	2	8	18	32	12	2	
	Re	75	2	8	18	32	13	2	
	Os	76	2	8	18	32	14	2	
	Ir	77	2	8	18	32	15	2	
	Pt	78	2	8	18	32	17	1	
	Au	79	2	8	18	32	18	1	
	Hg	80	2	8	18	32	18	2	
	Tl	81	2	8	18	32	18	3	
	Pb	82	2	8	18	32	18	4	
	Bi	83	2	8	18	32	18	5	
	Po	84	2	8	18	32	18	6	
	At	85	2	8	18	32	18	7	
7	Rn	86	2	8	18	32	18	8	
	Fr	87	2	8	18	32	18	8	1
	Ra	88	2	8	18	32	18	8	2
	Ac	89	2	8	18	32	18	9	2
	Th	90	2	8	18	32	19	9	2

TABLE 3.1 (*cont.*)

Period	Element	Atomic no.	K Shell 1	L Shell 2	M Shell 3	N Shell 4	O Shell 5	P Shell 6	Q Shell 7
	Pa	91	2	8	18	32	20	9	2
	U	92	2	8	18	32	21	9	2
	Np	93	2	8	18	32	22	9	2
	Pu	94	2	8	18	32	23	9	2
	Am	95	2	8	18	32	25	8	2
	Cm	96	2	8	18	32	25	9	2
	Bk	97	2	8	18	32	26	9	2
	Cf	98	2	8	18	32	27	9	2
	Es	99	2	8	18	32	28	9	2
	Fm	100	2	8	18	32	29	9	2
	Md	101	2	8	18	32	30	9	2
	No	102	2	8	18	32	31	9	2

physicist because they are all naturally radioactive and because they include the fuel used in nuclear reactors.

Characteristic X-rays

Some virtues of the solar system type of atomic model, in which electrons rotate about the nucleus in certain radii corresponding to unique energy levels, are the simple explanations that it allows for transfer of energy to matter by excitation and ionization, for the photoelectric effect, and for the origin of certain X-rays called characteristic X-rays. It was pointed out that optical and ultraviolet spectra of elements are due to excitation of outer electrons to levels up to several electron volts, and that spectral lines represent energy differences between excited states. As more and more electron shells are added the energy differences between the principal levels increases greatly. For the high atomic numbered elements, they reach tens of thousands of electron volts. In the case of lead, for example, the energy difference between the K and L shells is 72,000 eV. If this K electron is struck by a photon whose energy exceeds 87.95 keV, the binding energy of the K electron, the electron is ejected from the atom, and leaves an empty slot in the K shell, as shown schematically in Fig. 3.4. Instantaneously, one of the outer electrons falls down into the vacant slot left by the photoelectron. When this happens, a photon is emitted whose energy is equal to the difference between the initial and final energy levels, in accordance with equation (3.4A). For the lead atom, when an electron falls from the L to the K levels, the emitted photon has a quantum energy of 72,000 eV. A photon of such high energy is an X-ray. When produced in this manner, the photon is called a *characteristic X-ray* because the energy differences between electron orbits are unique for the different atoms, and the X-rays that represent these differences are "characteristic" of the elements in which they originate. This process is repeated until all the inner electron orbits are re-filled. It is possible, of course, that the first transition is from the *M* level, or even from the outermost electronic orbit. The most likely origin of the first electronic transition, however, is the *L* shell. When this happens, the resulting X-ray is called a K_α photon; if an electron falls from the *M* level to the *K* level, we have a K_β photon. When the vacancy in

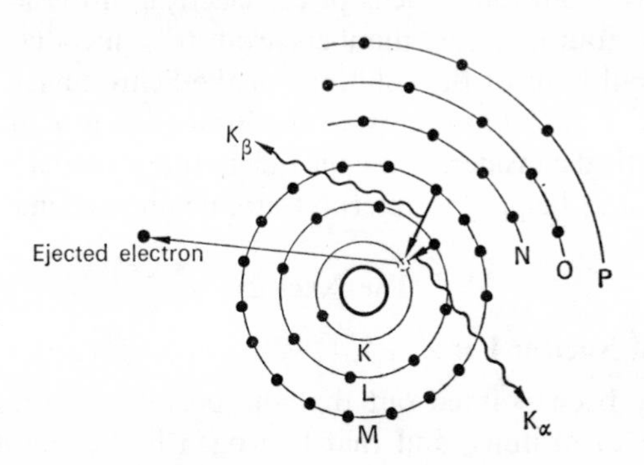

FIG. 3.4. Schematic representation of the origin of characteristic X-rays.

the *L* orbit is filled by an electron that falls from the *M* level, we have an L_α X-ray; if the *L* vacancy is filled by an electron originally in the *N* level, then an L_β X-ray results, and so on. These characteristic X-rays are sometimes called *fluorescent radiation*, since they are emitted when matter is irradiated with X-rays. Characteristic radiation is useful as a tool to the analytical chemist for identifying unknown elements. Obviously, characteristic radiation is of importance to the health physicist who must consider the fluorescent radiation that may be produced in radiation absorbers and in certain other cases where inner electrons are ejected from high-atomic-numbered elements.

The Wave Mechanics Atomic Model

The atomic model described above is sufficiently useful to explain most phenomena encountered in health physics. However, for the study of atomic physics, a more abstract concept of the atom was proposed by the Austrian physicist Schrödinger (for which he and the British physicist Dirac shared the Nobel prize in 1933). Instead of working with particulate electrons as Bohr had done, he treated them as de Broglie waves, and developed the branch of physics known as "wave mechanics". Starting with the de Broglie equation for the associated electron wave, Schrödinger derived a general differential equation that must be satisfied by an electron within an atom. The present-day atomic theory consists of solutions of this equation subject to certain conditions. A number of different solutions, corresponding to different energy levels, is possible. However, whereas Bohr pictured an atom with electrons at precisely determined distances from the nucleus, the Schrödinger wave equation gives the probability of finding an electron at any given distance from the nucleus. The two atomic pictures coincide to the extent that the most probable radius for the hydrogen electron is exactly the same as the first Bohr radius. Similarly, the second Bohr radius corresponds to the most probable distance from the nucleus of the electron in the first excited state. Furthermore, the four quantum numbers arbitrarily introduced into the Bohr atom falls naturally out of the solutions of the Schrödinger wave equation. Although the wave model has replaced the Bohr system of atomic mechanics for highly theoretical considerations, the older atomic model is still considered a very useful tool in helping to interpret atomic phenomena.

The Nucleus

The Neutron and Nuclear Force

It has already been pointed out that the positive charges in the atomic nucleus are due to protons, and that hydrogen is the simplest nucleus—it consists of only a single proton. If succeeding nuclei merely were multiples of the proton, then the mass numbers of the nuclei, if a mass number of 1 is assigned to the proton, should be equal to the atomic numbers of the nuclei. This was not found to be the case. Except for hydrogen, the nuclear mass numbers were found to be about twice as great as the corresponding atomic numbers, and to become relatively greater as the atomic numbers increased. Furthermore, it was necessary to account for the stability of the nucleus in the face of the repulsive coulombic forces among the nuclear protons. A simple calculation shows that the gravitational force of attraction is insufficient to overcome the repulsive electrical forces. Both these problems were solved by the discovery, in 1932, by the British physicist Chadwick, of the third basic building block in nature: the neutron. (Chadwick won the Nobel prize in 1935 for this discovery.) This particle, whose mass is about the same as that of a proton, 1.67474×10^{-24} g, is electrically neutral. Its presence in the nucleus accounts for the difference between the atomic number and the atomic mass number; it also supplies the cohesive force that holds the nucleus together. This force is called the nuclear force. It is thought to act over an extremely short range—about 2 to 3×10^{-13} cm. By analogy to the ordinary case of charged particles, it may be assumed that the neutron and the proton carry certain nuclear charges and that force fields due to these nuclear charges are established around the nucleons (particles within the nucleus are called nucleons). Nuclear forces are all attractive, and the interaction between the nuclear force fields supplies the cohesive forces which overcome the repulsive electrical forces. However, since the range of the nuclear force is much shorter than the range of the electrical force, neutrons can interact only with those nucleons to which they are immediately adjacent, while protons interact with each other even though remotely located within the nucleus. For this reason, the number of neutrons must increase more rapidly than the number of protons.

Isotopes

It has been found that, for any particular element, the number of neutrons within the nucleus is not constant. Oxygen, for example, consists of three nuclear species; one whose nucleus has eight neutrons, one of nine neutrons, and one of ten neutrons. In these three cases, of course, the nucleus contains eight protons. The atomic mass numbers of these three species are 16, 17, and 18 respectively. These three nuclear species of the same element are called *isotopes* of oxygen. Isotopes of an element are atoms that contain the same number of positive nuclear charges and have the same extra-nuclear electronic structure, but differ in the number of neutrons. Most elements contain several isotopes. The atomic weight of an element is the weighted average of the weights of the different isotopes of which the element is composed. Isotopes cannot be distinguished chemically, since they have the same electronic structure and therefore undergo the same chemical reactions. An isotope is identified by writing the chemical symbol with a subscript to the left giving the atomic number and a superscript giving the atomic mass number, or the total number of nucleons. Thus the three isotopes of oxygen may be written as ${}^{16}_{8}O$, ${}^{17}_{8}O$, and ${}^{18}_{8}O$. Since the atomic number is synonomous with the chemical symbol, however, the subscript is usually omitted and the isotope is written as ${}^{16}O$. It should be pointed out that not all isotopes are equally abundant. In the case of oxygen 99.975% of the naturally occurring atoms are ${}^{16}O$, while ${}^{17}O$ and ${}^{18}O$ include 0.037% and 0.204% respectively. In other elements the distribution of isotopes may be quite different. Chlorine, for example, consists of two naturally occurring isotopes, ${}^{35}Cl$ and ${}^{37}Cl$, whose respective abundances are 75.4% and 24.6%.

The Atomic Mass Unit

Atomic masses may be given either in grams or in relative numbers called atomic mass units. Since one mole of any substance contains 6.03×10^{23} molecules (Avogadro's number), and since the weight in grams of one mole is equal numerically to its molecular weight, the weight of a single atom can easily be computed. In the case of ${}^{16}O$, for example, 1 mole (22.4 liters at normal temperature and pressure) weighs 32.000 g because the oxygen molecule is diatomic. One atom, therefore, weighs

$$\frac{32.0000 \text{ g/mole}}{6.027 \times 10^{23} \text{ molecules/mole} \times 2 \text{ atoms/molecule}} = 2.65656 \times 10^{-24} \text{ g.}$$

Because the weight of one mole of oxygen was found to be a whole number, oxygen was chosen as the reference standard for the system of relative weights known as atomic weights. (Actually the atomic weight of an element is the weighted average of all the isotopic weights. Because of the preponderance of the ^{16}O isotope, however, and the limited sensitivity of the determination of the atomic weight of oxygen by the early chemists, the discrepancy due to the presence of the other two isotopes was overlooked when the chemical atomic weights were determined. This introduces a very slight error, which is corrected in the physical scale of atomic weights. The physical scale is based only on the weight of ^{16}O.) Since oxygen was assigned an atomic weight of 16.0000, one atomic weight unit is

$$1 \text{ amu} = \frac{2.65656 \times 10^{-24} \text{ g}}{16} = 1.66035 \times 10^{-24} \text{ g}.$$

On this basis, the weight of a neutron, m_n, is 1.008987 amu, that of a proton, m_p, is 1.007593 amu, and the atomic weight of an electron is 0.000552 amu. The energy equivalent of one atomic mass unit is

$$E = \frac{mc^2 \text{ ergs}}{1.6 \times 10^{-6} \text{ ergs/MeV}}$$

$$= 931 \text{ MeV}.$$

Binding Energy

At this point, it is interesting to compare the sum of the weights of the constituent parts of an isotope, W, with the measured isotopic weight M. For the case of ^{17}O, whose atomic mass is 17.004533 amu, we have

$$W = Zm_p + (A - Z)m_n + Zm_e, \tag{3.15}$$

where Z is the atomic number and A is the atomic mass number, which is equal to the number of nucleons within the nucleus.

$$W = 8(1.007593) + (17 - 8)(1.008987) + 8(0.000552)$$

$$= 17.146043 \text{ amu}.$$

The weight of the sum of the parts is seen to be much greater than the actual weight of the entire atom. This is true not only for ^{17}O, but for all nuclei. The difference between the atomic weight and the sum of the weights of the parts is called the mass defect, and is defined by δ in equation (3.16):

$$\delta = W - M. \tag{3.16}$$

The mass decrement represents the mass equivalent of the work that must be done in order to separate the nucleus into its individual component nucleons, and is therefore called the binding energy. In energy units, the binding energy, BE, is

$$BE = (W - M) \text{ amu} \times 931 \frac{\text{MeV}}{\text{amu}}. \qquad (3.17)$$

Quite clearly, the binding energy is a measure of the cohesiveness of a nucleus. Since the total binding energy of a nucleus depends on the number of nucleons within the nucleus, a more useful measure of the cohesiveness is the binding energy per nucleon, E_b, as given below:

$$E_b = \frac{931\,(W - M)}{A} \frac{\text{MeV}}{\text{nucleon}}, \qquad (3.18)$$

where A, the atomic mass number, represents the number of nucleons within the nucleus. For the case of ^{17}O, the binding energy is 131.7 MeV, and the binding energy per nucleon is 7.74 MeV.

The binding energy per nucleon is very low for the low atomic numbered elements, but rises rapidly to a very broad peak at binding energies in excess of 8 MeV per nucleon, and then decreases very slowly until a value of 7.58 MeV per nucleon is reached for ^{238}U. Figure 3.5, in which the binding energy per nucleon is plotted against the number of nucleons in the various isotopes, shows that, with a very few exceptions, there is a systematic variation of binding energy per nucleon with the number of nucleons within the nucleus.

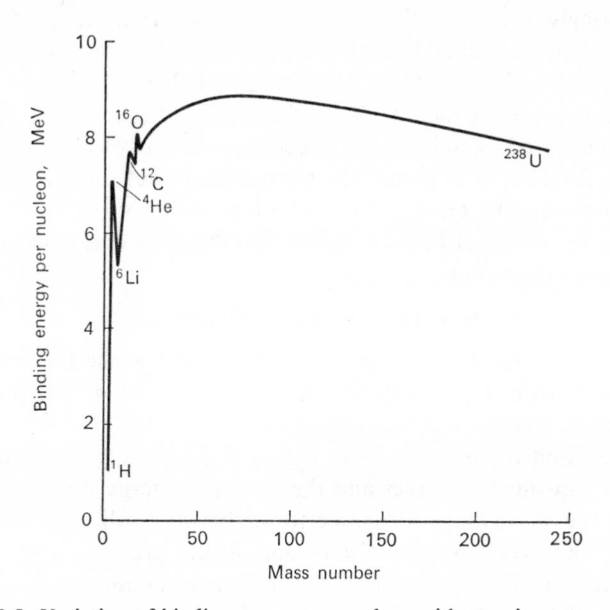

FIG. 3.5. Variation of binding energy per nucleon with atomic mass number.

The most notable departures from the smooth curve are the isotopes ^{4}He, ^{12}C, and ^{16}O. Each of these isotopes lies above the curve, indicating that they are very strongly bound. The ^{12}C and ^{16}O isotopes, as well as ^{20}Ne, which has more binding energy per nucleon than either of the isotopes that flank it, may be thought of as containing three, four, and five sub-units of ^{4}He respectively. The exceptional binding energies in these nuclei, together with the fact that ^{4}He nuclei, as alpha particles, are emitted in certain modes of radioactive disintegration, suggest that nucleons tend to form stable sub-groups of two protons and two neutrons within the nucleus.

The fact that the binding energy curve, Fig. 3.5, has the shape that it does, explains why it is possible to release energy by splitting the very heavy elements and by fusing two very light elements. Since the binding energy per nucleon is greater for nuclei in the center of the curve than for nuclei at both extremes, any change in nuclear structure that drives the nucleons towards the center of the curve must release the energy difference between the final and the initial states.

Nuclear Models

(a) *Liquid drop*

Although the nuclear building blocks seem to be well known, no definite structure for the nucleus has yet been established. However, two different nuclear models have been postulated. According to one of these, the nucleus is thought to be a homogenous mixture of nucleons in which all the nucleons interact strongly with each other. As a result, the internal energy of the nucleus is about equally distributed among the constituent nucleons, while surface tension forces tend to keep the nucleus spherical. This is analogous to a drop of liquid, and hence is called the *liquid drop model* of the nucleus. This model, which was proposed by Bohr and Wheeler, is particularly successful in explaining nuclear fission and in permitting the calculation of the atomic masses of isotopes whose atomic mass is very difficult to measure.

From the preceding discussion, we see that the mass of a nucleus of atomic number A and atomic mass number Z is

$$M = (A - Z)\,m_n + Zm_p - BE. \qquad (3.19)$$

Furthermore, the binding energy per nucleon, and hence the total binding energy, is seen from Fig. 3.5 to be a function of Z and A. The nuclear drop model permits a semi-empirical equation to be formulated that relates the nuclear mass and binding energy to A and Z. According to the liquid drop model, the intra-nuclear forces and the potential energy due to these forces are due to the short-range attractive forces between adjacent nucleons, the long-range repulsive Coulomb forces among the protons, and the surface tension effect, in which nucleons on the surface of the nucleus are less tightly bound than those in the nuclear interior. The binding energy due to these forces is modified according to whether the numbers of neutrons and protons are even or odd. On the basis of this reasoning, the following equation was fitted to the experimental data relating nuclear mass, in atomic mass units, with A and Z:

$$M = 0.99389\,A - 0.00081Z + 0.014A^{2/3} + 0.083\,\frac{(A/2 - Z)^2}{A} + + 0.000627\,\frac{Z^2}{A^{1/3}} + \Delta, \tag{3.20}$$

where: $\Delta = 0$ for odd A,

$\Delta = -\,0.036/A^{3/4}$ for even A, even Z,

$\Delta = +\,0.036/A^{3/4}$ for even A, odd Z.

(b) *Shell model*

The alternate nuclear model is called the *shell model.* According to this model of the nucleus, the various nucleons exist in certain energy levels within the nucleus, and interact weakly among themselves. Many observations and experimental data lend support to such a nuclear structure. Among the stable isotopes, the "even–even" nuclei, that is, nuclei with even numbers of protons and neutrons, are most numerous, with a total of 162 isotopes. Even–odd nuclei, in which either the protons or neutrons are even in number, and the other one is odd, are second in abundance with a total of 108 isotopes. Odd–odd nuclei are the fewest in number; only four such stable isotopes are found in nature. Furthermore, so-called "magic numbers" have been found to recur among the stable isotopes. These magic numbers include 2, 8, 20, 50, 82, and 126. Isotopes containing these numbers of protons or neutrons or both are most abundant in nature, suggesting unusual stability in their structures. Nuclei containing these magic numbers are relatively inert in a nuclear sense, that is, they do not react easily when bombarded with neutrons. This is analogous to the case of chemically inert elements that have filled electron energy levels. All these observed facts are compatible with an energy level model of the nucleus similar to the electronic energy level model of the atom. Each nucleon in a nucleus is identified by its own set of four quantum numbers, as in the case of the extra-nuclear electrons. By application of the Pauli exclusion principle to nucleons, it is possible to construct energy levels which contain successively 2, 8, 20, 50, 82, and 126 nucleons.

As in the case of the extra-nuclear electrons, nucleons too may be excited by raising them to higher energy levels. When this occurs, the nucleon falls back into its ground state, and emits a photon whose energy is equal to the energy difference between the excited and ground states. This is the same type phenomenon as in the case of optical and characteristic X-ray spectra.

The photon in this case is called a *gamma-ray*. Because nuclear energy levels are usually much further apart than electronic energy levels, gamma-rays are usually (though not necessarily) more energetic than X-ray photons. It should be emphasized that from the practical health physics point of view X-rays and gamma-rays are identical. They differ only in their place of origin—X-rays in the extra-nuclear structure and gamma-rays within the nucleus. Once produced, it is impossible to distinguish between X-rays and gamma-rays.

Nuclear Stability

If a plot is made of the number of protons versus the number of neutrons for the stable isotopes, the curve shown in Fig. 3.6 is obtained. The stable isotopes lie within a relatively narrow range, indicating that the neutron to proton ratio must lie within certain limits if a nucleus is to be stable. Most radioactive nuclei lie outside this range of stability. The plot also shows that the slope of the curve, which initially has a value of unity, gradually increases as the atomic number increases, thereby showing the continuously increasing ratio of neutrons to protons.

Since all nuclear forces are attractive, it may appear surprising to find unstable nuclei with an excessive number of neutrons. This apparent anomaly may be explained simply in terms of the shell model of the nucleus. According to the Pauli exclusion principle, like nucleons may be grouped in pairs,

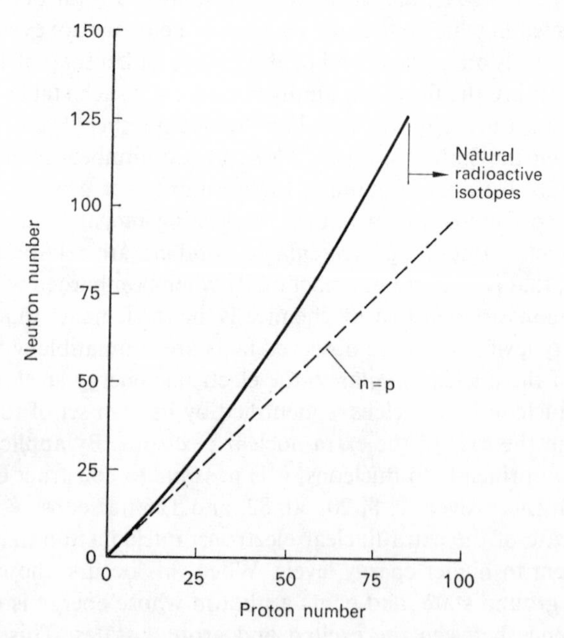

FIG. 3.6. Nuclear stability curve. The line represents the best fit to the neutron-proton coordinates of stable isotopes.

with each pair having all quantum numbers the same, except the spin quantum number. Since nuclei with completely filled energy levels are more stable than those with unfilled inner levels, additional neutrons in the case of nuclei with unfilled proton levels but filled neutron levels results in unstable nuclei. To achieve stability, the nucleus may undergo an internal rearrangement in which the additional neutron transforms itself into a proton by emitting an electron. The new proton then pairs off with a proton in one of the unfilled proton levels. As an example of this possible mechanism, consider the consequences of the addition of a neutron to ${}^{31}_{15}P$. This is the stable isotope of phosphorous that occurs naturally. According to the shell model, the fifteen protons inside the nucleus may be distributed among seven pairs plus one, while the neutrons may be paired off into eight groups. If now an additional neutron is added to the nucleus to make ${}^{32}_{15}P$, then the additional neutron may go into another energy level. This condition, however, is unstable. The additional neutron may therefore become a proton and an electron, with the electron being ejected from the nucleus, and the proton pairing off with the single proton, thereby forming stable ${}^{32}_{16}S$. This internal nuclear transformation is called a *radioactive transformation*, and the entire process is called a *radioactive disintegration* or a *radioactive decay*.

Problems

1. What is the closest approach that a 5.3-MeV alpha particle can make to a gold nucleus?

2. Calculate the number of atoms per cm^3 of lead, given that the density of lead is 11.3 g/cm^3 and its atomic weight is 207.21.

3. A μ^- meson has a charge of -4.8×10^{-10} sC and a mass 207 times that of a resting electron. If a proton should capture a μ^- to form a "mesic" atom, calculate

(a) the radius of the first Bohr orbit, and

(b) the ionization potential.

4. Calculate the ionization potential of a singly ionized ^{4}He atom.

5. Calculate the current due to the hydrogen electron in the ground state of hydrogen.

6. Calculate the ratio of the velocity of a hydrogen electron in the ground state to the velocity of light.

7. Calculate the Rydberg constant for deuterium.

8. What is the uncertainty in the momentum of a proton inside a nucleus of ^{27}Al? What is the kinetic energy of this proton?

9. A sodium ion is neutralized by capturing a 1-eV electron. What is the wavelength of the emitted radiation if the ionization potential of Na is 5.41 volts?

10. (a) How much energy would be released if one gram deuterium were fused to form helium according to the equation ${}^2H + {}^2H\ {}^4He + Q$? (b) How much energy is necessary to drive the two deuterium nuclei together?

11. The density of beryllium, atomic number 4, is 1.84 g/cm^3, and the density of lead, atomic number 82, is 11.3 g/cm^3. Calculate the density of a ^{9}Be and a ^{208}Pb nucleus.

12. Determine the electronic shell configuration for aluminium, atomic number 13.

13. What is the difference in mass between the hydrogen atom and the sum of the masses of a proton and an electron? Express the answer in energy equivalent (eV) of the mass difference.

14. If the heat of vaporization of water is 540 calories per gram at atmospheric pressure, what is the binding energy of a water molecule?

15. The ionization potential of He is 24.5 eV.

(a) What is the minimum velocity with which an electron is moving before it can ionize an unexcited He atom?

(b) What is the maximum wavelength of a photon in order that it ionize the He atom?

16. In a certain 25-watt mercury-vapor ultraviolet lamp, 0.1% of the electrical energy input appears as U.V. radiation of wavelength 2537 Å. What is the photon emission rate, per second, from this lamp?

Suggested References

1. GLASSTONE, S.: *Sourcebook on Atomic Energy*, D. Van Nostrand, Princeton, 1958.
2. LAPP, R. E. and ANDREWS, H. L.: *Nuclear Radiation Physics*, Prentice Hall, Englewood Cliffs, 1963.
3. SEMAT, H.: *Introduction to Atomic and Nuclear Physics*, Rinehart & Co., New York, 1964.
4. EVANS, R. D.: *The Atomic Nucleus*, McGraw-Hill, New York, 1955.
5. CUNNINGHAME, J. G.: *Introduction to the Atomic Nucleus*, Elsevier, Amsterdam, 1964.
6. BORN, M.: *Atomic Physics*, Hafner, New York, 5th edition.
7. KAPLAN, I.: *Nuclear Physics*, Addison-Wesley Pub. Co., Reading, 1962.
8. HALLIDAY, D.: *Introductory Nuclear Physics*, John Wiley & Sons, New York, 1955.
9. RICHTMYER, F. K., KENNARD, E. H. and LAURITSEN, T.: *Introduction to Modern Physics*, McGraw-Hill, New York, 1955.
10. FEYNMAN, R. P., LEIGHTON, R. B. and SANDS, M.: *The Feynman Lectures on Physics*, Vols. I, II, and III, Addison-Wesley, Reading, 1965.

CHAPTER 4

RADIOACTIVITY

Radioactivity and Decay Mechanisms

Radioactivity may be defined as spontaneous nuclear changes that result in the formation of new elements. These changes are accomplished by one of several different mechanisms, including alpha particle emission, beta particle and positron emission, and orbital electron capture. Each of these reactions may or may not be accompanied by gamma radiation. Radioactivity and radioactive properties of nuclides are determined by nuclear considerations only, and are independent of the chemical and physical states of the radioisotope. Radioactive properties of radioisotopes therefore cannot be changed by any means, and are unique to the respective radionuclides. The exact mode of radioactive decay depends on two factors: the particular type of nuclear instability—that is, whether the neutron to proton ratio is either too high or too low for the particular nuclide under consideration, and on the mass–energy relationship among the parent nucleus, daughter nucleus, and the emitted particle.

Alpha Emission

An alpha particle is a highly energetic helium nucleus that is emitted from the nucleus of the radioactive isotope when the neutron to proton ratio is too low. It is a positively charged, massive particle, consisting of an assembly of two protons and two neutrons. Since atomic numbers and mass numbers are conserved in alpha transitions, it follows that the result of alpha emission is a daughter whose atomic number is two less than that of the parent, and whose atomic mass number is four less than that of the parent. In the case of ^{210}Po, for example, the reaction is

$$^{210}_{84}Po \longrightarrow {}^{4}_{2}He + {}^{206}_{82}Pb.$$

In this example, ^{210}Po has a neutron to proton ratio of 126 to 84, or 1.5 to 1. After decaying by alpha particle emission, a stable daughter nucleus, $^{206}_{82}Pb$, is formed whose neutron to proton ratio is 1.51 to 1. With one exception, $^{147}_{62}Sm$, naturally occurring alpha emitters are found only among elements of atomic number greater than 82. The explanation for this is two-fold: first is the fact that the electrostatic repulsive forces in the heavy nuclei increase much more rapidly than the cohesive nuclear forces, and the magnitude of the electrostatic forces, consequently, may closely approach or even exceed that of the nuclear force; the second part of the explanation is concerned with the fact that the emitted particle must have sufficient energy to overcome the high potential barrier at the surface of the nucleus resulting from the presence of the positively charged nucleons. This potential barrier may be graphically represented by the curve in Fig. 4.1. The inside of the nucleus, because of the negative potential there, may be thought of as a potential well that is surrounded by a wall whose height is about 25 MeV for an alpha particle inside a high atomic numbered nucleus. According to quantum mechanical theory, an alpha particle may escape from the potential well by tunneling through the potential barrier. For alpha emission to be observed from the high atomic numbered naturally occurring elements, theoretical considera-

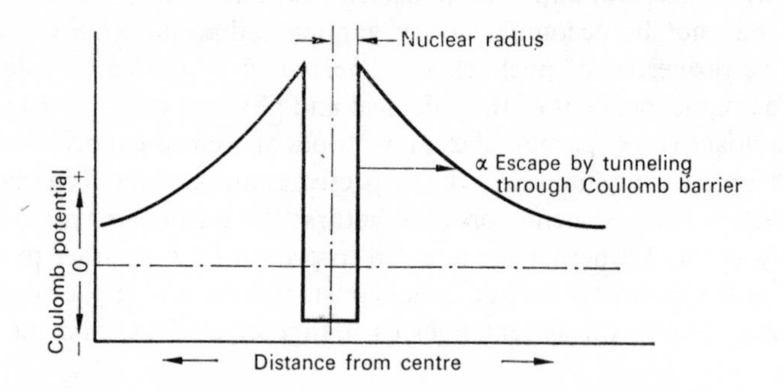

FIG. 4.1. Potential inside and in the vicinity of a nucleus.

tions demand that an alpha particle have a kinetic energy greater than 3.8 MeV. This condition is verified by the experimental finding that the lowest energy alpha particle from the high atomic numbered elements is 3.93 MeV. This alpha particle originates in ^{232}Th. (Samarium-147 emits an alpha particle whose energy is only 2.18 MeV. This low energy is consistent, however, with the theoretical calculations mentioned above if the low atomic number, 62, of samarium is considered.) The question regarding the source of this kinetic energy naturally arises. This energy results from the net decrease in mass following the formation of the alpha particle. Generally, for alpha emission to occur, the following conservation equation must be satisfied:

$$M_p = M_d + M_\alpha + 2M_e + Q, \qquad (4.1)$$

where M_p, M_d, M_α, and M_e are respectively equal to the masses of the parent, the daughter, the emitted alpha particle, and the two orbital electrons that are lost during the transition to the lower atomic numbered daughter, while Q is

FIG. 4.2. Tracks in a Wilson cloud chamber of alpha particles from thorium C (^{212}Bi), energy = 6.05 MeV, and thorium C (^{212}Po), energy = 8.78 MeV. (E. Rutherford, J. Chadwick, and C. D. Ellis, *Radiations from Radioactive Substances*, Macmillan, New York, 1930.)

the total energy release associated with the radioactive disintegration. In the case of the decay of ^{210}Po, for example, we have, from equation (4.1),

$$\begin{aligned} Q &= M_{Po} - M_{Pb} - M_{\alpha} - 2M_e \\ &= 210.04850 - 206.03883 - 4.00277 - 2 \times 0.00055 \\ &= 0.0058 \text{ atomic mass units.} \end{aligned}$$

In energy units,

$$\begin{aligned} Q &= 0.0058 \text{ amu} \times 931 \frac{\text{MeV}}{\text{amu}} \\ &= 5.4 \text{ MeV.} \end{aligned}$$

This Q value represents the total energy associated with the disintegration of ^{210}Po. Since no gamma-ray is emitted in this transition, the total released energy appears as kinetic energy, and is divided between the alpha particle and the daughter, which recoils after the alpha particle is emitted. The exact energy division between the alpha and recoil nucleus depends on the mass of the daughter, and may be calculated by application of the laws of conservation of energy and momentum. If M and m are the masses, respectively, of the recoil nucleus and the alpha particle, and if V and v are their velocities, then

$$Q = \tfrac{1}{2} MV^2 + \tfrac{1}{2} mv^2. \tag{4.2}$$

We have, according to the law of conservation of momentum,

$$MV = mv, \tag{4.3}$$

or,

$$V = \frac{mv}{M}.$$

When the value for V from equation (4.3) is substituted into equation (4.2), we have

$$Q = \tfrac{1}{2} M \frac{m^2v^2}{M^2} + \tfrac{1}{2} mv^2. \tag{4.4}$$

If we let E represent the kinetic energy of the alpha particle, $\tfrac{1}{2} mv^2$, then equation (4.4) may be rewritten as

$$Q = E\left(\frac{m}{M} + 1\right),$$

or

$$E = \frac{Q}{1 + m/M}. \tag{4.5}$$

According to equation (4.5), the kinetic energy of the alpha particle emitted in the decay of ^{210}Po is:

$$E = \frac{5.4}{(1 + 4/206)}$$
$$= 5.3 \text{ MeV}.$$

The kinetic energy of the recoil nucleus, therefore, is 0.1 MeV.

Alpha particles are essentially monoenergetic. However, alpha-particle spectrograms do show discrete energy groupings, with small energy differences among the different groups. These small differences are attributed to differences in the energy level of the daughter nucleus. That is, a nucleus that emits one of the lower energy alpha particles is left in an excited state, while the nucleus that emits the highest energy alpha particle for any particular isotope is usually left in the "ground" state. A nucleus left in an excited state emits its energy of excitation in the form of a gamma-ray. It should be pointed out that most of the alpha particles are usually emitted with the maximum energy. Very few nuclei, consequently, are left in excited states, and gamma radiation, therefore, accompanies only a small fraction of the alpha-rays. Radium may be cited as an example of an alpha emitter with a complex spectrum. In the overwhelming majority of disintegrations of ^{226}Ra, 94.3 %, alphas are emitted with a kinetic energy of 4.777 MeV. The balance of the alpha particles, 5.7 %, have kinetic energies of only 4.591 MeV. In that instance where a lower energy alpha is emitted, the daughter nucleus is left in an excited state, and rids itself of its energy of excitation by emitting a gamma-ray photon whose energy is equal to the difference between the energies of the two alpha particles: 4.777 − 4.591 = 0.186 MeV. (About 35 % of these gamma-ray photons are internally converted: see section below on internal conversion.) The ^{226}Ra spectrum is the least intricate of all the complex alpha spectra. Most alpha emitters have several groups of alphas, and therefore more gammas. All alpha spectra, however, show the same consistent relationship among the various nuclear energy levels.

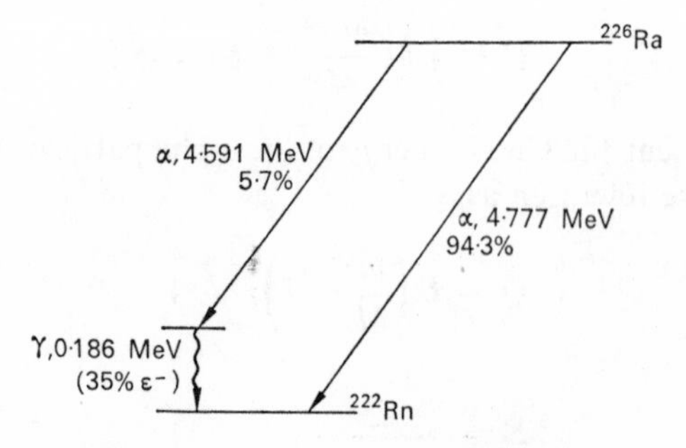

FIG. 4.3. Radium-226 decay scheme.

Alpha particles are extremely limited in their ability to penetrate matter. The dead outer layer of skin is sufficiently thick to absorb all alpha radiations from radioactive materials. As a consequence, alpha radiation from sources outside the body do not constitute a radiation hazard. In the case of internally deposited alpha-emitting isotopes, however, the shielding effect of the dead outer layer of skin is absent, and the energy of the alpha radiation is dissipated in living tissue. For this reason, and others to be discussed in Chapter 7, alpha radiation is highly toxic when it irradiates the inside of the body from internally deposited radioisotopes.

Beta Emission

A beta particle is an ordinary electron that is ejected from the nucleus of a beta-unstable radioactive atom. The particle has a single negative electrical charge (4.8×10^{-10} sC) and a very small mass (0.00055 atomic mass units). Since theoretical considerations preclude the independent existence of an intra-nuclear electron, it is postulated that the beta particle is formed at the instant of emission by the decay of a neutron into a proton and an electron according to the equation

$$_0n^1 \rightarrow {_1H^1} + {_{-1}e^0}. \tag{4.6}$$

This transformation shows that beta decay occurs among those isotopes that have a surplus of neutrons. For beta emission to be energetically possible, the exact nuclear mass of the parent must be greater than the sum of the exact masses of the daughter nucleus plus the beta particle.

$$M_p = M_d + M_e + Q. \tag{4.7}$$

This restriction, of course, is analogous to the corresponding restriction on alpha emitters. Because a unit negative charge is lost during beta decay, and because the mass of the beta particle is very much less than 1 atomic mass unit, the daughter nucleus is one atomic number higher than its parent but retains the same atomic mass number as the parent. For example, radioactive phosphorous decays to stable sulfur according to the equation

$$^{32}_{15}\mathrm{P} \rightarrow {^{32}_{16}\mathrm{S}} + {_{-1}e^0} + 1.71 \text{ MeV}.$$

The energy of disintegration, in this instance 1.71 MeV, is the energy equivalent of the difference in mass between the ^{32}P nucleus and the sum of the ^{32}S nucleus plus the beta particle, and appears as kinetic energy of the beta particle. If neutral atomic masses are used to complete the mass-energy equation, then, of course, the mass of the electron shown in the right hand side of equation (4.7) is not considered, since it is implicitly included in the extra-nuclear electronic structure of the ^{32}S. The mass difference is

$$31.98403 = 31.98224 + Q,$$

$$Q = 0.00179 \text{ am}\mu,$$

and the energy equivalent of the mass difference is

$$0.00179 \text{ amu} \times 931 \frac{\text{MeV}}{\text{amu}} = 1.71 \text{ MeV}.$$

Examination of equation (4.5) shows that in the case of beta emission, an extremely small part of the energy of the reaction is dissipated by the recoil nucleus, since m/M, where m is now the mass of the beta particle, and M is the mass of the daughter nucleus, is very small. In the example given above,

$$\frac{m}{M} = \frac{0.00055}{32} = 0.000017,$$

and Q is only 1.000017 times greater than the kinetic energy of the beta particle.

On the basis of the above analysis, one might expect beta particles to be monoenergetic, as in the case of alpha radiation. This expectation is not confirmed by experiment. Instead, beta particles are found to be emitted with

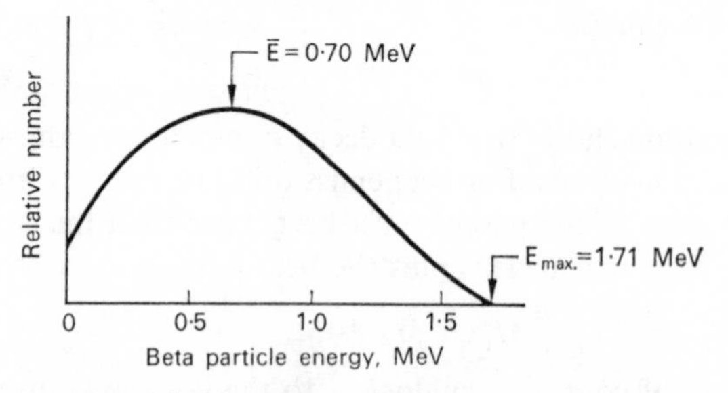

FIG. 4.4. Phosphorous-32 beta-ray spectrum.

a continuous energy distribution ranging from zero to the theoretically expected value based on mass-energy considerations for the particular beta transition. In the case of ^{32}P, for example, although the maximum energy of the beta particle may be 1.71 MeV, most of the beta-rays have considerably smaller kinetic energies, as shown in Fig. 4.4. The average energy of a ^{32}P beta particle is 0.7 MeV, or about 41 % of the maximum energy. Generally, the average energy of the beta radiation from most beta-active radioisotopes is about 30–40 % of the maximum energy. Unless otherwise specified, when the energy of a beta emitter is given, it is the maximum energy.

The fact that beta radiation is emitted with a continuous energy distribution up to a definite maximum seems to violate the established energy-mass conservation laws. To prevent violation of the conservation laws, it was postulated that the beta particle is accompanied by another particle, called a neutrino, whose energy is equal to the difference between the kinetic energy of the accompanying beta particle and the maximum energy of the spectral distribution. The neutrino, as postulated, has no electrical charge and a vanishingly small mass. Because of these two characteristics, detection of the neutrino is extremely difficult. Recent experimental work, however, has confirmed the validity of the neutrino hypothesis. Equation (4.6) should therefore be modified to

$$_0n^1 \rightarrow {}_1H^1 + {}_{-1}e^0 + \nu, \tag{4.8}$$

where ν represents the neutrino.

Phosphorous-32, like several other beta emitters including ^{3}H, ^{14}C, ^{90}Sr, and ^{90}Y, emits no gamma-rays. These isotopes are known as pure beta emitters. The opposite of a pure beta emitter is a beta–gamma emitter. In this case, the beta particle is followed (instantaneously) by a gamma-ray. The explanation for the gamma-ray here is the same as that for the alpha-ray. The daughter nucleus, after the emission of a beta-ray, is left in an excited condition, and rids itself of the energy of excitation by the emission of a gamma-ray. Mercury-203 may be given as an example. It emits a 0.21-MeV beta-ray and a 0.279-MeV gamma photon, as seen in the decay scheme shown in Fig. 4.5.

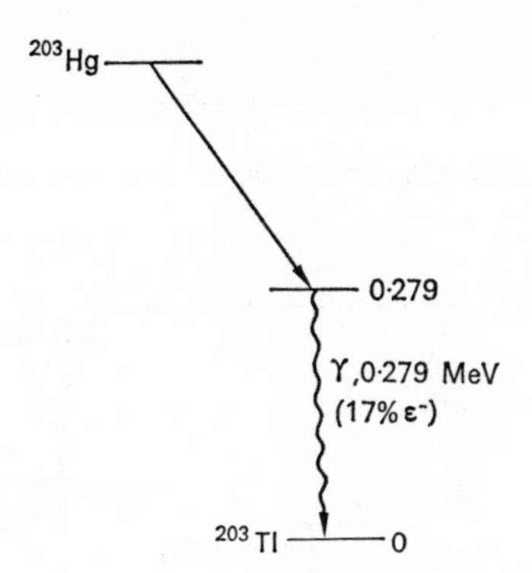

FIG. 4.5. Decay scheme of ^{203}Hg.

Both illustrations given above (^{32}P and ^{203}Hg) are for beta emitters with simple spectra, that is, for emitters with only one group of beta-rays. Complex beta emitters are those isotopes whose beta ray spectra contains more than one distinct group of beta rays. Potassium-42, for example, in about 82% of its disintegrations, decays to stable ^{42}Ca by emission of a beta particle from a group whose maximum energy is 3.55 MeV and in 18% of its disintegrations by emitting a 2.04-MeV beta. In this case, however, the excited ^{42}Ca immediately emits a gamma-ray photon whose energy is 1.53 MeV. A commonly used isotope which has an even more complex beta-gamma spectrum is ^{131}I. This isotope decays to stable ^{131}Xe by emission of a beta particle. In 85% of the disintegrations, however, the beta particle is a member of a group whose maximum energy is 0.6 MeV, while the remainder of beta particles belong to a group whose maximum energy is 0.315 MeV. In both instances, the xenon daughter nucleus is left in an excited state, and rids itself of its energy of excitation by the emission of gamma radiation. In the case of the nucleus resulting from the emission of the lower energy beta particle, a gamma photon whose energy is 0.638 MeV is emitted. The nucleus in the lower energy level, the one resulting from the emission of the 0.6-MeV beta particle, rids itself of its excitation energy by two competing gamma-ray transitions. About 93%

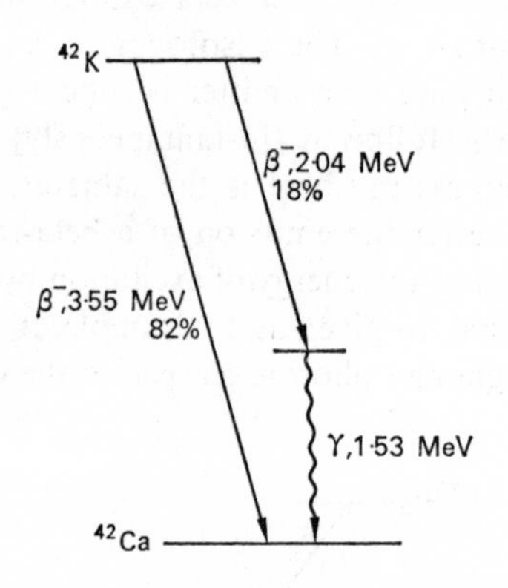

FIG. 4.6. Potassium-42 decay scheme.

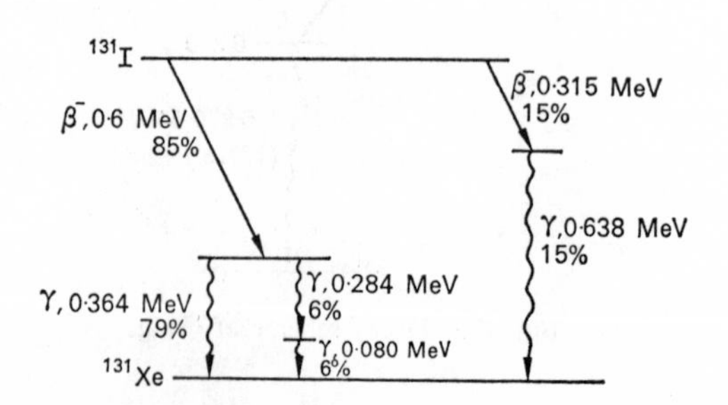

FIG. 4.7. Iodine-131 decay scheme.

of these nuclei (corresponding to 79% of the ^{131}I disintegrations) emit 0.364-MeV gamma-rays and the balance of the excited nuclei emit two gamma photons in cascade, one of 0.284 MeV and one of 0.080 MeV. The decay scheme for ^{131}I is shown in Fig. 4.7.

Beta radiation, because of its ability to penetrate tissue to varying depths, depending on the energy of the beta particle, may be an external radiation hazard. The exact degree of hazard, of course, depends on the beta-emitting isotope, and must be evaluated in every case. Generally, however, beta-rays whose energies are less than 200 keV and therefore have very limited penetrability, such as those from tritium, ^{35}S, and ^{14}C, are not considered as external radiation hazards. It should be noted, however, that beta-rays give rise to highly penetrating X-rays called *bremsstrahlung* when they are stopped by shielding. (This interaction will be more fully discussed later, in Chapter 5.) Unless shielding is properly designed, and proper precautionary measures adopted, beta radiation may indirectly result in an external radiation hazard through the production of bremsstrahlung. Any beta-emitting isotope, of course, is potentially hazardous when it is deposited in the body in amounts exceeding those thought to be safe.

Positron Emission

In those instances where the neutron to proton ratio is too low and alpha emission is not energetically possible, the nucleus may, under certain conditions, attain stability by emitting a positron. A positron is a beta particle whose charge is positive. In all other respects it is the same as the negative beta particle, or an ordinary electron. Its mass is 0.000548 atomic mass units and its charge is $+4.8 \times 10^{-10}$ sV. Because of the fact that the nucleus loses a positive charge when a positron is emitted, the daughter product is one atomic number less than the parent. The mass number of the daughter remains unchanged, as in all nuclear transitions involving electrons. In the case of ^{22}Na, for example, we have

$$^{22}_{11}\mathrm{Na} \rightarrow {}^{22}_{10}\mathrm{Ne} + {}_{1}e^{0} + \nu. \tag{4.9}$$

Whereas negative electrons occur freely in nature, positrons have only a transitory existence. They occur in nature only as the result of the interaction between cosmic rays and the atmosphere, and disappear in a matter of microseconds after formation. The manner of disappearance is of interest and great importance to the health physicist. The positron combines with an electron, and the two particles are annihilated, giving rise to two gamma-ray photons whose energies are equal to the mass equivalent of the positron and electron. This interaction will be discussed more fully in Chapter 5. The positron is not thought to exist independently within the nucleus. Rather, it is believed that the positron results from a transformation, within the nucleus, of a proton into a neutron according to the reaction

$${}_{1}H^{1} \rightarrow {}_{0}n^{1} + {}_{1}e^{0} + \nu. \tag{4.10}$$

For positron emission, the following conservation equation must be satisfied:

$$M_p = M_d + M_e + Q, \tag{4.11}$$

where M_p, M_d, and M_e are the masses of the parent nucleus, daughter nucleus, and positron respectively, the Q is the mass equivalent of the energy of the reaction. Since the daughter is one atomic number less than the parent, it must also lose an orbital electron immediately after the nuclear transition. In terms of atomic masses, therefore, the conservation equation is

$$M'_p = M'_d + 2M_e + Q. \tag{4.12}$$

Sodium-22, a useful isotope for bio-medical research, disintegrates by two competing mechanisms, positron emission and *K* capture (which is discussed in the following section), according to the decay scheme shown below in Fig. 4.8. Positrons are emitted in 89.8% of the disintegrations, while the

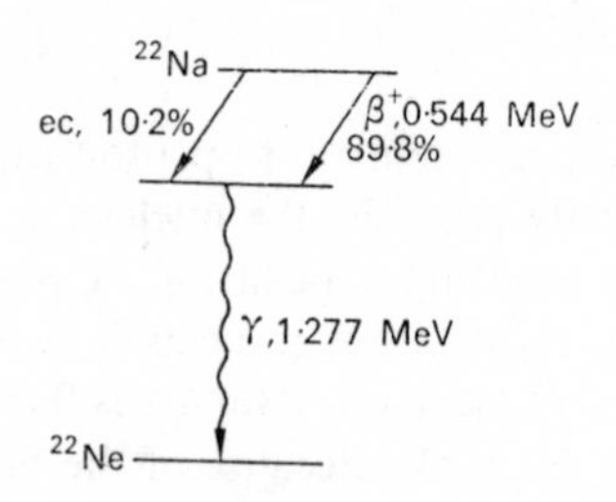

FIG. 4.8. Sodium-22 decay scheme.

competing decay mode, *K* capture, occurs in 10.2% of the nuclear transformations. Both modes of decay result in ^{22}Ne, which is in an excited state; the excitation energy instantly appears as a 1.277 MeV gamma-ray. The exact atomic mass of the neon may be calculated from the positron decay data with the aid of equation (4.12).

$$\begin{aligned} M\,(^{22}\text{Ne}) &= M\,(^{22}\text{Na}) - 2M_e - Q_m \\ &= 22.001404 - 2 \times 0.000548 - \frac{(0.544 + 1.277)\ \text{MeV}}{931\ \text{MeV/amu}} \\ &= 21.998352\ \text{amu}. \end{aligned}$$

Since positrons are electrons, the radiation hazard from the positrons themselves is very similar to the hazard from beta particles. The gamma radiation resulting from the annihilation of the positron, however, makes all positron emitting isotopes potential external radiation hazards.

Orbital Electron Capture

Equation (4.12) shows that, if a neutron-deficient atom is to attain stability by positron emission, it must exceed the weight of its daughter by at least two electron masses. If this requirement cannot be met, then the neutron deficiency is overcome by the process known as orbital electron capture or, alternatively, as *K* capture. In this radioactive transformation, one of the extra-nuclear electrons is captured by the nucleus, and unites with an intra-nuclear proton to form a neutron according to the equation

$$_{-1}e^0 + {}_1H^1 \rightarrow {}_0n^1 + \nu. \tag{4.13}$$

Since the electrons in the K shell are much closer to the nucleus than those in any other shell, the probability that the captured orbital electron will be from the K shell is much greater than that for any other shell, hence the alternate (and most common name) for this mechanism. In the case of K capture, as in positron emission, the atomic number of the daughter is one less than that of the parent, while the atomic mass number remains unchanged. The energy conservation requirements for K capture are much less rigorous than for positron emission. It is merely required that the following conservation equation be satisfied:

$$M_p + M_e = M_d + \phi + Q, \tag{4.14}$$

where M_p and M_d are the exact atomic masses of the parent and daughter, M_e is the mass of the captured electron, ϕ is the binding energy of the captured electron, and Q is the energy of the reaction.

Equation (4.14) may be illustrated by the K capture mode of decay of ^{22}Na. The binding energy, ϕ, of the sodium K electron is 1.08 keV. The energy of decay, Q, may therefore by calculated as follows:

$$Q = M\,(^{22}\text{Na}) + M_e - M\,(^{22}\text{Ne}) - \phi$$

$$= 22.001404 + 0.000548 - 21.998352 - \frac{0.00108 \text{ MeV}}{931 \text{ MeV/amu}}$$

$$= 0.003600 \text{ amu}.$$

In terms of MeV, we have

$$Q = 0.0036 \text{ amu} \times 931 \frac{\text{MeV}}{\text{amu}}$$

$$= 3.352 \text{ MeV}.$$

Since a 1.277 MeV gamma-ray is emitted, we are left with an excess of 3.352 − 1.277 = 2.075 MeV. The recoil energy associated with the emission of the gamma-ray photon is insignificantly small. The excess energy, therefore, must be carried away by a neutrino. Although the example given above is for a specific reaction, it is nevertheless typical of all reactions involving K capture; a neutrino is always emitted when an orbital electron is captured. It is thus seen that in all types of radioactive decay involving either the capture or emission of an electron, a neutrino must be emitted in order to conserve energy. In contrast to positron and negatron (ordinary beta) decay, however, in which the nuetrino carries off the difference between the actual kinetic energy of the particle and the maximum observed kinetic energy, and therefore has a continuous energy distribution, the neutrino in orbital electron capture is necessarily monoenergetic.

Whenever an atom decays by orbital electron capture, an X-ray characteristic of the daughter element is emitted as an electron from an outer orbit falls into the energy level occupied by the electron which had been captured. That characteristic X-rays of the daughter should be observed follows from the fact that the X-ray photon is emitted after the nucleus captures the orbital electron and is thereby transformed into the daughter. These low energy characteristic X-rays must be considered by the health physicist when he computes absorbed radiation doses from internally deposited isotopes which decay by orbital electron capture.

Gamma-rays

Gamma-rays are monochromatic electromagnetic radiations that are emitted from nuclei of excited atoms following radioactive disintegration; they provide a mechanism for ridding excited nuclei of their excitation energy. Since the health physicist is concerned with all radiations which come from radioactive substances, and since X-rays are indistinguishable from gamma-rays, characteristic X-rays that arise in the extra-nuclear structure of many isotopes must be considered by the health physicist when he evaluates radiation hazards. However, because of the low energy of characteristic X-rays, they are of importance mainly in the case of internally deposited radioisotopes. Annihilation radiation, the gamma-rays resulting from the mutual annihilation of positrons and negatrons, are usually associated, for health physics purposes, with the isotope that emits the positrons. When considering the radiation hazard from ^{22}Na, for example, two photons from the annihilation process, which are not shown on the decay scheme, must be considered together with the 1.277-MeV gamma photon that is shown on the decay scheme. The general rule in health physics, therefore, is to automatically associate positron emission with gamma radiation in all problems involving shielding, dosimetry, and radiation hazard evaluation.

Internal Conversion

Internal conversion is an alternative mechanism by means of which an excited nucleus of a gamma-emitting isotope may rid itself of the excitation energy. It is an interaction in which a tightly bound electron interacts with its nucleus, absorbs the excitation energy from the nucleus, and is ejected from the atom. Internally converted electrons appear in monoenergetic groups. The kinetic energy of the converted electron is always found to be equal to the difference between the energy of the gamma-ray photon emitted by the radioisotope and the binding energy of the converted electron of the daughter element. Since electrons in the L energy level of high atomic-numbered elements are also tightly bound, internal conversion in those elements results in two groups of electrons which differ in energies by the difference between the binding energies of the K and L levels. Because of these experimental findings, internal conversion may be thought of as an internal photoelectric effect, that is, an interaction in which the gamma photon collides with the tightly bound electron and transfers all of its energy to the electron. The energy of the photon is divided between work done to overcome the binding energy of the electron and kinetic energy imparted to the electron. Mathematically, this may be expressed by the equation

$$E_\gamma = E_e + \phi, \tag{4.15}$$

where $E\gamma$ is the energy of the gamma ray, E_e is the kinetic energy of the conversion electron, and ϕ is the binding energy of the electron. Since conversion electrons are monoenergetic, they appear as line spectra superimposed on the continuous beta-ray spectra of the isotope. An interesting example of internal conversion is given by ^{137}Cs. This isotope decays by beta emission to an excited state of ^{137}Ba. The ^{137}Ba then emits a 0.661-MeV photon which undergoes internal conversion in 11% of the transitions. The internal conversion coefficient α, which is defined as the ratio of the number of conversion electrons per gamma-ray photon,

$$\alpha = \frac{N_e}{N_\gamma} \tag{4.16}$$

is equal to 0.11 in this instance. The conversion electrons, which seem to come from the ^{137}Cs, are found to be superimposed on the beta spectrum of the cesium, as shown below in Fig. 4.9. After internal conversion, characteristic X-rays are emitted as outer orbital electrons fill the vacancies left in the deeper energy levels by the conversion electrons. These characteristic X-rays may themselves be absorbed by an internal photoelectric effect on the atom

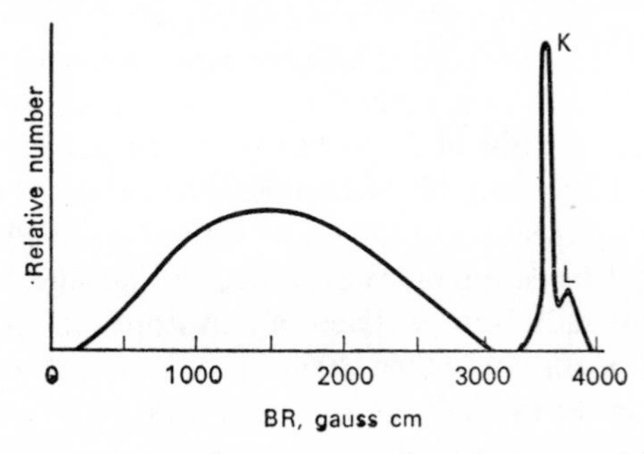

FIG. 4.9. Cesium-137 beta spectrum, showing conversion electrons from the K and L energy levels.

from which they were emitted, a process that is of the same nature as internal conversion. The ejected electrons from this process are called Auger electrons, and they possess very little kinetic energy.

Kinetics of Decay

Half-life

Different isotopes disintegrate at different rates, and each isotope has its own characteristic decay rate. For example, when the activity of ^{32}P is measured daily over a period of about 3 months, and the percent of the initial activity is plotted as a function of time, the curve shown in Fig. 4.10 is obtained. The data show that one-half of the ^{32}P is gone in 14.3 days, half of

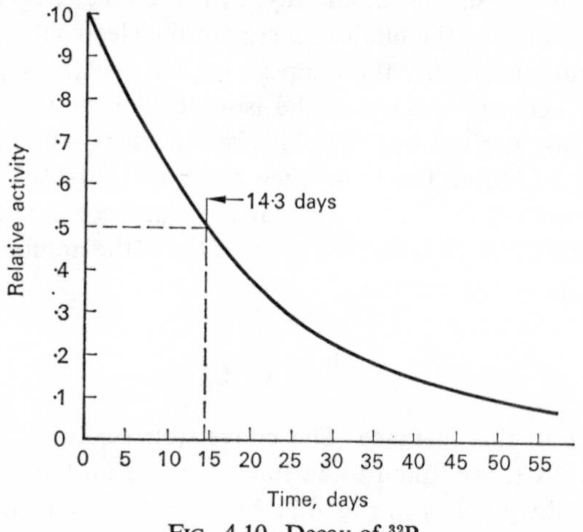

Fig. 4.10. Decay of ^{32}P.

the remainder in another 14.3 days, half of what is left during the following 14.3 days, and so on.

If a similar series of measurements were made on ^{131}I, it would be observed that the iodine would disappear at a faster rate. One-half would decay after 8 days, and three-fourths of the initial activity would have decayed after only 16 days, while seven-eighths of the iodine would be gone after 24 days.

The time required for any given radioisotope to decay to one-half of its original quantity is a measure of the speed with which the isotope undergoes radioactive decay. This period of time is called the half-life, and is characteristic of the particular radioisotope. Each radioisotope has its own unique rate of decay, and no operation, either chemical or physical, is known that will change the decay rate; the half-life of a radioisotope is an unalterable property of the isotope. Half-lives of radioisotopes range from microseconds to billions of years.

From the definition of the half-life, it follows that the fraction of a radio-isotope remaining after n half-lives is given by the relationship

$$\frac{A}{A_0} = \frac{1}{2^n}, \tag{4.17}$$

where A_0 is the original quantity of activity, and A is the activity left after n half-lives.

Example 4.1

Cobalt-60, a gamma-emitting radioisotope whose half-life is 5.3 years, is used as a radiation source for radiographing pipe welds. Because of the radio-active decay, the exposure time for a radiograph will be increased annually. Calculate the correction factor to be applied to the exposure time in order to account for the decrease in the strength of the source. Equation (4.17) may be rewritten as

$$\frac{A_0}{A} = 2^n.$$

By taking the logarithm of each side of the equation, we have

$$\log \frac{A_0}{A} = n \log 2,$$

where n, the number of ^{60}Co half-lives in one year, is $1/5.3 = 0.189$.

$$\log \frac{A_0}{A} = 0.189 \times 0.301,$$

$$\frac{A_0}{A} = \text{antilog } 0.0569$$

$$= 1.14.$$

The ratio of the initial quantity of cobalt to the quantity remaining after 1 year is 1.14. The exposure time after 1 year, therefore, must be increased by 14%. It should be noted that this ratio is independent of the actual amount of activity at the beginning and end of the year. After the second year, the ratio of the cobalt at the beginning of the second year to that at the end will be 1.14. The same correction factor, 1.14, therefore, is applied every year to the exposure time for the previous year.

If the decay data for any isotope are plotted on semi-logarithmic paper, with the activity measurements recorded on the logarithmic axis and time on the linear axis, a straight line results. If time is measured in units of half-lives, the generally useful curve shown in Fig. 4.11 results.

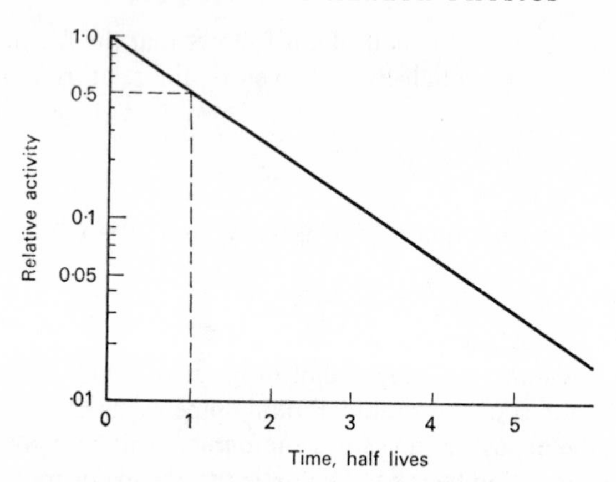

FIG. 4.11. Generalized semi-logarithmic plot of radioactive decay.

The illustrative example given above could have been solved graphically with the aid of this curve. The ordinate at which the time in units of half-life, 0.189, intersects the curve shows that 87.7% of the original activity is left. The correction factor, therefore, is the reciprocal of 0.877:

$$\text{Correction factor} = \frac{1}{0.877} = 1.14.$$

That fact that the graph of activity vs. time, when drawn on semi-logarithmic paper, is a straight line tells us that the quantity of activity left after any time interval is given by the following equation:

$$A = A_0 e^{-\lambda t}, \tag{4.18}$$

where A_0 is the initial quantity of activity, A is the amount left after time t, λ is the decay constant, and e is the base of the system of natural logarithms. The decay constant is the fraction of the activity that decays per unit time, and is defined as

$$\lim_{\Delta t \to 0} \frac{\Delta N/N}{\Delta t} = -\lambda, \tag{4.19}$$

where N is a number of radioactive atoms, and ΔN is the number of these atoms that disintegrate during a time interval Δt. The fraction $\Delta N/N$ is the fractional decrease in the number of radioactive atoms during the time interval Δt. A negative sign is given to λ to indicate that the quantity N is decreasing. For a short-lived radioisotope, λ may be determined from the slope of an experimentally determined decay curve. For long-lived isotopes, the decay constant may be determined by measuring N, then by counting the number of disintegrations per second, and then calculating the numerical value of λ from equation (4.19).

Example 4.2

One microgram radium is found to emit 3.7×10^4 alpha particles per second. If each of these alphas represents a radioactive transformation of radium, what is the decay constant for radium?

In this case, ΔN is 3.7×10^4, Δt is one second, and N, the number of radium atoms per microgram, may be calculated as follows:

$$N = \frac{6.03 \times 10^{23} \text{ atoms/mole}}{A \text{ g/mole}} \times W \text{ g}, \tag{4.20}$$

where A is the atomic weight and W the weight of the radium sample.

$$N = \frac{6.03 \times 10^{23} \times 10^{-6}}{2.26 \times 10^2} = 2.66 \times 10^{15} \text{ atoms.}$$

The decay constant, therefore, is

$$\lambda = \frac{\Delta N/N}{\Delta t} = \frac{3.7 \times 10^4 \text{ atoms}/2.66 \times 10^{15} \text{ atoms}}{1 \text{ sec}} = 1.39 \times 10^{-11} \text{ sec}^{-1}.$$

On a per year basis, the decay rate is

$$\lambda = 1.39 \times 10^{-11} \text{ sec}^{-1} \times 8.64 \times 10^4 \frac{\text{sec}}{\text{day}} \times 3.65 \times 10^2 \frac{\text{day}}{\text{yr}}$$

$$= 4.38 \times 10^{-4} \text{ yr}^{-1}.$$

This value of λ may be used in equation (4.18) to compute the amount of radium left after any given time period.

Example 4.3

What percent of a given amount of radium will decay during a period of 1000 years?

The fraction remaining after 1000 years is given by

$$\frac{A}{A_0} = e^{-\lambda t} = e^{-4.38 \times 10^{-4} \text{ yr}^{-1} \times 10^3 \text{ yr}} = e^{-0.438}.$$

This equation may be easily solved with a log-log slide-rule, with tables of values for powers of e, or with logarithms.

$$\frac{A}{A_0} = 0.658 \text{ or } 65.8\%.$$

The percent that decayed away during the 1000-year period, therefore, is

$$100 - 65.8 = 34.2\%.$$

The quantitative relationship between half-life, T, and decay constant, λ, may be found by setting A/A_0 in equation (4.18) equal to $\frac{1}{2}$, and solving the equation for t. In this case, of course, the time is the half-life.

$$\frac{A}{A_0} = \frac{1}{2} = e^{-\lambda T},$$
$$T = \frac{0.693}{\lambda}. \tag{4.21}$$

Example 4.4

Given that the decay constant for ^{226}Ra is 4.38×10^{-4} per year, calculate the half-life for radium.

$$T = \frac{0.693}{\lambda}$$
$$= \frac{0.693}{4.38} \times 10^{-4}\ \text{yr}^{-1}$$
$$= 1580 \text{ years.}$$

Average Life

Although the half-life of an isotope is a unique, reproducible characteristic of that isotope, it is nevertheless a statistical property, and is valid only because of the very large number of atoms involved. (One microgram radium contains 2.79×10^{15} atoms.) Any particular atom of a radioisotope may disintegrate at any time, from zero to infinity, after it is observed. For some applications, such as in the case of dosimetry of internally deposited radioisotopes (to be discussed in Chapter 6), it is convenient to use the average life of the radioisotope. The average life is defined simply as the sum of the lifetimes of the individual atoms divided by the total number of atoms originally present.

The instantaneous disintegration rate of a quantity of radioisotope containing N atoms is λN. During the time interval between t and $t + dt$, the total number of disintegrations is $\lambda N dt$. Each of the atoms that decayed during this interval, however, had existed for a total lifetime t since the beginning of observation on them. The sum of the lifetimes, therefore, of all the atoms that decayed during the time interval between t and $t + dt$, after having survived since time $t = 0$, is $t\lambda N dt$. The average life of the radioactive species, τ, is

$$\tau = \frac{1}{N_0} \int_0^\infty t\, \lambda\, N\, dt, \tag{4.22}$$

where N_0 is the number of radioactive atoms in existence at time $t = 0$. Since

$$N = N_0 \, e^{-\lambda t},$$

we have

$$\tau = \frac{1}{N_0} \int_0^\infty t \, \lambda \, N_0 \, e^{-\lambda t} \, dt. \tag{4.23}$$

This expression, when integrated by parts, shows the value for the mean life of a radioisotope to be

$$\tau = \frac{1}{\lambda}. \tag{4.24}$$

If the expression for the decay constant in terms of the half-life of the radioisotope,

$$\lambda = \frac{0.693}{T},$$

is substituted into equation (4.22), the relationship between the half-life and the mean life is found to be

$$\tau = \frac{T}{0.693} = 1.45 \, T. \tag{4.25}$$

The Curie

Uranium-238 and its daughter ^{234}Th each contain about the same number of atoms per gram; approximately 2.5×10^{21}. Their half-lives, however, are greatly different; ^{238}U has a half-life of 4.5×10^{9} years while ^{234}Th has a half-life of 24.1 days (or 6.63×10^{-2} years). Thorium-234, therefore, is decaying 6.8×10^{10} times faster than ^{238}U. Another example of greatly different rates of decay that may be cited is ^{35}S and ^{32}P. These two radioisotopes, which have about the same number of atoms per gram, have half-lives of 87 and 14.3 days respectively. The radiophosphorous, therefore, is decaying about 6 times faster than the ^{35}S. When radioisotopes are used, the radiations are the center of interest. In this context, therefore, $\frac{1}{6}$ of a gram of ^{32}P is about equivalent to 1 g of ^{35}S in radioactivity, while 15 micromicrograms of ^{234}Th is about equivalent in activity to 1 g of ^{238}U. Obviously, therefore, when interest is centered on radioactivity, the gram is not a very useful unit of quantity. To be meaningful, the unit for quantity of radioactivity must be based on activity. Such a unit is called the *curie* (symbolized by Ci) and is defined as follows:

> The curie is the activity of that quantity of radioactive material in which the number of disintegrations per second is 3.7×10^{10}.

It should be emphasized that, although the curie is defined in terms of a number of disintegrating atoms per second, it is not a measure of rate of decay. *The curie is a measure only of quantity* of radioactive material. The phrase "disintegrations per second" as used in the definition of the curie is not synonymous with number of particles emitted by the radioactive isotope. In the case of a simple pure beta emitter, for example, 1 curie, or 3.7×10^{10} disintegrations per second, does in fact result in 3.7×10^{10} beta particles per

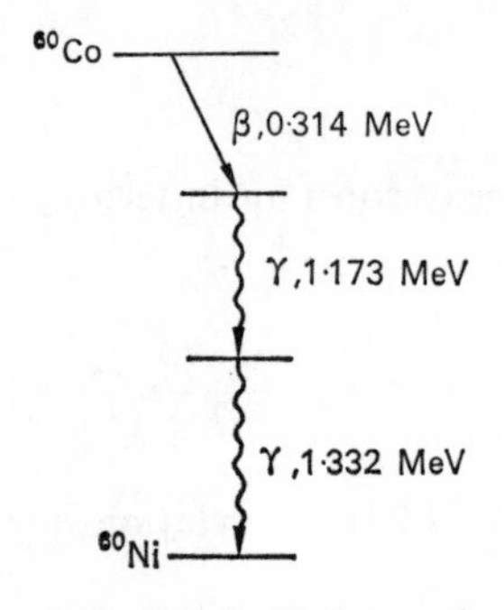

FIG. 4.12. Cobalt-60 decay scheme.

second. In the case of a more complex radioactive isotope, however, such as ^{60}Co, Fig. 4.12, each disintegration releases 1 beta particle and 2 gamma photons; the total number of radiations, therefore, is $3 \times 3.7 \times 10^{10}$, or 11.1×10^{10} per second per curie ^{60}Co. In the case of ^{42}K, Fig. 4.6, on the other hand, 20% of the beta decays are accompanied by a single quantum of gamma radiation. The total number of emissions from 1 curie ^{42}K, therefore, is

$$3.7 \times 10^{10} + 0.2 \times 3.7 \times 10^{10} = 4.44 \times 10^{10} \text{ per sec.}$$

For health physics, as well as for many other purposes, the curie is a very large quantity of activity. Submultiples of the curie, as listed below, therefore are used:

$$1 \text{ millicurie (mCi)} = 10^{-3} \text{ Ci}$$

$$1 \text{ microcurie } (\mu\text{Ci}) = 10^{-6} \text{ Ci}$$

$$1 \text{ nanocurie (nCi)} = 10^{-9} \text{ Ci}$$

$$1 \text{ picocurie (pCi)} = 10^{-12} \text{ Ci}$$

Multiples of the curie that are frequently used are the kilocurie and the megacurie. These quantities are generally not abbreviated.

Specific Activity

Note that the curie, although used as a unit of quantity, does not mention anything about the mass or volume of the radioactive material in which the specified number of disintegrations per second occur. The concentration of radioactivity, or relationship between the mass of radioactive material and the activity, is called the specific activity. Specific activity is the number of curies per unit mass or volume. The specific activity of a carrier free radioisotope, that is, a radioisotope that is not mixed with any other isotope of the same element, may be calculated as follows:

If λ is the decay constant in units of reciprocal seconds, then the number of disintegrating atoms per second among an aggregation of N atoms is simply

$$\frac{\text{disintegrations}}{\text{second}} = \lambda N.$$

If the radioisotope under consideration weighs 1 g, then, according to equation (4.20), the number of atoms is simply equal to

$$N = \frac{6.03 \times 10^{23} \text{ atoms/mole}}{A \text{ g/mole}},$$

where A is the atomic weight of the isotope. The activity per unit time, therefore, is

$$\lambda N = \frac{\lambda \times 6.03 \times 10^{23}}{A} \frac{\text{dis}}{\text{sec/g}}. \tag{4.26}$$

Equation (4.26) gives the desired relationship between activity and weight of an isotope. The unit for activity in the equation may be converted from disintegrations per second to curies by application of the fact that there are 3.7×10^{10} disintegrations per second per curie:

$$\text{Specific activity} = \frac{\lambda \times 6.03 \times 10^{23}/A \text{ dps/g}}{3.7 \times 10^{10} \text{ dps/curie}}.$$

$$\text{Specific activity} = 1.63 \times 10^{13} \frac{\lambda}{A} \frac{\text{curies}}{\text{gram}}. \tag{4.27}$$

Note that the decay constant, λ, in equation (4.27), must be in reciprocal seconds. Equation (4.27) may be rewritten in terms of half-life rather than decay constant.

$$\text{S.A.} = \frac{1.63 \times 10^{13}}{A} \times \frac{0.693}{T} = \frac{1.13 \times 10^{13}}{A \times T} \frac{\text{curies}}{\text{g}}. \tag{4.28}$$

A more convenient form for the equation for specific activity may be derived by making use of the fact that the specific activity of ^{226}Ra is 1 curie per gram. This value for the specific activity of radium is no coincidence. When radium had been the radioactive material of greatest interest to scientists, the curie was defined in terms of the activity of 1 g of radium. Since the value for the specific activity of radium was an experimentally determined quantity, it naturally varied within the limits of the experimental errors of the investigators who were making the measurements. Since there was no theoretical reason for using the precise value for the specific activity of radium as the standard for the curie, the curie was redefined in terms independent of radium. The activity chosen as the standard, 3.7×10^{10} disintegrations per second, however, is extremely close to the best estimate of the activity of 1 g of radium. For practical purposes, therefore, radium has a specific activity of 1 curie per gram. It should be pointed out that another unit for quantity of radioactive material was proposed, but was not officially adopted. This other unit is called the rutherford (rd), and is defined as a quantity of radioactive material in which the activity is 1×10^6 disintegrations per second. Although the rutherford was not officially adopted, it nevertheless is sometimes used instead of the curie to denote quantity of radioactivity.

Since the specific activity of ^{226}Ra is 1 curie per gram, equation (4.29) may be used to specify the ratio of activity of any radioisotope to that of radium.

$$\frac{X \text{ Ci/g}}{1 \text{ Ci/g}} = \frac{1.13 \times 10^{13}/A_x T_x}{1.13 \times 10^{13}/A_{Ra} T_{Ra}}$$

or

$$\text{S.A.}_x = \frac{A_{Ra} \times T_{Ra}}{A_x \times T_x} \frac{\text{Ci}}{\text{g}}. \qquad (4.29)$$

In equation (4.29), A_{Ra}, the atomic weight of ^{226}Ra, is 226, A_x is the atomic weight of the radioisotope whose specific activity is being calculated, and T_{Ra} and T_x are the half-lives of radium and the isotope X. The only restriction on equation (4.29) is that both half-lives be in the same units of time.

Example 4.5

Calculate the specific activities of ^{14}C and ^{35}S, given that their half-lives are 5600 years and 87 days respectively.

For ^{14}C:

$$\text{S.A.} = \frac{2.26 \times 10^2 \times 1.65 \times 10^3 \text{ years}}{14 \times 5.6 \times 10^3 \text{ years}} = 4.75 \frac{\text{Ci}}{\text{g}}.$$

For ^{35}S:

$$S.A. = \frac{2.26 \times 10^2 \times 1.65 \times 10^3 \text{ years} \times 3.65 \times 10^2 \text{ days/year}}{35 \times 87 \text{ days}}$$

$$= 44{,}600 \frac{\text{Ci}}{\text{g}}.$$

The specific activities calculated above are for the carrier-free isotopes of ^{14}C and ^{35}S. Very frequently, especially when radioisotopes are used to label compounds, the isotope is not carrier free, but rather it constitutes an extremely small fraction, either by weight or number of atoms, of the element that is labeled. In such cases, it is customary to refer to the specific activity either of the element or the compound which is labeled. Generally, the exact meaning of specific activity is clear from the context.

Example 4.6

A solution of $Hg(NO_3)_2$ tagged with ^{203}Hg has a specific activity of 4 μCi/ml. If the concentration of Hg in the solution is 5 mg/ml,

(a) what fraction of the Hg in the $Hg(NO_3)_2$ is ^{203}Hg?
(b) What is the specific activity of the $Hg(NO_3)_2$?

(a) The specific activity of the Hg is

$$\frac{4 \ \mu\text{Ci/ml}}{5 \text{ mg/ml Hg}} = 0.8 \ \frac{\mu\text{Ci}}{\text{mg Hg}},$$

and the specific activity of the carrier-free ^{203}Hg is

$$\frac{226 \times 1650 \text{ years} \times 365 \text{ days/year}}{203 \times 46.5 \text{ days}} = 14{,}400 \ \frac{\text{Ci}}{\text{g}};$$

the fraction tagged is:

$$\frac{8 \times 10^{-4} \text{ Ci/g Hg}}{1.44 \times 10^{4} \text{ Ci/g } ^{203}\text{Hg}} = 5.55 \times 10^{-8} \ \frac{\text{g } ^{203}\text{Hg}}{\text{g Hg}}.$$

(b) Since an infinitesimally small fraction of the mercury is tagged with ^{203}Hg, it may be assumed that the formula weight of the tagged $Hg(NO_3)_2$ is 324.63, and that the concentration of $Hg(NO_3)_2$ is:

$$\frac{324.63 \text{ mg Hg (NO}_3)_2}{200.61 \text{ mg Hg}} \times \frac{5 \text{ mg Hg}}{\text{ml}} = 8.1 \ \frac{\text{mg Hg (NO}_3)_2}{\text{ml}}.$$

The specific activity, therefore, of the $Hg(NO_3)_2$ is:

$$\frac{4 \ \mu\text{Ci/ml}}{8.1 \text{ mg Hg (NO}_3)_2\text{/ml}} = 0.495 \ \frac{\mu\text{Ci}}{\text{mg Hg (NO}_3)_2}.$$

Example 4.7

Can commercially available ^{14}C-tagged absolute ethanol, CH_3—C^*H_2—OH, whose specific activity is 1 mCi/mole, be used in an experiment that requires a minimum specific activity of 10^7 disintegrations per minute per milliliter? The density of the alcohol is 0.789 g/cm^3.

The specific activity of ^{14}C is 4.75 Ci/g. One millicurie ^{14}C, therefore, weighs

$$\frac{10^{-3}\text{ Ci}}{4.75\text{ Ci/g}} = 2.1 \times 10^{-4}\text{ g},$$

and the number of radioactive atoms represented by 0.21 mg ^{14}C is

$$\frac{6.03 \times 10^{23}\text{ atoms/mole}}{14\text{ g/mole}} \times 2.1 \times 10^{-4}\text{ g} = 9.05 \times 10^{18}\text{ atoms.}$$

Since one mole contains Avogadro's number of molecules, and each tagged molecule contains only one carbon atom, there are

$$\frac{9.05 \times 10^{18}}{6.03 \times 10^{23}} = 15\text{ per million}$$

ethanol molecules that are tagged. For all practical purposes, therefore, the additional mass due to the isotopic carbon may be neglected when calculating the molecular weight of the labeled ethanol, and the accepted molecular weight of ethanol, 46.078, may be used to compute the activity per cubic centimeter of the alcohol:

$$1\,\frac{\text{mCi}}{\text{mole}} \times \frac{1\text{ mole}}{46.078\text{ g}} \times 2.22 \times 10^9\,\frac{\text{dis/min}}{\text{mCi}} \times 0.789\,\frac{\text{g}}{\text{cm}^3}$$

$$= 3.74 \times 10^7\,\frac{\text{dis/min}}{\text{cm}^3}.$$

The commercially available ethanol may be used.

Naturally Occurring Radioactivity

The naturally occurring radioactive substance which Becquerel discovered in 1896 was a mixture of several isotopes which were later found to be related to each other. They were members of long series of isotopes of various elements, all of which were radioactive but the last. Uranium, the most abundant of the radioactive elements in this mixture, consists of three different isotopes: about 99.3% of naturally occurring uranium is ^{238}U, about 0.7% is ^{235}U, and a trace quantity (about 5×10^{-6}%) is ^{234}U. The ^{238}U and ^{234}U belong to one family, the uranium series, while the ^{235}U isotope of uranium is the first member of another series called the actinium series. The most abundant of all naturally occurring radioisotopes, ^{232}Th, is the first member of still another ong chain of successive radioisotopes. All of the isotopes that are members of adioactive series are found in the upper portion of the periodic table; the owest atomic number in these groups is 81, while the lowest mass number s 207. All the radioactive series have several common characteristics. First s the fact that the first member of each series is very long lived, with half-lives hat may be measured in geological time units. That the first member of each nust be very long lived is obvious, since, if the time since the creation of the world is considered, relatively short-lived isotopes would have decayed away during the several billions of years that the earth is believed to be in existence. This point is well illustrated by the artificially produced series of isotopes called the neptunium series. In this case, the first member is the transuranium element ^{241}Pu which is produced in the laboratory by neutron irradiation of reactor produced ^{239}Pu. The half-life of ^{241}Pu, however, is only 13 years. Because of the short half-life, even a period of a century is long enough to permit most of the ^{241}Pu to decay away. Even the longest-lived member of this series, ^{237}Np, whose half-life is 2.2×10^6 years, is sufficiently short to have completely disappeared if it had been created at the same time as all the other elements of the earth.

A second characteristic common to all three naturally occurring series is that each has a gaseous member, and furthermore that the radioactive gas in each case is a different isotope of the element radon. In the case of the uranium series, the gas, $^{222}_{83}$Rn, is called radon; in the thorium series the gas, $^{220}_{83}$Rn, is called thoron, while in the actinium series it is called actinon, $^{219}_{83}$Rn. It should be noted that the artificial neptunium series has no gaseous member. The existence of the radioactive gasses in the three chains is one of the chief reasons for the presence of naturally occurring environmental radioactivity. The radon gas diffuses out of the earth into the air, and the radioactive radon daughters, which are solids under ordinary conditions, attach themselves to atmospheric dust. Atmospheric concentrations of radioactivity from this source vary widely around the earth, and are dependent on the local concentrations of uranium and thorium in the earth. Although the probable atmospheric radon concentration is in the order of 5×10^{-11} μCi/ml, concentrations 10 times greater are not uncommon. Since the radioactive radon daughters are found on the surface of atmospheric particulates, and since air-borne particulates are washed out of the atmosphere by rain, it is reasonable to expect increased background radiation during periods of rain. This phenomenon is in fact observed, and must be considered by health physicists and others when interpreting routine monitoring data. Fallout caused by rain in Upton, N.Y., is shown in Fig. 4.13, a set of curves giving the beta and gamma activity during rainy and dry periods. When the ground is covered with snow, however, a decrease in airborne radioactivity occurs because of the filtering action of the snow blanket on the effusing radon and its daughters. Increased environmental radioactivity from this source also occurs during temperature inversions, when vertical mixing of the air and consequent dilution of rado and its daughters temporarily ceases. Because of this naturally occurrin airborne activity, certain correction factors must be applied when computin the atmospheric concentration, from dust samples, of air-borne radioativ contaminants. These corrections will be discussed in more detail in Chapter 13

A third common characteristic among the three natural radioactive serie is that the end product in each case is lead. In the case of the uranium series the final member is stable ^{206}Pb, in the actinium series it is ^{207}Pb, and in th thorium series it is ^{208}Pb. The artificial neptunium series differs in this charac teristic too from the natural series; the terminal member is stable bismuth ^{209}Bi.

These four radioactive chains, the three naturally occurring ones and th

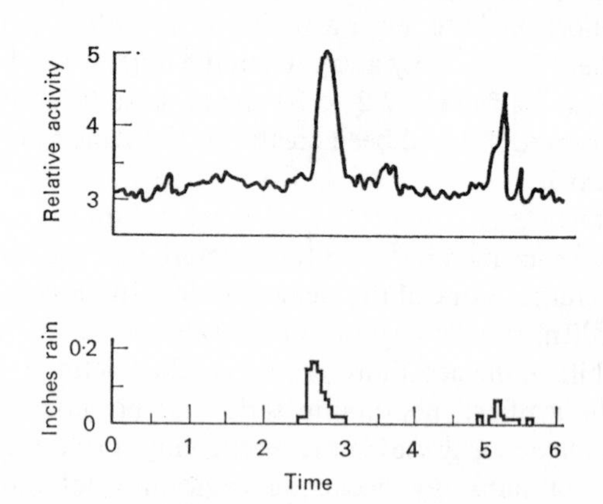

FIG. 4.13. Washout of atmospheric radioactivity by rain (M. M. Weiss, *Area Survey Manual*, BNL 344 (T-61), 15 June, 1955).

artificially produced neptunium series, are often designated as the $4n$, $4n+1$, $4n+2$ and $4n+3$ series. These identification numbers refer to the divisibility of the mass numbers of each of the series by 4. The atomic mass number of ^{232}Th, the first member of the thorium series, is exactly divisible by 4. Since all disintegrations in the series are accomplished by the emission of either an alpha particle of 4 atomic mass units, or a beta particle of 0 atomic mass units, it follows that the mass numbers of all members of the thorium series are exactly divisible by 4. The uranium series, whose first member is ^{238}U, consists of isotopes whose mass numbers are divisible by 4, and leave a remainder of 2 ($238 \div 4 = 59 + 2/4$). This series, therefore, is called the $4n+2$ series. The actinium series, whose first member is ^{235}U (actinouranium), is the $4n+3$ series. The "missing" series, $4n+1$, is the artificially produced neptunium series, which begins with ^{241}Pu.

Table 4.1. Thorium Series (4*n*)

Nuclide	Half-life	Energy, MeV		
		Alpha[a]	Beta	Gamma (photons dis.)[b]
$^{232}_{90}$Th	1.39 10^{10} y	3.98		
$^{228}_{88}$Ra (MsThl)	6.7 y		0.01	
$^{228}_{89}$Ac (MsTh2)	6.13 h		Complex decay scheme. Most intense beta group is 1.11 MeV	1.59 (n.v.) 0.966 (0.2) 0.908 (0.25)
$^{228}_{90}$Th (RdTh)	1.91 y	5.421		0.084 (0.016)
$^{224}_{88}$Ra (ThX)	3.64 d	5.681		0.241 (0.038)
$^{220}_{86}$ Em (Tn)	52 s	6.278		0.542 (0.0002)
$^{216}_{84}$ Po (ThA)	0.158 s	6.774		
$^{212}_{82}$Pb (ThB)	10.64 h		0.35, 0.59	0.239 (0.40)
$^{212}_{83}$Bi (ThC)	60.5 m	6.086 (33.7%)[c]	2.25 (66.3%)[c]	0.04 (0.034 branch)
$^{212}_{84}$Po (ThC′)	3.04×10^{-7} s	8.776		
$^{208}_{81}$Tl (ThC″)	3.1 m		1.80, 1.29, 1.52	2.615 (0.997)
$^{208}_{82}$Pb (ThD)	Stable			

[a] Only the highest energy alpha is given. Complete information on alpha energies may be obtained from Sullivan's *Trilinear Chart of Nuclides*, Government Printing Office, Washington, D.C., 1957.

[b] Only the most prominent gamma photons are listed. For the complete gamma ray information, consult T. P. Kohman, "Natural radioactivity" in *Radiation Hygiene Handbook*, pp. 6–6 to 6–13, H. Blatz, editor, McGraw-Hill, 1959.

[c] Indicates branching. The percentage enclosed in the parentheses gives the proportional decay by the indicated mode.

TABLE 4.2. NEPTUNIUM SERIES ($4n + 1$)

Nuclide	Half-life	Energy, MeV		
		Alpha[a]	Beta	Gamma (photons dis.)[b]
$^{241}_{94}Pu$	13.2 y		0.02	
$^{241}_{95}Am$	462 y	5.496		0.060 (0.4)
$^{237}_{93}Np$	2.2×10^6 y	4.77		
$^{233}_{91}Pa$	27.4 d		0.26, 0.15, 0.57	0.31 (very strong)[d]
$^{233}_{92}U$	1.62×10^5 y	4.823		0.09 (0.02) 0.056 (0.02) 0.042 (0.15)
$^{229}_{90}Th$	7.34×10^3 y	5.02		
$^{225}_{88}Ra$	14.8 d		0.32	
$^{225}_{89}Ac$	10.0 d	5.80		
$^{221}_{87}Fr$	4.8 m	6.30		0.216 (1)
$^{217}_{85}At$	0.018 s	7.02		
$^{213}_{83}Bi$	47 m	5.86 (2%)[c]	1.39 (98%)[c]	
$^{213}_{84}Po$	4.2×10^{-6} s	8.336		
$^{209}_{81}Tl$	2.2 m		2.3	0.12 (weak)[d]
$^{209}_{82}Pb$	3.32 h		0.635	
$^{209}_{83}Bi$	Stable			

[a],[b],[c] See footnotes under Table 4.1.
[d] Exact intensity not known.

TABLE 4.3. URANIUM SERIES ($4n + 2$)

Nuclide	Half-life	Energy, MeV		
		Alpha[a]	Beta	Gamma (photons dis.)[b]
${}^{38}_{2}$U	4.51×10^{9} y	4.18		
${}^{34}_{0}$Th (UX_1)	24.10 d		0.193, 0.103	0.092 (0.04) 0.063 (0.03)
${}^{34m}_{1}$Pa (UX_2)	1.175 m		2.31	1.0 (0.015) 0.76 (0.0063), I.T.
${}^{34g}_{1}$Pa (UZ)	6.66 h		0.5	Many weak
${}^{34}_{2}$U (UII)	2.48×10^{5} y	4.763		
${}^{30}_{0}$Th (I_o)	8.0×10^{4} y	4.685		0.068 (0.0059)
${}^{26}_{8}$Ra	1,622 y	4.777		
${}^{22}_{6}$Em (Rn)	3.825 d	5.486		0.51 (very weak)
${}^{8}_{4}$Po (RaA)	3.05 m	5.998 (99.978%)[c]	Energy not known (0.022%)[c]	0.186 (0.030)
${}^{8}_{5}$At (RaA′)	2 s	6.63 (99.9%)[c]	Energy not known (0.1%)[c]	
${}^{8}_{6}$Em (RaA″)	0.019 s	7.127		
${}^{4}_{2}$Pb (RaB)	26.8 m		0.65	0.352 (0.036) 0.295 (0.020) 0.242 (0.07)
${}^{4}_{3}$Bi (RaC)	19.7 m	5.505 (0.04%)[c]	1.65, 3.7 (99.96%)[c]	0.609 (0.295) 1.12 (0.131)
${}^{4}_{4}$Po (RaC′)	1.64×10^{-4} s	7.680		
${}^{0}_{1}$Tl (RaC″)	1.32 m		1.96	2.36 (1) 0.783 (1) 0.297 (1)
${}^{0}_{2}$Pb (RaD)	19.4 y		0.017	0.0467 (0.045)
${}^{0}_{3}$Bi (RaE)	5.00 d		1.17	
${}^{0}_{4}$Po (RaF)	138.40 d	5.298		0.802 (0.000012)
${}^{6}_{2}$Pb (RaG)	Stable			

[a],[b],[c] See footnotes under Table 4.1.

TABLE 4.4. ACTINIUM SERIES ($4n + 3$)

Nuclide	Half-life	Energy, MeV		
		Alpha[a]	Beta	Gamma (photons dis.)[b]
$^{235}_{92}$U	7.13×10^{8} y	4.39		0.18 (0.7)
$^{231}_{90}$Th (UY)	25.64 h		0.094, 0.302, 0.216	0.022 (0.7) 0.0085 (0.4) 0.061 (0.16)
$^{231}_{91}$Pa	3.43×10^{4} y	5.049		0.33 (0.05) 0.027 (0.05) 0.012 (0.01)
$^{227}_{89}$Ac	21.8 y	4.94 (1.2%)[c]	0.0455 (98.8%)[c]	
$^{227}_{90}$Th (RdAc)	18.4 d	6.03		0.24 (0.2) 0.05 (0.15)
$^{223}_{87}$Fr (AcK)	21 m		1.15	0.05 (0.40) 0.08 (0.24)
$^{223}_{88}$Ra (AcX)	11.68 d	5.750		0.270 (0.10) 0.155 (0.055)
$^{219}_{86}$Em (An)	3.92 s	6.824		0.267 (0.086) 0.392 (0.048)
$^{215}_{84}$Po (AcA)	1.83×10^{-3} s	7.635		
$^{211}_{82}$Pb (AcB)	36.1 m		1.14, 0.5	Complex spectrum, 0.065 to 0.829 MeV
$^{211}_{83}$Bi (AcC)	2.16 m	6.619 (99.68%)[c]	Energy not known (0.32)%[c]	0.35 (0.14)
$^{211}_{84}$Po (AcC′)	0.52 s	7.434		0.88 (0.005) 0.56 (0.005)
$^{207}_{81}$Tl (AcC″)	4.78 m		1.47	0.87 (0.005)
$^{207}_{82}$PB	Stable			

[a],[b],[c] See footnotes under Table 4.1.

Radioactive isotopes that are found in nature are not restricted to the thorium, uranium, and actinium series. Several of the elements among the lower atomic numbered members of the periodic table also have radioactive isotopes. The most important of these low-atomic-numbered natural emitters are listed below in Table 4.5.

TABLE 4.5. SOME LOW-ATOMIC-NUMBERED NATURALLY OCCURRING RADIOISOTOPES

Nuclide	Isotopic abundance (%)	Half-life (years)	Principal radiations	
			Particles	Gamma
^{40}K	0.0119	1.3×10^9	1.35 MeV	1.46 MeV
^{87}Rb	27.85	5×10^{10}	0.275 MeV	None
^{138}La	0.089	1.1×10^{11}	1.0 MeV	0.80, 1.43 MeV
^{147}Sm	15.07	1.3×10^{11}	2.18 MeV	None
^{176}Lu	2.6	3×10^{10}	0.43 MeV	0.20, 0.31 MeV
^{187}Re	62.93	5×10^{10}	0.043 MeV	None

Of these naturally radioactive isotopes, ^{40}K, by virtue of the widespread distribution of potassium in the environment (the average concentration of potassium in crustal rocks is about 27 g/kg and in the ocean is about 380 mg/l., and in plants and animals, including man (the average concentration of potassium in man is about 1.7 g/kg), is the most important from the health physics point of view. Estimates of body burden of many radioisotopes, from which the degree of exposure to environmental contaminants may be inferred, is made from radiochemical analysis of urine from persons suspected of overexposure. Potassium, whose concentration in urine is about 1.5 g/l. may interfere with the determination of the suspected contaminant unless special care is taken to remove the potassium from the urine or unless allowance is made for the ^{40}K activity. That this interfering activity must be considered is clearly shown by a comparison of the ^{40}K activity in urine with that of certain isotopic concentrations which are thought to be indicative of a significant body burden.

Example 4.8

The specific activity of urine, with respect to ^{40}K, is:

$$\frac{\text{dis/min}}{\text{liter}} = \frac{1600 \times 226}{1.3 \times 10^9 \times 40} \frac{\text{Ci}}{\text{g } ^{40}\text{K}} \times 1.19 \times 10^{-4} \frac{\text{g } ^{40}\text{K}}{\text{g K}} \times \frac{1.5 \text{ g K}}{\text{liter}}$$

$$\times 2.22 \times 10^{12} \frac{\text{dis/min}}{\text{Ci}} = 2730 \frac{\text{dis/min}}{\text{liter}}$$

A gross beta activity in excess of 200 disintegrations per minute per liter of urine following possible exposure to mixed fission products is considered indicative of internal deposition of the fission products. This activity is less than 10% of that due to the naturally occurring potassium.

Another naturally occurring radioisotope of importance is ^{14}C. It may appear surprising that this radioisotope is found in our environment, since its half-life is only 5600 years. It is not a "natural" isotope in the same sense as the very long-lived isotopes listed in Table 4.5; ^{14}C was not created at the beginning of time, as those are thought to have been. The production of ^{14}C is still going on. This isotope of carbon is the result of a nuclear transformation induced by the cosmic-ray bombardment of ^{14}N. The environmental burden of ^{14}C before the advent of nuclear bombs was about 1900 lb in the air, 3 tons in plants, and about 56 tons in the ocean. Since the specific activity of ^{14}C is 4.75 Ci/g, or about 2150 Ci/lb, the ^{14}C content of each of these environmental compartments was about 4, 13, and 240 megacuries respectively. It should be pointed out that testing of nuclear weapons has resulted in a significant increase in the atmospheric level of radiocarbon. It is estimated that about 2.8 megacuries ^{14}C have been injected into the air by all weapon tests which have been conducted until 1960. The atmospheric radiocarbon exists as $^{14}CO_2$. It is therefore inhaled by all animals, and utilized by plants in the process of photosynthesis. Because only living plants continue to incorporate ^{14}C along with non-radioactive carbon, it is possible to determine the age of organic matter by measuring the specific activity of the carbon present. If it is assumed that the rate of production of ^{14}C, as well as its concentration in the air, has remained constant during the past several tens of thousands of years, then a simple correction of specific activity data for half-life permits the estimation of the age of ancient samples of organic matter.

Example 4.9

If 2 g of carbon from a piece of wood found in an ancient temple is analyzed and found to have an activity of 10 disintegrations per minute per gram, what is the age of the wood if the current specific activity of ^{14}C in carbon is assumed to have been constant at 15 dis/min per gram:

Fraction of the original ^{14}C that remains today is, according to equation (4.18),

$$\frac{A}{A_0} = \frac{10}{15} = e^{-\lambda t}.$$

Since the half-life for ^{14}C is 5600 years,

$$\lambda = \frac{0.693}{5600} \text{ yr} = 1.238 \times 10^{-4} \text{ yr}^{-1},$$

$$\frac{10}{15} = e^{-1.238 \times 10^{-4} t},$$

$$t = 3.28 \times 10^3 \text{ years}.$$

Serial Decay

In addition to the four chains of radioactive isotopes described above, there are a number of other groups of sequentially decaying isotopes which are important to the health physicist and the radiobiologist. Most of these series are associated with nuclear fission, and the first member of each series is a fission fragment. One of the most widely known fission products, for example, ^{90}Sr, is the middle member of a five-member series that starts with ^{90}Ks, and finally terminates with stable ^{90}Zr according to the following sequence:

$$^{90}_{36}\text{Kr} \xrightarrow[33\text{ R}]{\beta} {}^{90}_{37}\text{Rb} \xrightarrow[2.74\text{ m}]{\beta} {}^{90}_{38}\text{Sr} \xrightarrow[19.9\text{ y}]{\beta} {}^{90}_{39}\text{Y} \xrightarrow[64.2\text{ h}]{\beta} {}^{90}_{40}\text{Zr}$$

The quantitative relationship among the various members of the series is of great significance, and must be considered when dealing with any of the group's members. Intuitively, it can be seen that any amount of ^{90}Kr will, in a time period of 10–15 min, have decayed to such a degree that for practical purposes, the ^{90}Kr may be assumed to have completely disintegrated. Rubidium-90, the ^{90}Kr daughter, because of its 2.74-min half-life, will suffer the same fate after about an hour. Essentially, all the ^{90}Kr is, as a result, converted into ^{90}Sr within about an hour after its formation. The buildup of ^{90}Sr is therefore very rapid. The half-life of ^{90}Sr is 20 years, and its decay, therefore, is very slow. The ^{90}Y daughter of ^{90}Sr, with a half-life of 64.2 hours, decays rapidly to stable ^{90}Zr. If initially pure ^{90}Sr is prepared, its radioactive disintegration will result in an accumulation of ^{90}Y. Because the ^{90}Y decays very much faster than ^{90}Sr, however, a point is soon reached at which the instantaneous amount of ^{90}Sr that decays is equal to that of ^{90}Y. Under these conditions, the ^{90}Y is said to be in secular equilibrium. The quantitative relationship between isotopes in secular equilibrium may be derived in the following manner for the general case

$$A \xrightarrow{\lambda_A} B \xrightarrow{\lambda_B} C,$$

where the half-life of isotope A is very much greater than that of isotope B. The decay constant of A, λ_A, is therefore much smaller than λ_B, the decay constant for isotope B. Isotope C is stable and does not disintegrate. Because of the very long half-life of A relative to B, the rate of formation of B may be considered to be constant, and equal to K. Under these conditions, the net rate of change of isotope B with respect to time, if N_B is the number of atoms of isotope B in existence at any time t after an initial number N_{B0}, is given by:

rate of change = rate of formation − rate of decay,

$$\frac{dN_B}{dt} = K - \lambda_B N_B, \tag{4.30}$$

or,

$$\int_{N_{B_0}}^{N_B} \frac{dN_B}{K - \lambda_B N_B} = \int_0^t dt. \tag{4.31}$$

The integrand can be changed to the form

$$\int_a^b \frac{dv}{v} = \ln v \Big|_a^b \tag{4.32}$$

if it is multiplied by $-\lambda_B$, and if the entire integral is multiplied by $-1/\lambda_B$ in order to keep the value of the integral unchanged. Equation (4.31) therefore may be solved to yield

$$\ln \left(\frac{K - \lambda_B N_B}{K - \lambda_B N_{B_0}} \right) = - \lambda_B t. \tag{4.33}$$

Equation (4.33) may be written in the exponential form

$$\frac{K - \lambda_B N_B}{K - \lambda_B N_{B_0}} = e^{-\lambda_B t} \tag{4.34}$$

and then solved for N_B:

$$N_B = \frac{K}{\lambda_B} (1 - e^{-\lambda_B t}) + N_{B_0} e^{-\lambda_B t}. \tag{4.35}$$

If we start with pure A, that is, if $N_{B0} = 0$, then equation (4.35) reduces to

$$N_B = \frac{K}{\lambda_B} (1 - e^{-\lambda_B t}). \tag{4.36}$$

The rate of formation of B from A is equal to the rate of decay of A. K, therefore, is simply equal to $\lambda_A N_A$. An alternative way of expressing equation (4.36), therefore, is:

$$N_B = \frac{\lambda_A N_A}{\lambda_B} (1 - e^{-\lambda_B t}). \tag{4.37}$$

Note that the quantity of both the parent, isotope A, and the daughter, isotope B, is given in the same units, namely, λN, or disintegrating atoms per unit time. This is a reasonable unit, since each parent atom that decays is transformed into a daughter. Any other unit that implicitly states this fact is equally usable in equation (4.37). If λN represents disintegrating atoms per second, then division of both sides of the equation by the proper factor to convert the activity to curies, or multiples thereof is permissible, and converts equation (4.37) into a slightly more usable form. For example, if the activity of the parent is given in millicuries, then the activity of the daughter must also be in units of millicuries, and equation (4.37) may be written as

$$Q_B = Q_A (1 - e^{-\lambda_B t}), \tag{4.38}$$

where Q_A and Q_B are the respective activities in millicuries of the parent and daughter.

Example 4.10

If we have 500 mg radium, how much ^{222}Rn will be collected after 1 day, after 3.8 days, after 10 days, and after 100 days?

Since the specific activity of radium is 1 Ci/g, 500 mg = 500 mCi. The half-life of radon is 3.8 days; its decay constant therefore is:

$$\lambda_{Rn} = \frac{0.693}{(T_{\frac{1}{2}})_{Rn}} = \frac{0.693}{3.8 \text{ days}},$$

$$\lambda_{Rn} = 0.1825 \text{ day}^{-1}.$$

From equation (4.38), we have:

$$Q_{Rn} = Q_{Ra} (1 - e^{-\lambda_{Rn} t}),$$

$$Q_{Rn} = 500 (1 - e^{-0.1825 \text{ day}^{-1} \times t \text{ days}}).$$

Substituting the respective time in days for t in the equation above gives 83.5 mCi Rn after 1 day, 250 mCi after 3.8 days, 419.5 mCi after 10 days, and 500 mCi after 100 days. Equation (4.38), as well as the illustrative example given above, shows a buildup of radon from 0 to a maximum activity that is equal to that of the parent from which it was derived. This buildup of daughter activity may be shown graphically by plotting equation (4.38). A generally

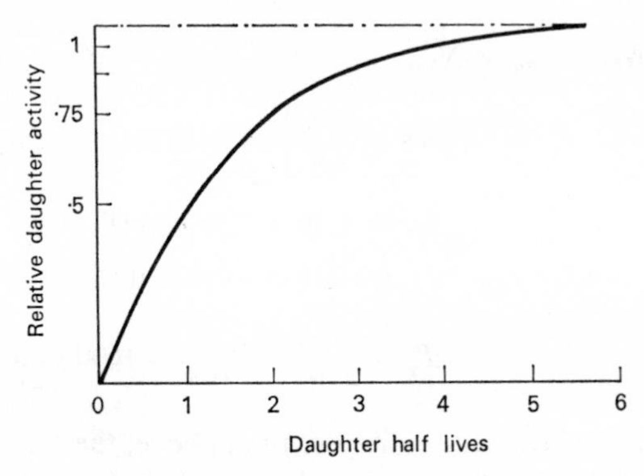

FIG. 4.14. Secular equilibrium: build-up of a very short-lived daughter from a long-lived parent. The activity of the parent remains constant.

useful curve showing the buildup of daughter activity under conditions of secular equilibrium may be obtained if t is plotted in units of daughter half-life, as shown in Fig. 4.14. As time increases, $e^{-\lambda t}$ decreases, and Q_B approaches Q_A. For practical purposes, equilibrium may be considered established after 7 daughter half-lives. At equilibrium, it should be noted that

$$\lambda_A N_A = \lambda_B N_B. \tag{4.39}$$

Equation (4.39) tells us that, at equilibrium, the activity of the parent is equal to that of the daughter, and that the ratio of the decay constants of the parent and daughter are in the inverse ratio of the equilibrium concentrations of the parent and daughter.

Example 4.11

Deduce the decay constant and half-life of ^{226}Ra if the radon gas in secular equilibrium with 1 g Ra exerts a partial pressure of 4.8×10^{-4} mm Hg in a one-liter flask, and if the half-life of radon is 3.8 days.

The amount of radon in equilibrium with the radium is computed by:

$$\frac{\text{moles/liter}}{4.8 \times 10^{-4}\ \text{mm}} = \frac{1\ \text{mole/22 liters}}{760\ \text{mm}}$$
$$= 2.88 \times 10^{-8}\ \text{moles,}$$

and the number of atoms of radon is

$$6.03 \times 10^{23}\ \frac{\text{atoms}}{\text{mole}} \times 2.88 \times 10^{-8}\ \text{moles} = 1.736 \times 10^{16}\ \text{atoms.}$$

Since the radon is in equilibrium with 1 g radium, and since there are

$$\frac{1\ \text{g}}{226\ \text{g/mole}} = 4.42 \times 10^{-3}\ \text{moles radium}$$

$$\lambda_{Ra} \times N_{Ra} = \lambda_{Rn} \times N_{Rn},$$

$$\lambda_{Ra} \times 4.42 \times 10^{-3}\ \text{moles} = \frac{0.693}{3.8\ \text{days}} \times 2.88 \times 10^{-8}\ \text{moles}$$

$$\lambda_{Ra} = 1.19 \times 10^{-6}\ \text{day}^{-1}$$
$$= 4.35 \times 10^{-4}\ \text{year}^{-1},$$

$$T_{\frac{1}{2}} = \frac{0.693}{4.35 \times 10^{-4}} = 1600\ \text{years.}$$

In the case of secular equilibrium discussed above, the quantity of parent remains substantially constant during the period that it is being observed. Since it is required, for secular equilibrium, that the half-life of the parent be very much longer than that of the daughter, it is evident that secular equilibrium is a special case of a more general situation in which the half-life of the parent may be of any conceivable magnitude, and no restrictions are applied to the relative magnitudes of the decay constants of the parent and daughter. For the general case, where the parent activity is not relatively constant,

$$A \xrightarrow{\lambda_A} B \xrightarrow{\lambda_B} C,$$

the time rate of change of the number of atoms of species B is given by the differential equation

$$\frac{dN_B}{dt} = \lambda_A N_A - \lambda_B N_B. \tag{4.40}$$

In this equation, $\lambda_A N_A$ is the rate of decay of species A, and is exactly equal to the rate of formation of species B. The rate of decay of isotope B is $\lambda_B N_B$, and the difference between these two rates at any time is the instantaneous rate of growth of species B at that time.

Since we have, from equation (4.18),

$$N_A = N_{A_0} e^{-\lambda_A t}, \tag{4.41}$$

equation (4.40) may be rewritten, after substituting the expression above for N_A and transposing $\lambda_B N_B$, as

$$\frac{dN_B}{dt} + \lambda_B N_B = \lambda_A N_{A_0} e^{-\lambda_A t}. \tag{4.42}$$

Equation (4.42) is a first-order linear differential equation of the form

$$\frac{dy}{dx} + P(x)y = Q(x), \tag{4.43}$$

and may be integrable by multiplying both sides of the equation by

$$e^{\int P dx} = e^{\int \lambda_B dt} = e^{\lambda_B t},$$

and the solution to equation (4.43) is

$$y e^{\int P dx} = \int e^{\int P dx} . Q dx. \tag{4.44}$$

Since N_B, λ_B, and $\lambda_A N_{A_0} e^{-\lambda_A t}$ from equation (4.42) are represented in equation (4.44) by y, P, and Q, respectively, the solution of equation (4.42) is

$$N_B e^{\lambda_B t} = \int e^{\lambda_B t} \lambda_A N_{A_0} e^{-\lambda_A t} dt + C \tag{4.45}$$

or, if the two exponentials are combined, we have

$$N_B e^{\lambda_B t} = \int \lambda_A N_{A_0} e^{(\lambda_B - \lambda_A)t} dt + C. \tag{4.46}$$

If the integrand in equation (4.46) is multiplied by the integrating factor $\lambda_B - \lambda_A$, then equation (4.46) is in the form

$$\int e^v dv = e^v + C, \tag{4.47}$$

and the solution is

$$N_B e^{\lambda_B t} = \frac{1}{\lambda_B - \lambda_A} \lambda_A N_{A_0} e^{(\lambda_B - \lambda_A)t} + C. \tag{4.48}$$

The constant C may be evaluated by applying the boundary conditions

$$N_B = 0 \text{ when } t = 0.$$

$$0 = \frac{1}{\lambda_B - \lambda_A} \cdot \lambda_A N_{A_0} + C,$$

$$C = -\frac{\lambda_A N_{A_0}}{\lambda_B - \lambda_A}. \tag{4.49}$$

If the value for C, from equation (4.49), is substituted into equation (4.48), the solution for N_B is found to be

$$N_B = \frac{\lambda_A N_{A_0}}{\lambda_B - \lambda_A} (e^{-\lambda_A t} - e^{-\lambda_B t}). \tag{4.50}$$

For the case in which the half-life of the parent is very much greater than that of the daughter, that is, when $\lambda_A \ll \lambda_B$, equation (4.50) approaches the condition of secular equilibrium, the limiting case described by equation (4.39). Two other general cases should be considered. The case where the parent half-life is slightly greater than that of the daughter, $\lambda_A < \lambda_B$, and the case in which the parent half-life is less than that of the daughter, $\lambda_B < \lambda_A$. In the former case, where the half-life of the daughter is slightly smaller than that of the parent, the daughter activity, if the parent is initially pure and free of any daughter activity, starts from zero, rises to a maximum, and then seems to decay with the same half-life as the parent. When this occurs, the daughter is disintegrating at the same rate as it is being produced, and the two isotopes are said to be in a state of transient equilibrium. The quantitative relationships prevailing during transient equilibrium may be inferred from equation (4.50). If both sides of that equation are multiplied by λ_B, we have an explicit expression for the activity of the daughter:

$$\lambda_B N_B = \frac{\lambda_B \lambda_A N_{A_0}}{\lambda_B - \lambda_A} (e^{-\lambda_A t} - e^{-\lambda_B t}). \tag{4.51}$$

Since λ_B is greater than λ_A, then, after a sufficiently long period of time, $e^{-\lambda_B t}$ will become much smaller than $e^{-\lambda_A t}$. Under this condition, equation (4.51) may be rewritten as

$$\lambda_B N_B = \frac{\lambda_B \lambda_A N_{A_0}}{\lambda_B - \lambda_A} e^{-\lambda_A t}. \tag{4.52}$$

By application of equation (4.41) the mathematical expression for transient equilibrium, equation (4.52), may be rewritten as

$$\lambda_B N_B = \frac{\lambda_B \lambda_A N_A}{\lambda_B - \lambda_A}, \tag{4.53}$$

or in terms of activity units (curies, etc.)

$$Q_B = \frac{\lambda_B}{\lambda_B - \lambda_A} Q_A. \tag{4.54}$$

An example of this equilibrium that is of importance to the health physicist is the ThB to ThC to ThC′ and ThC″ chain which occurs near the end of the thorium series. In this sequence of disintegrations, ThB (^{212}Pb), whose half-life is 10.6 hr, decays by beta emission to 60.5 min ThC (^{212}Bi), which then branches, 35.4% of the disintegrations going by alpha emission to ThC″ (^{208}Te) and 64.6% of the transitions being accomplished by beta emission to form ThC′ (^{212}Po). The ThC′ and ThC″ half-lives are very short, 3×10^{-7} sec and 3.1 min respectively, and both decay to stable ^{208}Pb. Since ThB is a

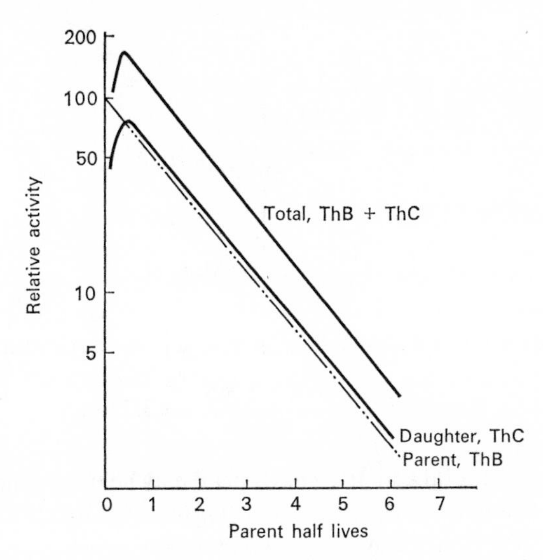

FIG. 4.15. Transient equilibrium: growth and decay of 60.5-min ThC from 10.6-hr ThB.

naturally occurring atmospheric isotope, correction for its activity and its daughter activity must be made before data on air samples can be accurately interpreted. The growth and decay of ThC is shown in Fig. 4.15. This curve graphically emphasizes the fact that, at transient equilibrium, the daughter seems to decay at the same rate as the parent.

Figure 4.15 also shows that the daughter activity, which starts from zero, rises to a maximum value and then decreases. It is also seen that the total activity, daughter plus parent, reaches a maximum value that does not coincide in time with that of the daughter. The time after isolation of the parent that the daughter reaches its maximum activity may be computed by differentiating equation (4.51), the daughter activity, setting the derivative equal to zero, and then solving for t_{md}.

$$\lambda_B N_B = \frac{\lambda_B \lambda_A N_{A_0}}{\lambda_B - \lambda_A} (e^{-\lambda_A t} - e^{-\lambda_B t}), \tag{4.51}$$

$$\frac{d(\lambda_B N_B)}{dt} = \frac{\lambda_B \lambda_A N_{A_0}}{\lambda_B - \lambda_A} (-\lambda_A e^{-\lambda_A t} + \lambda_B e^{-\lambda_B t}) = 0,$$

$$\lambda_A e^{-\lambda_A t} = \lambda_B e^{-\lambda_B t},$$

$$\frac{\lambda_B}{\lambda_A} = \frac{e^{-\lambda_A t}}{e^{-\lambda_B t}} = e^{(\lambda_B - \lambda_A) t},$$

$$\ln \frac{\lambda_B}{\lambda_A} = (\lambda_B - \lambda_A) t,$$

$$t = t_{md} = \frac{\ln \lambda_B/\lambda_A}{\lambda_B - \lambda_A} = 2.3 \frac{\log \lambda_B/\lambda_A}{\lambda_B - \lambda_A}. \tag{4.55}$$

For the case illustrated in Fig. 4.15,

$$\lambda_A = \frac{0.693}{10.6} = 0.065 \text{ hr}^{-1},$$

and

$$\lambda_B = \frac{0.693}{1.01} = 0.686 \text{ hr}^{-1}.$$

The time at which the daughter, ThC, will reach its maximum activity is:

$$t_{md} = \frac{2.3 \log 0.686/0.065}{0.686 - 0.065} = 3.78 \text{ hr}.$$

The time when the total activity is at its peak may be determined in a similar manner. In this case, the total actitivy, $A(t)$, must be maximized. The total activity at any time is given by

$$A(t) = \lambda_A N_A + \lambda_B N_B. \tag{4.56}$$

Substituting equations (4.41) and (4.50) for N_A and N_B, respectively, in equation (4.56), we have

$$A(t) = \lambda_A N_{A_0} e^{-\lambda_A t} + \frac{\lambda_B \lambda_A N_{A_0}}{\lambda_B - \lambda_A} (e^{-\lambda_A t} - e^{-\lambda_B t}), \tag{4.57}$$

and, differentiating yields:

$$\frac{dA(t)}{dt} = - \lambda_A^2 N_{A_0} e^{-\lambda_A t} + \frac{\lambda_B \lambda_A N_{A_0}}{\lambda_B - \lambda_A} (- \lambda_A e^{-\lambda_A t} + \lambda_B e^{\lambda_B t}) = 0. \tag{4.58}$$

Expanding and collecting terms, we have

$$-\lambda_A^2 N_{A_0} e^{-\lambda_A t} \left(1 + \frac{\lambda_B}{\lambda_B - \lambda_A}\right) + \frac{\lambda_A \lambda_B^2}{\lambda_B - \lambda_A} N_{A_0} e^{-\lambda_B t} = 0,$$

and solving for t_{mt}, the time when the total activity is a maximum, we get

$$t = t_{mt} = \frac{1}{\lambda_B - \lambda_A} \ln \left(\frac{\lambda_B^2}{2 \lambda_A \lambda_B - \lambda_A^2}\right) = \frac{2.3}{\lambda_B - \lambda_A} \log \left(\frac{\lambda_B^2}{2 \lambda_A \lambda_B - \lambda_A^2}\right). \tag{4.59}$$

Referring once more to the example in Fig. 4.15, the maximum total activity is found to occur at

$$t_{mt} = \frac{2.3}{0.686 - 0.065} \log \left(\frac{(0.686)^2}{2 \times 0.065 \times 0.686 - 0.065^2}\right)$$

$$= 2.78 \text{ hr}$$

after the initial purification of the parent. It should be noted that the maximum total activity occurs earlier than that of the daughter alone.

The time when the parent and daughter isotopes may be considered to be equilibrated depends on their respective half-lives. The shorter the half-life

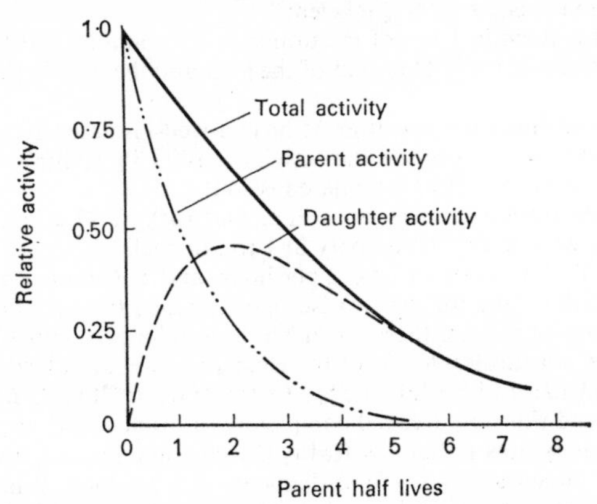

FIG. 4.16. No equilibrium: growth and decay of 24 min ^{146}Pr from 14 min ^{146}Ce.

of the daughter, relative to the parent, the more rapidly will equilibrium be attained.

In the case where the half-life of the daughter exceeds that of the parent, no equilibrium is possible. The daughter activity reaches a maximum, at a time which can be calculated from equation (4.55), and then reaches a point where it decays at its own characteristic rate. The parent, in the meantime, because of its shorter half-life, decays away. The total activity in this case does not increase to a maximum; it decreases continuously. Equation (4.56), which gives the time after isolation of a parent at which the total activity is maximum, cannot be solved for the case in which the half-life of the daughter exceeds that of the parent, or if $\lambda_B < \lambda_A$. These points are all illustrated in Fig. 4.16, which shows the course in time of the growth and decay of ^{146}Pr, whose half-life is 24.4 min, from the fission product ^{146}Ce, whose half-life is only 13.9 min. Solution of equation (4.55) shows the ^{146}Pr activity to reach its peak 26.2 min after isolation of ^{146}Ce.

Problems

1. Carbon-14 is a pure beta emitter that decays to ^{14}N. If the exact atomic masses of the parent and daughter are 14.007687 and 14.007520 atomic mass units, calculate the kinetic energy of the most energetic beta particle.

2. If 25 μCi ^{131}I is needed for a diagnostic test, and if 3 days elapse between shipment of the radioiodine and its use in the test, how many μCi must be shipped?

3. The gamma radiation from 1 ml of a solution containing 0.01 μCi ^{198}Au and 0.005 μCi ^{131}I is counted daily with a scintillation detector, over a 16-day period. Assume equal detection efficiency, 10%, for all the quantum energies involved. What will be the relative counting rates of the ^{131}I and ^{198}Au at time $t = 0$, $t = 3$ days, $t = 8$ days, $t = 16$ days? Plot the daily total counting rates that one would expect during the observation period.

4. The decay constant for ^{235}U is 9.72×10^{-10} per year. Compute the number of disintegrations per second in a 500 mg sample of ^{235}U.

5. Five millicuries ^{210}Po are necessary for a certain ionization source. How many grams ^{210}Po does this represent?

6. If uranium ore contains 10% U_3O_8, how many tons are necessary to produce 1 g radium if the extraction process is 90% efficient?

7. How much ^{234}U is there in 1 ton of the uranium ore containing 10% U_3O_8?

8. Compare the activity of the ^{234}U to that of the ^{235}U and the ^{238}U in the ore of problems 6 and 7.

9. What will be the temperature rise after 24 hr in a well-insulated 100-ml aqueous solution containing 1 g $Na^{35}SO_4$, whose specific activity is 100 Ci/g sulfur?

10. Show that ^{22}Na cannot decay by alpha emission.

11. The mean concentration of potassium in crystal rocks is 27 g/kg. If ^{40}K constitutes 0.012% of potassium, what is the ^{40}K activity in 1 ton of rock?

12. ThB decays to ThC at a rate of 6.54% per hour, and ThC decays at a rate of 1.15% per hour. How long will it take for the two isotopes to reach their equilibrium state?

13. How many grams of ^{90}Y are there when ^{90}Y is equilibrated with 10 mg ^{90}Sr?

14. Radiogenic lead constitutes 98.5% of the element as found in lead ore. The isotopic constitution of lead in nature is: ^{204}Pb, 1.5%; ^{206}Pb, 23.6%; ^{207}Pb, 22.6%; ^{208}Pb, 52.3%. How much uranium and thorium decayed completely to produce 985 mg radiogenic lead?

15. One hundred milligrams radium as $RaBr_2$ (specific gravity = 5.79) is in a platinum capsule whose inside dimensions are 1 mm diameter $\times$ 3 cm long. What will be the gas pressure, at body temperature, inside the capsule 100 years after manufacture if it originally contained air at atmospheric pressure at room temperature (27°C)?

16. A solution of ^{203}Hg is received with the following assay: 1 mCi/ml on 1 March, 1967 at 8.00 a.m. It is desired to make a solution whose activity will be 100 μCi/ml on 1 April, 1967. Calculate the dilution factor to give the desired activity. $T_{\frac{1}{2}}$ $^{203}Hg = 46$ days.

17. In a mixture of two radioisotopes, 99% of the activity is due to ^{24}Na and 1% is due to ^{32}P. At what subsequent time will the two activities be equal?

18. A nucleus, originally at rest, decays by emitting a beta particle whose momentum is 1×10^{-16} g cm/sec. At a right angle to the direction of the beta particle, a neutrino whose momentum is 0.2×10^{-16} g cm/sec is emitted. In what direction, relative to the direction of the beta particle does the nucleus recoil?

19. Barium-140 decays to ^{140}La with a half-life of 12.8 days, and the ^{140}La decays to stable ^{140}Ce with a half-life of 40.5 hours. A radio chemist, after precipitating ^{140}Ba, wishes to wait until he has a maximum amount of ^{140}La before separating the ^{140}La from the ^{140}Ba. (a) How long must he wait? (b) If he started with 25 mCi ^{140}Ba, how many micrograms ^{140}La will he collect?

20. Strontium-90 is to be used as a heat source for generating electrical energy in a satellite. (a) How many curies of ^{90}Sr are required to generate 50 watts of electrical power, if the conversion efficiency from heat to electricity is 30%? (b) Weight of the isotopic heat source is an important factor in design of the power source. If weight is to be kept at a minimum, and if the source is to generate 50 watts after 1 year of operation, would there be an advantage to using ^{210}Po?

Suggested References

1. GLASSTONE, S.: *Sourcebook on Atomic Energy*, D. Van Nostrand, Princeton, 1958.
2. LAPP, R. E. and ANDREWS, H. L.: *Nuclear Radiation Physics*, Prentice-Hall, Englewood Cliffs, 1963.
3. SEMAT, H.: *Introduction to Atomic and Nuclear Physics*, Rinehart & Co., New York, 1964.
4. EVANS, R. D.: *The Atomic Nucleus*, McGraw-Hill, New York, 1955.
5. CUNNINGHAME, J. G.: *Introduction to the Atomic Nucleus*, Elsevier, Amsterdam, 1964.
6. BORN, M.: *Atomic Physics*, Hafner, New York, 5th edition.
7. KAPLAN, I.: *Nuclear Physics*, Addison-Wesley Pub. Co., Reading, 1962.
8. HALLIDAY, D.: *Introductory Nuclear Physics*, John Wiley & Sons, New York, 1955.
9. RICHTMYER, F. K., KENNARD, E. H., and LAURITSEN, T.: *Introduction to Modern Physics*, McGraw-Hill, New York, 1955.
10. CORK, J. A.: *Radioactivity and Nuclear Physics*, D. Van Nostrand, Princeton, 1957.
11. BLATZ, H. (Ed.): *Radiation Hygiene Handbook*, McGraw-Hill, New York, 1959.
12. ETHERINGTON, H. (Ed.): *Nuclear Engineering Handbook*, McGraw-Hill, New York, 1958.
13. *Radiation Health Handbook*, U.S. Department of Health, Education and Welfare, Public Health Service, Washington, 1960.

CHAPTER 5

INTERACTION OF RADIATION WITH MATTER

Introduction

In order for the health physicist to understand the physical basis for radiation dosimetry and the theory of radiation shielding, he must understand the mechanisms by which the various radiations interact with matter. In most instances, these interactions involve a transfer of energy from the radiation to the matter with which it interacts. Matter consists of atomic nuclei and extra-nuclear electrons. Radiation may interact with either or both of these constituents of matter. The probability of occurrence of any particular category of interaction, and hence the penetrating power of the several radiations, depends on the type and energy of the radiation as well as on the nature of the absorbing medium. In all instances, excitation and ionization of the absorber atoms results from their interaction with the radiation. Ultimately, the energy transferred either to tissue or to a radiation shield is dissipated as heat.

Beta-rays

Range–Energy Relationship

The attenuation of beta-rays by any given absorber may be measured by interposing successively thicker absorbers between a beta-ray source and a suitable beta-ray detector, such as a Geiger counter, as shown in Fig. 5.1, and counting the beta particles that penetrate the absorbers. When this is done with a pure beta emitter, it is found that the beta-particle counting rate decreases rapidly at first, and then slowly as the absorber thickness increases. Eventually, a thickness of absorber is reached that stops all the beta particles; the Geiger counter then registers only background counts due to environmental radiation. If semi-log paper is used to plot the data, and if the counting rate is plotted on the logarithmic axis while absorber thickness is plotted on the linear axis, the data approximate a straight line, as shown in Fig. 5.2. The end point in the absorption curve, where no further decrease in the counting rate is observed, is called the *range* of the beta-rays in the material of which the absorbers are made. As a rough rule of thumb, a useful relationship

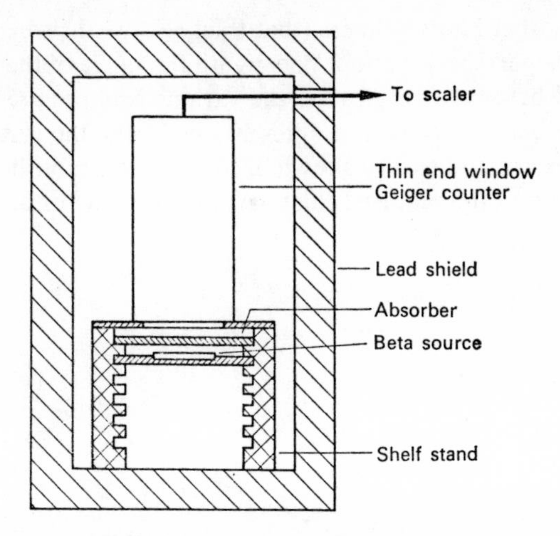

FIG. 5.1. Experimental arrangement for absorption measurements on beta particles.

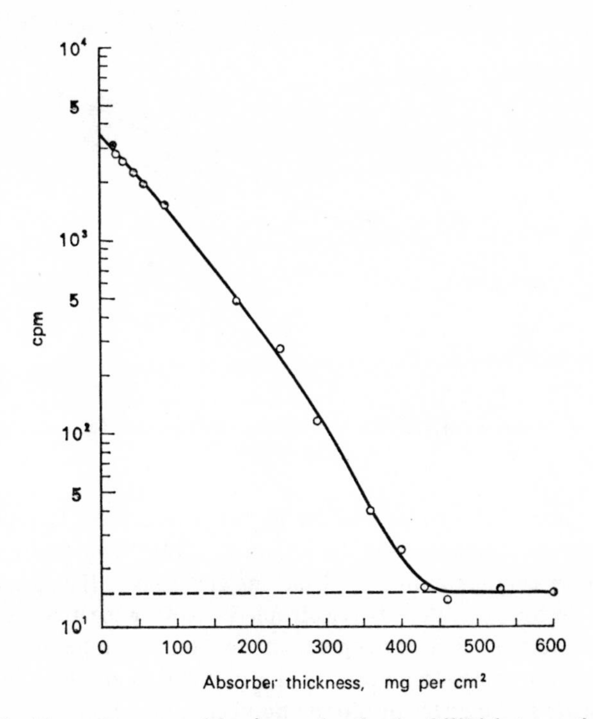

FIG. 5.2. Absorption curve (aluminum absorbers) of ^{210}Bi beta particles, 1.17 MeV. The broken line represents the mean background counting rate.

is that the absorber half-thickness (that thickness of absorber that stops one-half of the beta particles) is about one-eighth the range of the beta-rays. Since the maximum beta-ray energies for the various isotopes are known, then by measuring the beta-ray ranges in different absorbers, the systematic relationship between range and energy shown in Fig. 5.3. is established. Inspection of Fig. 5.3 shows that the required thickness of absorber for any given beta-ray

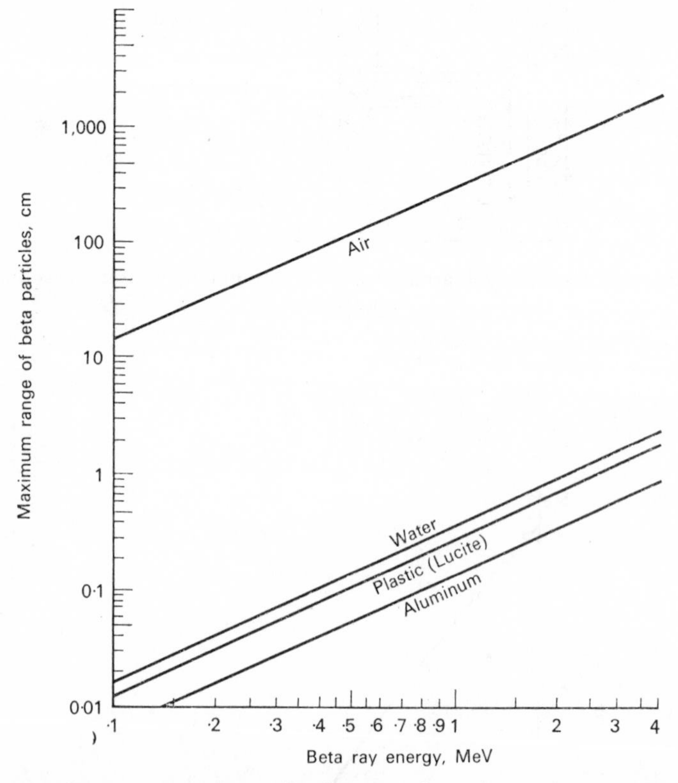

FIG. 5.3. Range–energy curves for beta-rays in various substances. (Adapted from *Radiological Health Handbook*, Office of Technical Services, Washington, 1960.)

energy decreases as the density of the absorber increases. Detailed analyses of experimental data show that the ability to absorb energy from beta-rays depends mainly on the number of absorbing electrons in the path of the beta-ray, that is, on the areal density (electrons per cm^2) of electrons in the absorber; and, to a very much lesser degree, on the atomic number of the absorber. For practical purposes, therefore, in the calculation of shielding thickness against beta-rays, the effect of atomic number is neglected. Areal density of electrons is approximately proportional to the product of the density of the absorber material and the linear thickness of the absorber, thus giving rise to the unit of thickness called the *density thickness*. Mathematically, density thickness, t_d, is defined as

$$t_d \text{ g/cm}^2 = \rho \text{ g/cm}^3 \times t_l \text{cm}. \tag{5.1}$$

The units of density and thickness in equation (5.1), of course, need not be grams and centimeters, they may be any consistent set of units. Use of the density thickness unit, such as g/cm² or mg/cm² for absorber materials makes it possible to specify such absorbers independently of the absorber material. (It should be pointed out, that, for reasons to be given later, beta-ray shields are almost always made from low atomic-numbered materials.) For example, the density of aluminum is 2.7 g/cm³. From equation (5.1), a sheet of aluminum 1 cm thick, therefore, has a density thickness of

$$t_d = 2.7 \text{ gm/cm}^3 \times 1 \text{ cm} = 2.7 \text{ gm/cm}^2.$$

If a sheet of Plexiglass, whose density is 1.18 g/cm³ is to have a beta-ray absorbing quality very nearly equal to that of the 1-cm thick sheet of aluminum, i.e. 2.7 g/cm², its linear thickness is found, from equation (5.1), to be

$$t_l = \frac{t_d}{\rho} = \frac{2.7 \text{ g/cm}^2}{1.18 \text{ g/cm}^3} = 2.39 \text{ cm}.$$

Another practical advantage of using this system of thickness measurement is that it allows the addition of thicknesses of different materials in a radiologically meaningful way. A universal curve of beta-ray range (in units of density

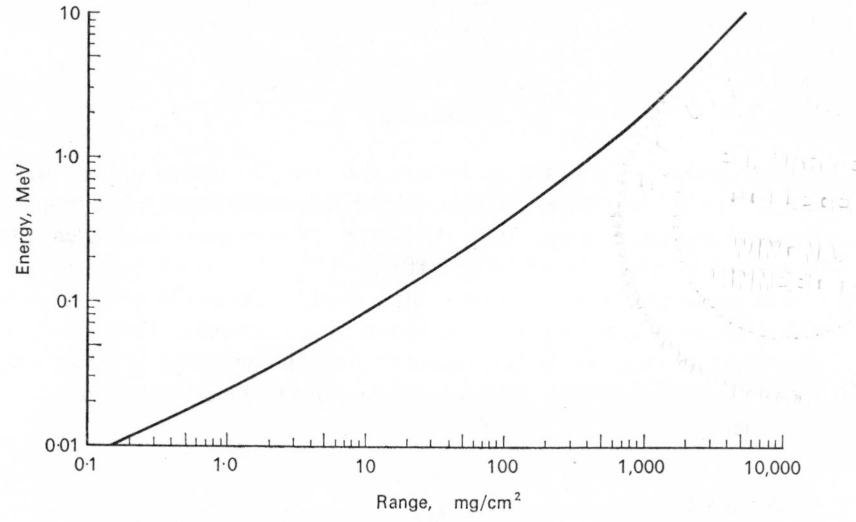

FIG. 5.4. Range–energy curve for beta particles. The range is expressed in units of density thickness. (From *Radiation Health Handbook*, Office of Technical Services, Washington, 1960.)

thickness) versus energy is given in Fig. 5.4. This curve, which is based on experimental range–energy measurements, is fitted by the following equations:

$$R = 412E^{1.265-0.0954 \ln E} \tag{5.2}$$

for $0.01 \leq E \leq 2.5$ MeV,

$$\ln E = 6.63 - 3.2376\,(10.2146 - \ln R)^{\frac{1}{2}} \tag{5.3}$$

for $R \leq 1200$,

$$R = 530\,E - 106 \tag{5.4}$$

for $E > 2.5$ MeV, $R > 1200$,

where R = range, mg/cm²

E = maximum beta-ray energy, MeV.

Example 5.1

What must be the minimum thickness of a shield made of (a) Plexiglass, and (b) aluminum in order that no beta-rays from a ^{90}Sr source pass through?

Strontium-90 emits a 0.54-MeV beta particle. However, its daughter, ^{90}Y, emits a beta particle whose maximum energy is 2.27 MeV. Since ^{90}Y beta particles always accompany ^{90}Sr beta-rays, the shield must be thick enough to stop these more energetic betas. From Fig. 5.4, the range of a 2.27-MeV beta particle is found to be 1.1 g/cm². The density of Plexiglass is 1.18 g/cm³. From equation (5.1), the required thickness is found to be

$$t_l = \frac{t_d}{\rho} = \frac{1.1 \text{ g/cm}^2}{1.18 \text{ g/cm}^3} = 0.932 \text{ cm.}$$

Plexiglass may suffer radiation damage and crack if exposed to very intense radiation for a long period of time. Under these conditions, aluminum is a better choice for a shield. Since the density of aluminum is 2.7 g/cm³, the required thickness is found to be 0.41 cm.

The range–energy relationship is often used by the health physicist as an aid in identifying an unknown beta-emitting contaminant. This is done by measuring the range of the beta radiation, then finding energy of the beta-ray, then looking up in a table of isotopes the isotope that emits a beta particle of that energy.

Example 5.2

Using the counting set-up shown in Fig. 5.5, the range of an unknown beta particle was found to be 0.111 mm aluminum. No measurable decay of the isotope was observed during a period of 1 month, and no other radiation was emitted from the isotope.

(a) What was the energy of the beta particle?
(b) What is the isotope?

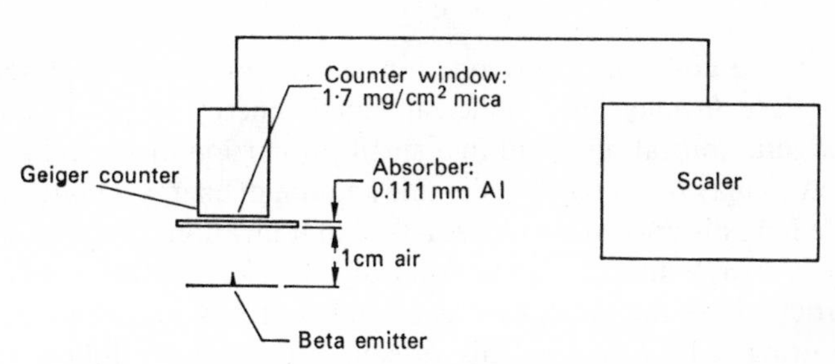

FIG. 5.5. Measuring the range of an unknown beta particle to identify the isotope.

The total range of the beta-ray is

$$\text{Range} = 1.7\ \frac{\text{mg}}{\text{cm}^2}\ \text{mica} + 1\ \text{cm air} + 0.111\ \text{mm aluminum}.$$

These different absorbing media may be added together if their thicknesses are expressed in density thickness. The density of air is 1.293 mg/cm³ at STP. With equation (5.1) the density thicknesses of the air and aluminum are computed, and the range of the unknown beta particle is found to be

$$\text{Range} = 1.7\ \frac{\text{mg}}{\text{cm}^2} + 1.29\ \frac{\text{mg}}{\text{cm}^2} + 30\ \frac{\text{mg}}{\text{cm}^2} = 32.99\ \frac{\text{mg}}{\text{cm}^2}.$$

In Fig. 5.4, the energy corresponding to this range is seen to be about 0.17 MeV. The unknown isotope is therefore likely to be ^{14}C, a pure beta emitter whose maximum beta-ray energy is 0.155 MeV and whose half-life is 5700 years.

Mechanisms of Energy Loss

Ionization and excitation

Interaction between the electric fields of a beta particle and the orbital electrons of the absorbing medium leads to electronic excitation and ionization. Such interactions are inelastic collisions, analogous to that described in Example 2.13. The electron is held in the atom by electrical forces, and energy is lost by the beta particle in overcoming these forces. Since electrical forces act over long distances, the "collision" between a beta particle and an electron occurs without the two particles coming into actual contact—as in the case of the collision between like poles of two magnets. The amount of energy lost by the beta particle depends on its distance of approach to the electron and on its kinetic energy. If ϕ is the ionization potential of the absorbing medium and E_t is the energy lost by the beta particle during the collision, the kinetic energy of the ejected electron, E_k, is

$$E_k = E_t - \phi. \tag{5.5}$$

In many ionizing collisions, only one ion pair is produced. In other cases, the ejected electron may have sufficient kinetic energy to produce a small cluster of several ionizations; and in a small proportion of the collisions the ejected electron may receive a considerable amount of energy, enough to cause it to travel a long distance and to leave a trail of ionizations. Such an electron, whose kinetic energy may be on the order of 1000 eV, is called a *delta-ray*.

Beta particles have the same mass as orbital electrons, and hence are easily deflected during collisions. For this reason, beta particles follow tortuous paths as they pass through absorbing media. Figure 5.6 shows the path of a beta particle through a photographic emulsion. The ionizing events expose the film at the points of ionization, thereby making them visible after development of the film.

By using a cloud chamber or a film to visualize the ionizing events, and by counting the actual number of ionizations due to a single primary ionizing particle of known energy, it was learned that the average energy expended in the production of an ion pair is about two to three times greater than the ionization potential. The difference between the energy expended in ionizing collisions and the total energy lost by the ionizing particle is attributed to

TABLE 5.1. AVERAGE ENERGY LOST BY A BETA PARTICLE IN THE PRODUCTION OF AN ION PAIR

Gas	Ionization potential	Mean energy expenditure per ion pair
H_2	13.6 eV	36.6 eV
He	24.5	41.5
N_2	14.5	34.6
O_2	13.6	30.8
Ne	21.5	36.2
A	15.7	26.2
Kr	14.0	24.3
Xe	12.1	21.9
Air		33.7
CO_2	14.4	32.9
CH_4	14.5	27.3
C_2H_2	11.6	25.7
C_2H_4	12.2	26.3
C_2H_6	12.8	24.6

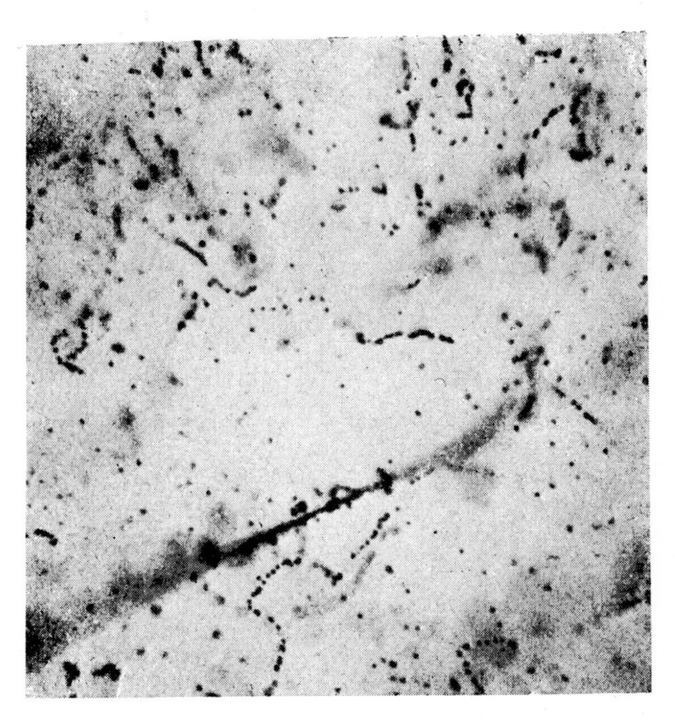

FIG. 5.6. Electron tracks in photographic emulsion. The tortuous lines are the electron tracks; the heavy line near the bottom was made by an oxygen nucleus in primary cosmic radiation. (From H. Yagoda, The tracks of nuclear particles, *Scientific American*, May 1956.)

electronic excitation. For oxygen and nitrogen, for example, the ionization potentials are 12.5 and 15.5 eV respectively, while the average energy expenditure per ion pair in air is 34 eV. Table 5.1 shows the ionization potential and mean energy expenditure, w, for several gases of practical importance.

Specific ionization. The linear rate of energy loss of a beta particle due to ionization and excitation, which is an important parameter in health physics instrument design and in the biological effects of radiation, is usually expressed by the *specific ionization*. Specific ionization is the number of ion pairs formed per unit distance travelled by the beta particle. Generally, the specific ionization is relatively high for low energy betas; it decreases rapidly as the beta particle energy increases, until a broad minimum is reached around 1 MeV.

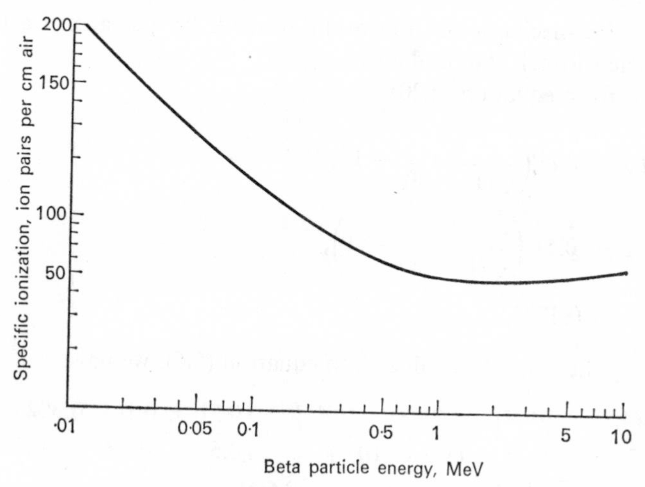

FIG. 5.7. Relationship between beta particle energy and specific ionization of air.

Further increase in beta energy results in slowly increasing specific ionization, as shown in Fig. 5.7.

The linear rate of energy loss due to excitation and ionization may be calculated from the equation,

$$\frac{dE}{dx} = \frac{2\pi q^4 NZ}{E_m \beta^2 (1.6 \times 10^{-6})^2} \left\{ \ln \left[\frac{E_m E_k \beta^2}{I^2(1-\beta^2)} \right] - \beta^2 \right\} \frac{\text{MeV}}{\text{cm}} \tag{5.6}$$

where q = charge on the electron, 4.8×10^{-10} sC,
N = number of absorber atoms per cm³,
Z = atomic number of the absorber,
NZ = number of absorber electrons per cm³ = 3.88×10^{20} for air at 0° and 76 cm,
E_m = energy equivalent of electron mass, 0.51 MeV,

E_k = kinetic energy of the beta particle, MeV,
β = v/c,
I = mean ionization and excitation potential of absorbing atoms, MeV,
I = 8.6×10^{-5} for air; for other substances, $I = 1.35 \times 10^{-5} Z$.

If the mean energy expended in the creation of an ion pair, w, is known, then the specific ionization may be calculated from the equation below:

$$\text{S.I.} = \frac{dE/dx \text{ eV/cm}}{w \text{ eV/ip}}. \tag{5.7}$$

Example 5.3

What is the specific ionization resulting from the passage of a 0.1 MeV beta particle through standard air?

β^2 is found from equation (2.20):

$$E_k = m_0c^2\left(\frac{1}{\sqrt{(1-\beta^2)}} - 1\right),$$

$$0.1 = 0.51\left(\frac{1}{\sqrt{(1-\beta^2)}} - 1\right),$$

$$\beta^2 = 0.3025.$$

Substituting the respective values into equation (5.6), we have

$$\frac{dE}{dx} = \frac{2\pi(4.8 \times 10^{-10})^4 \times 3.88 \times 10^{20}}{0.51 \times 0.3025 \times (1.6 \times 10^{-6})^2}\left\{\ln\left[\frac{0.51 \times 0.1 \times 0.3025}{8.6 \times 10^{-5}(1 - 0.3025)}\right] - 0.3025\right\}\frac{\text{MeV}}{\text{cm}},$$

$$\frac{dE}{dx} = 4.75 \times 10^{-3}\,\frac{\text{MeV}}{\text{cm}}.$$

For air, w = 34 eV/ip. The specific ionization, therefore, from equation (5.7), is

$$\text{S.I.} = \frac{4750 \text{ eV/cm}}{34 \text{ eV/ip}} = 140 \text{ ip/cm}.$$

Very often, the unit of length used in expressing rate of energy loss is density thickness, that is, in units of MeV/g/cm². This is called the mass stopping power, and is defined by the equation

$$S = \frac{dE/dx}{\rho}. \tag{5.8}$$

Since the density of standard air is 1.293×10^{-3} g/cm³, the mass rate of energy loss, or the mass stopping power, in Example 5.3 is

$$S = \frac{4.75 \times 10^{-3}\ \text{MeV/cm}}{1.293 \times 10^{-3}\ \text{g/cm}^3} = 3.67\ \frac{\text{MeV}}{\text{g/cm}^2}.$$

Linear energy transfer. The term specific ionization is used when attention is focused on the energy lost by the radiation. When attention is focused on the absorbing medium, as is the case in radiobiology and radiation effects, we are interested in the linear rate of energy absorption by the absorbing medium as the ionizing particle traverses the medium. As a measure of the rate of energy absorption, we use the *linear energy transfer*, abbreviated LET, which is defined by the equation

$$\text{LET} = \frac{dE_L}{dl}, \tag{5.9}$$

where dE_L is the average energy locally imparted to the absorbing medium by a charged particle of specified energy in traversing a distance of dl. In health physics and radiobiology, LET is usually expressed in units of keV per micron. As used in the definition above, the term "locally imparted" may refer either to a maximum distance from the track of the ionizing particle or to a maximum value of discrete energy loss by the particle beyond which losses are no longer considered local. In either case, LET refers to energy imparted within a limited volume of absorber.

Relative mass stopping power. The relative mass stopping power is used to compare quantitatively the energy absorptive power of different media. It will be shown later that the mass stopping power of different absorbers relative to that of air is important in the practice of health physics. Relative mass stopping power is defined by

$$\rho_m = \frac{S_{\text{medium}}}{S_{\text{air}}}. \tag{5.10}$$

Example 5.4

What is the relative (to air) mass stopping power of graphite, density = 2.25 g/cm³, for a 0.1-MeV beta particle?

The mass rate of energy loss in graphite is found by substituting the appropriate values into equations (5.6) and (5.8):

$$NZ = \frac{6.03 \times 10^{23}\ \text{atoms/mole} \times 2.25\ \text{g/cm}^3 \times 6\ \text{electrons/atom}}{12\ \text{g/mole}}$$

$$= 6.77 \times 10^{23}\ \text{electrons/cm}^3$$

$$I = 1.35 \times 10^{-5} \times 6 = 8.1 \times 10^{-5},$$

therefore

$$S_{\text{graphite}} = 3.85 \frac{\text{MeV}}{\text{g/cm}^2}.$$

For air, the mass stopping power is 3.67 MeV/g/cm². From equation (5.10), the relative mass stopping power of graphite for a 0.1-MeV electron is

$$\rho_m = \frac{3.85}{3.67} = 1.05.$$

Bremsstrahlung

Bremsstrahlung are X-rays that are emitted when high-speed charged particles suffer rapid acceleration. When a beta particle passes close to a nucleus, the strong attractive coulomb force causes the beta particle to deviate sharply from its original path. The change in direction is due to radial acceleration, and the beta particle, in accordance with classical theory, loses energy by electromagnetic radiation at a rate proportional to the square of the acceleration. This means that the bremsstrahlung photons have a continuous energy distribution that ranges downward from a theoretical maximum equal to the kinetic energy of the beta particle. The exact shape of the bremsstrahlung spectrum, as well as the intensity of bremsstrahlung resulting from beta-ray absorption in any given configuration of beta source and absorber, is difficult to calculate and relatively simple to measure. For example, the bremsstrahlung exposure rate at a distance of 10 cm from an aqueous solution of 100 mCi ^{32}P in a 25-ml volumetric flask is about 3 mR/hr; for 100 mCi ^{90}Sr in a small brass container, the bremsstrahlung exposure rate at a distance of 10 cm is about 100 mR/hr. (The mR will be formally introduced in the next chapter. At this point, it is sufficient to know that the mR is a unit for measuring radiation exposure.)

For purposes of *estimating* the bremsstrahlung hazard, the following approximate relationship may be used:

$$f = 3.5 \times 10^{-4} ZE, \tag{5.11}$$

where f = the fraction of the incident beta energy converted into photons,
Z = atomic number of the absorber,
E = maximum energy of the beta particle, MeV.

Because the likelihood of bremsstrahlung production increases with atomic number of the absorber, beta-ray shields are made with material of the minimum practicable atomic number. In practice, beta-ray shields of higher atomic number than 13, aluminum, are seldom if ever used.

Example 5.5

A very small source (physically) of 1 Ci of ^{32}P is inside a lead shield just thick enough to prevent any beta particles from emerging. What is the bremsstrahlung flux at a distance of 10 cm from the source?

Since Z for lead is 82, and the maximum energy of the ^{32}P beta particle is 1.71 MeV, we have, from equation (5.11), the fraction of the beta energy converted into photons

$$f = 3.5 \times 10^{-4} \times 82 \times 1.71 = 0.049.$$

Since the average beta-ray energy is about one-third of the maximum energy, the energy in the beta particles incident on the shield, E_β, is

$$E_\beta = \tfrac{1}{3}E \times 3.7 \times 10^{10} \text{ MeV/sec.}$$

For health physics purposes, it is assumed that all the bremsstrahlung photons are of the maximum energy. The flux, ϕ, of bremsstrahlung photons at a distance r cm from a beta source of 1 Ci therefore is:

$$\phi = \frac{fE_\beta}{4\pi r^2 E} \tag{5.12}$$

$$= \frac{0.049 \times \frac{1}{3} \times 1.71 \text{ MeV}/\beta \times 3.7 \times 10^{10}\ \beta/\text{sec}}{4\pi \times 10^2 \text{ cm}^2 \times 1.71 \text{ MeV/photon}}$$

$$= 4.8 \times 10^5 \text{ photons/cm}^2\text{/sec.}$$

Alpha-rays

Range–Energy Relationship

Alpha-rays are the least penetrating of the radiations. In air, even the most energetic alphas from radioactive substances travel only several centimeters, while in tissue, the range of alpha radiation is measured in microns ($1\mu = 10^{-4}$ cm). The term range, in the case of alpha particles, may have two different definitions: mean range and extrapolated range. The difference between these two ranges can be seen in the alpha-particle absorption curve, Fig. 5.8.

An alpha-particle absorption curve is flat because alpha radiation is essentially monoenergetic. Increasing thickness of absorbers serves merely to reduce the energy of the alphas that pass through the absorbers; the number of alphas is not reduced until the approximate range is reached. At this point, there is a sharp decrease in the number of alphas that pass through the absorber. Near the very end of the curve, absorption rate decreases due to straggling, or the combined effects of the statistical distribution of the "average" energy loss per ion and the scattering by the absorber nuclei. The mean range is the range most accurately determined, and corresponds to the range of the

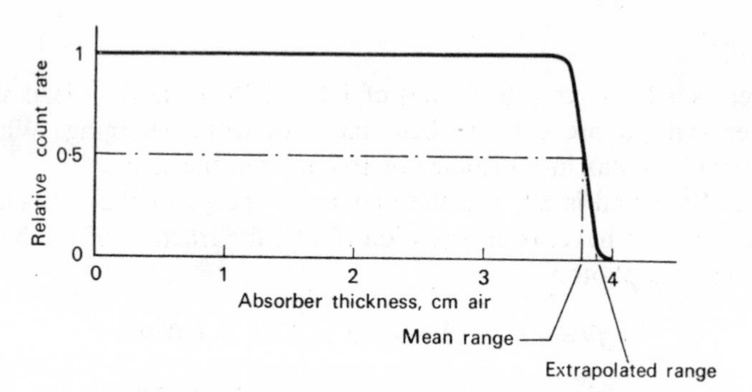

FIG. 5.8. Alpha-particle absorption curve.

"average" alpha particle. The extrapolated range is obtained by extrapolating the absorption curve to zero alpha particles transmitted.

Air is the most commonly used absorbing medium for specifying range–energy relationships of alpha particles. For energies less than 4 MeV, and for 4–8 MeV, the range in air at 0°C and 760 mm pressure is closely approximated (within 10%) by the equations

$$R, \text{cm} = 0.56E \text{ MeV} \qquad \text{for} \quad E < 4 \text{ MeV}, \tag{5.13}$$

$$R, \text{cm} = 1.24E \text{ MeV} - 2.62 \quad \text{for} \quad 4 < E < 8 \text{ MeV}. \tag{5.14}$$

The range of alpha particles in any other medium may be computed from the following relationship

$$R_m, \text{mg/cm}^2 = 0.56A^{1/3} R, \tag{5.15}$$

where A = atomic number of the medium, and R = range of the alpha particle in air, cm.

Example 5.6

What thickness of aluminum foil, density 2.7 g/cm³, is required to stop the alpha particles from ^{210}Po?

The energy of the ^{210}Po alpha particle is 5.3 MeV. From equation (5.14), the range of the alpha particle in air is

$$R = 1.24 \times 5.3 - 2.62 = 3.95 \text{ cm}.$$

Substituting this value for R into equation (5.15), and 27 for the atomic weight, A, we have

$$R_m = 0.56 \times 27^{1/3} \times 3.95 = 6.64 \text{ mg/cm}^2.$$

The thickness of this foil, in centimeters, is calculated from equation (5.1), and is found to be 0.00246 cm.

Because the effective atomic composition of tissue is not very much different from that of air, the following relationship may be used to calculate the range of alpha particles in tissue:

$$R_a \times \rho_a = R_t \times \rho_t, \tag{5.16}$$

where R_a and R_t = range in air and tissue, ρ_a and ρ_t = density of air and tissue.

Example 5.7

What is the range in tissue of a ^{210}Po alpha particle?

The range in air of this alpha was found in Example 5.6 to be 3.95 cm. Assuming tissue to have unit density, the range in tissue is, from equation (5.16),

$$R_t = \frac{3.95 \text{ cm} \times 1.293 \times 10^{-3} \text{ g/cm}^3}{1 \text{ g/cm}^3} = 5.1 \times 10^{-3} \text{cm}.$$

Energy Transfer

The major energy loss mechanism for alpha particles, and the only one considered significant in health physics is electronic excitation and ionization. In passing through air or soft tissue, an alpha particle loses, on the average, 35 eV per ion pair that it creates. Because of its high electrical charge and relatively low velocity due to its great mass, the specific ionization of an alpha particle is very high, on the order of tens of thousands of ion pairs per centimeter in air, Fig. 5.9.

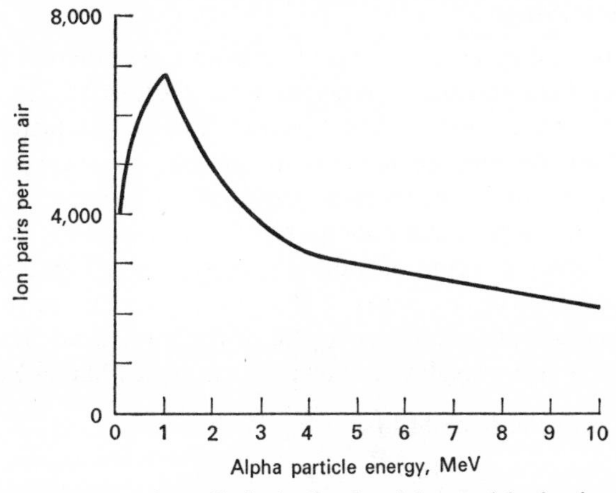

FIG. 5.9. Bragg curve of specific ionization by alpha particles in air at standard temperature and pressure.

The linear rate of energy loss for all charged particles more massive than an electron is

$$\frac{dE}{dx} = \frac{4\pi z^2 q^4 NZ}{Mv^2 \times 1.6 \times 10^{-6}} \left[\ln \frac{2Mv^2}{I} - \ln \left(1 - \frac{v^2}{c^2}\right) - \frac{v^2}{c^2}\right] \frac{\text{MeV}}{\text{cm}}, \tag{5.17}$$

where z = atomic number of the ionizing particle,
q = unit electrical charge, 4.8×10^{-10} sC,
zq = electrical charge on the ionizing particle,
M = rest mass of the ionizing particle, grams,
v = velocity of the ionizing particle, cm/sec,
N = number of absorber atoms per cm^3,
Z = atomic number of absorber,
NZ = number of absorber electrons per cm^3,
c = velocity of light, 3×10^{10} cm/sec,
I = mean excitation and ionization potential of absorber atoms; for air, $I = 1.38 \times 10^{-10}$ ergs, for other substances, $I = 2.16 \times 10^{-11} Z$.

For the case where the ionizing particle is an alpha particle, $z = 2$ and $M = 6.60 \times 10^{-24}$ g. Equation (5.17) is valid for ionizing particle velocities that are greater than the velocity of the orbital electrons of the absorber.

The mass stopping power of any substance for an alpha particle is defined in the same way as for a beta particle; the same thing is true for the relative mass stopping power of any absorber. Equations (5.8) and (5.10) are therefore used to define these two properties.

Gamma-rays

Exponential Absorption

The absorption of gamma radiation is qualitatively different from that of either alpha or beta radiation. Whereas both these particulate radiations have definite ranges in matter, and therefore can be completely absorbed, gamma radiation can only be reduced in intensity by increasingly thicker absorbers; it can not be completely absorbed. If gamma-ray absorption measurements are made under conditions of *good geometry*, that is, with a well-collimated, narrow beam of radiation, as shown in Fig. 5.10, and if the data are plotted on semi-log paper, a straight line results, as shown in Fig. 5.11, if the gamma-rays are monoenergetic. If the gamma-ray beam is heterochromatic, then a curve results, as shown by the dotted line in Fig. 5.11.

The equation of the straight line in Fig. 5.11 is

$$\ln I = -\mu t + \ln I_0 \tag{5.18A}$$

or

$$\ln I/I_0 = -\mu t. \tag{5.18B}$$

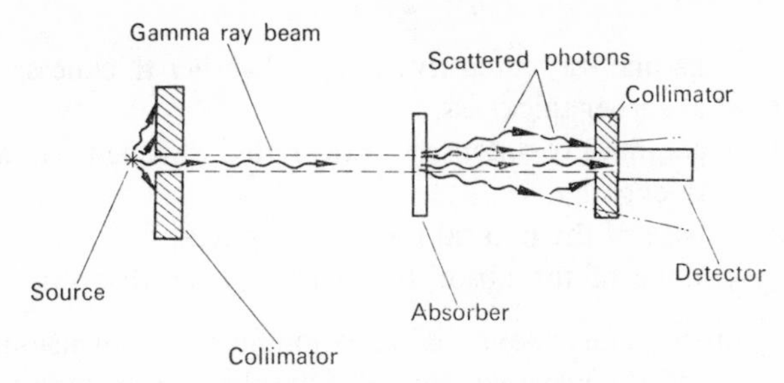

FIG. 5.10. Measuring attenuation of gamma-rays under conditions of good geometry. Ideally, the beam should be well collimated, and the source should be as far away as possible from the detector; the absorber should be midway between the source and the detector, and it should be thin enough so that the likelihood of a second interaction between a photon already scattered by the absorber and the absorber is negligible; and there should be no scattering material in the vicinity of the detector.

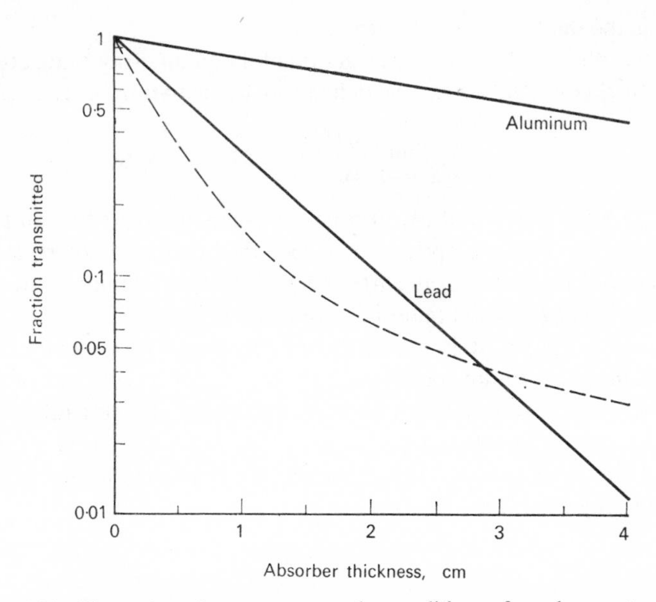

FIG. 5.11. Absorption of gamma-rays under conditions of good geometry. The solid lines are the absorption curves for a 0.662 MeV (monoenergetic) gamma-rays. The dotted line is the absorption curve for a heterochromatic beam.

Taking the anti-logs of both sides of equation (5.18), we have

$$I/I_0 = e^{-\mu t}, \tag{5.19}$$

where I_0 = gamma-ray intensity at zero absorber thickness,
t = absorber thickness,
I = gamma-ray intensity transmitted through an absorber of thickness t,
e = base of the natural logarithm system,
μ = slope of the absorption curve = the absorption coefficient.

Since the exponent in an exponential equation must be dimensionless, μ and t must be in reciprocal dimensions; that is, if the absorber thickness is measured in centimeters, then the absorption coefficient is called the *linear absorption coefficient*, μ_l, and must have dimensions of "per cm". If t is in g/cm^2, then the absorption coefficient is called the *mass absorption coefficient*, μ_m, and must have dimensions of $(\text{g/cm}^2)^{-1}$, or cm^2/g. The numerical relationship between μ_l and μ_m is given by the equation

$$\mu_l \text{ cm}^{-1} = \mu_m \text{ cm}^2/\text{g} \times \rho \text{ g/cm}^3, \tag{5.20}$$

where ρ is the density of the absorber.

The absorption coefficient is the fraction of the gamma-ray beam attenuated per unit thickness of absorber, as defined by the equation below:

$$\underset{\Delta t \to 0}{\text{limit}} \frac{\Delta I/I}{\Delta t} = -\mu, \tag{5.21}$$

where $\Delta I/I$ is the fraction of the gamma-ray beam attenuated by an absorber of thickness Δt. The absorption coefficient thus defined is often called the *total absorption coefficient* or the *attenuation coefficient*. Values of the attenuation coefficients for several materials are given in Table 5.2.

For some purposes, it is useful to use the *atomic absorption coefficient*, μ_a. The atomic absorption coefficient is the fraction of an incident gamma-ray beam that is absorbed by a single atom. Another way of saying the same thing is that the atomic absorption coefficient is the probability that an absorber atom will interact with one of the photons in the beam. The atomic absorption coefficient may be defined by the equation

$$\mu_a \text{ cm}^2 = \frac{\mu_l \text{ cm}^{-1}}{N \text{ atoms/cm}^3}, \tag{5.22}$$

where N = the number of absorber atoms per cm^3. Note that the dimensions of μ_a are cm^2, the units of area. For this reason, the atomic absorption coefficient is almost always referred to as the "cross section" of the absorber. The unit in which the cross section is specified is the *barn*.

$$1 \text{ barn} = 10^{-24} \text{ cm}^2.$$

The atomic absorption coefficient is also called the microscopic cross section, and is symbolized by σ, while the linear absorption coefficient is often called the *macroscopic cross section*, and is given the symbol Σ. This nomenclature is almost always used when dealing with neutrons. Equation (5.21) can thus be written as

$$\Sigma \text{ cm}^{-1} = \sigma \frac{\text{cm}^2}{\text{atom}} \times N \frac{\text{atoms}}{\text{cm}^3}. \tag{5.23}$$

Using the relationship given in equation (5.22), equation (5.19) may be re-written as

$$I/I_0 = e^{-\mu_a N t}, \tag{5.24A}$$

or

$$I/I_0 = e^{-\sigma N t}. \tag{5.24B}$$

The numerical values for μ_a have been published for many elements and for a wide range of quantum energies.†

With the aid of atomic cross sections, it is possible to compute the absorption coefficient of an alloy or a compound containing several different elements.

Example 5.8

Aluminum bronze, an alloy containing 90% Cu (atomic weight = 63.57) and 10% Al (atomic weight = 27) by weight, has a density of 7.6 g/cm³. What are the linear and mass absorption coefficients for 0.4 MeV gamma-rays, if the cross sections for Cu and Al for this quantum energy are 9.91 and 4.45 barns?

From equation (5.22) the linear absorption coefficient of aluminum bronze is

$$\mu_l = (\mu_a)_{Cu} \times N_{Cu} + (\mu_a)_{Al} \times N_{Al}.$$

The number of Cu atoms per cm³ in the alloy is

$$N_{Cu} = \frac{6.03 \times 10^{23} \text{ atoms/mole}}{63.57 \text{ g/mole}} (7.6 \times 0.9) \text{ g/cm}^3 = 6.49 \times 10^{22} \text{ atoms/cm}^3,$$

and for aluminum, N_{Al} is 1.7×10^{22} atoms/cm³. The linear absorption coefficient therefore is

$$\mu_l = 9.9 \times 10^{-24} \text{ cm}^2 \times 6.49 \times 10^{22} \text{ atoms/cm}^3 + 4.45 \times 10^{-24} \text{ cm}^2 \times \times 1.7 \times 10^{22} \text{ atoms/cm}^3,$$

$$\mu_l = 0.705 \text{ cm}^{-1}.$$

† GLADYS WHITE GROODSTEIN, *X-ray Attenuation Coefficients from 10 KeV to 100 MeV*. NBS Circular 583. U.S. Government Printing Office, Washington, 1957.

The mass absorption coefficient is, from equation (5.20)

$$\mu_m = \frac{0.705 \text{ cm}^{-1}}{7.6 \text{ g/cm}^3} = 0.0927 \text{ cm}^2/\text{g}.$$

The absorptive properties of matter vary systematically with the atomic number of the absorber and with the energy of the gamma radiation, as shown in Fig. 5.12. It should be noted, however, that in the region where the Comp-

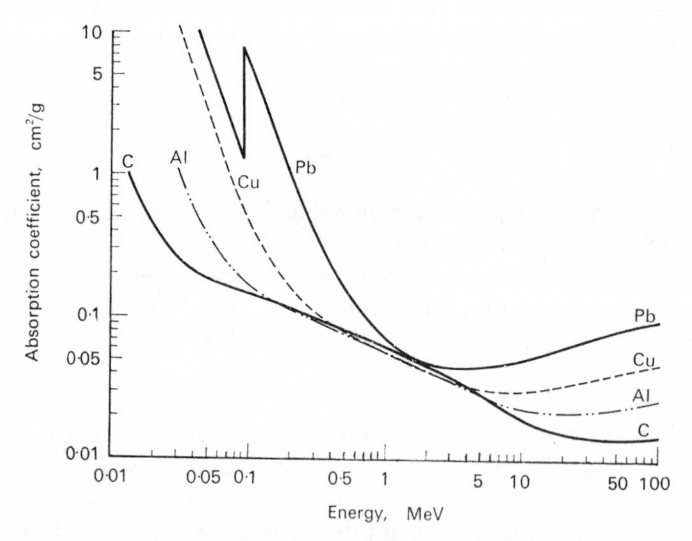

FIG. 5.12. Curves illustrating the systematic variation of absorption coefficient with atomic number of absorber and with quantum energy.

ton effect (this effect is more fully discussed below) predominates, the mass absorption coefficient is almost independent of the atomic number of the absorber.

Example 5.9

(a) Compute the thickness of aluminum and lead to transmit 10% of a narrow beam of 0.1 MeV gamma radiation.

From Table 5.2, μ_l for Al is 0.435 cm^{-1} and for lead it is 59.7. From equation (5.19) we have for aluminum

$$\frac{1}{10} = e^{-(0.435 \text{ cm}^{-1})(t \text{ cm})},$$

$$\ln 10 = 0.435t,$$

$$\frac{2.3}{0.435} = t = 5.3 \text{ cm}.$$

In a similar manner, we have for lead

$$\frac{1}{10} = e^{-(59.7 \text{ cm}^{-1})(t \text{ cm})},$$

$$t = 0.0385 \text{ cm}.$$

(b) Repeat part (a) for a 1.0 MeV gamma-ray. μ_l for Al = 0.166, μ_l for Pb = 0.771. For Al we have

$$\frac{1}{10} = e^{-(0.166 \text{ cm}^{-1})(t \text{ cm})},$$

$$t = 13.86 \text{ cm},$$

and for lead,

$$\frac{1}{10} = e^{-(0.771 \text{ cm}^{-1})(t \text{ cm})},$$

$$t = 2.97 \text{ cm}.$$

(c) Compare the density thicknesses of the Al and Pb in each part of the illustrative example above.

The density thickness for the Al in the case of the 0.1 MeV photon is, from equation (5.1)

$$t_{d\text{Al}} = 2.7 \text{ g/cm}^3 \times 5.3 \text{ cm} = 14.3 \text{ g/cm}^2,$$

and for lead,

$$t_{d\text{Pb}} = 11.34 \text{ g/cm}^3 \times 0.0385 \text{ cm} = 0.435 \text{ g/cm}^2.$$

In the case of the 1.0 MeV quantum energy, the density thickness for the Al is computed as 37.4 g/cm^2 and for lead it is 33.6 g/cm^2.

Example 5.9 shows that for the high-energy gamma-ray, lead is only a slightly better absorber, on a mass basis, than aluminum. For the low-energy photon, on the other hand, lead is a very much better absorber than aluminum. Generally, for energies between about 0.75 and 5 MeV almost all materials have, on a mass basis, about the same gamma-ray attenuating properties. To a first approximation in this energy range, therefore, shielding properties are approximately proportional to the density of the shielding material. For lower and higher quantum energies, absorbers of high atomic number are more effective than those of low atomic number. To understand this behaviour, we must examine the microscopic mechanisms of the interaction between gamma-rays and matter.

Absorption Mechanisms

For radiation-protection purposes, four major mechanisms for the absorption of gamma-ray energy are considered significant. Two of these mechanisms, photoelectric absorption and Compton scattering, which involve interactions only with the orbital electrons of the absorber, predominate in the case where the quantum energy of the photons does not greatly exceed 1.02 MeV, the energy equivalent of the rest mass of two electrons. In the case of higher energy photons, pair production, which is a direct conversion of electromagnetic energy into mass, occurs. These three gamma-ray absorption mechanisms result in the emission of electrons from the absorber. Very high energy photons, $E \gg 2m_0c^2$, may also be absorbed into the nuclei of the absorber atoms, and then initiate nuclear reactions which result in the emission, from the excited nuclei, of other radiations.

Pair production

A photon whose energy exceeds 1.02 MeV may, as it passes near a nucleus, spontaneously disappear, and its energy may reappear as a positron and an electron, as pictured in Fig. 5.13. Each of these two particles has a mass of m_0c^2, or 0.51 MeV, and the total kinetic energy is nearly equal to $hf - 2m_0c^2$. This transformation of energy into mass must take place near a particle, such as a nucleus, in order that momentum be conserved. The kinetic energy of the recoiling nucleus is very small. For practical purposes, therefore, all the photon energy in excess of that needed to supply the mass of the pair appears as kinetic energy of the pair. This same phenomenon may also occur in the

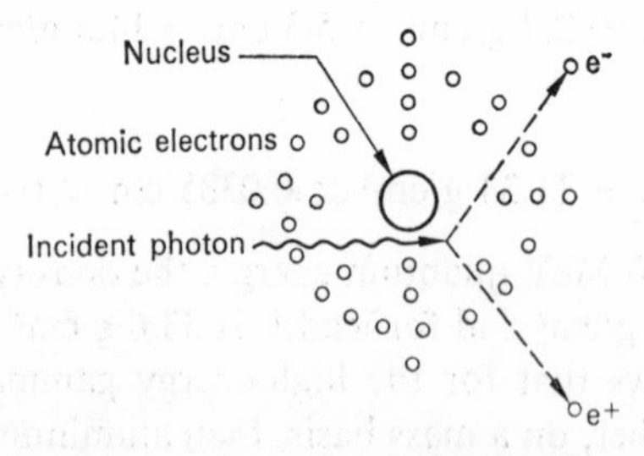

FIG. 5.13. Schematic representation of pair production. The positron-electron pair is generally projected in the forward direction (relative to the direction of the photon). The degree of forward projection increases with increasing photon energy.

vicinity of an electron, but the probability of occurrence near a nucleus is very much greater. Furthermore, the threshold energy for pair production near an electron is $4m_0c^2$. This higher threshold energy is necessary because the recoil electron, which conserves momentum, must be projected back with a very high velocity, since its mass is the same as that of each of the newly created particles. The cross section, or probability of the production of a positron–electron pair, is approximately proportional to $Z^2 + Z$, and is therefore important for high-atomic-numbered absorbers. The cross section increases slowly with increasing energy between the threshold of 1.02 MeV and about 5 MeV. For higher energies, the cross section is proportional to the logarithm of the quantum energy. It is this increasing cross section that accounts for the increasing absorption coefficient for high energy photons.

After production of a pair, the positron and electron are projected in a forward direction (relative to the direction of the photon) and lose their kinetic energy by excitation, ionization, and bremsstrahlung, as any other high-energy electron. When the positron has expended all of its kinetic energy, it combines with an electron to produce two quanta of 0.51 MeV each of annihilation radiations. Thus, a 10 MeV photon may, in passing through a lead absorber, be converted into a positron–electron pair in which each particle has about $4\frac{1}{2}$ MeV of kinetic energy. This kinetic energy is then dissipated in the same manner as beta particles. The positron then is annihilated by combining with an electron in the absorber, and two photons of 0.51 MeV each may emerge from the absorber (or they may undergo Compton scattering or photoelectric absorption). The net result of the pair production interaction in this case was the conversion of a single 10 MeV photon into two photons of 0.51 MeV each, and the dissipation of 8.98 MeV of energy.

Compton scattering

Compton scattering is an elastic collision between a photon and a "free" electron (a "free" electron is one whose binding energy to an atom is very much less than the energy of the photon), as shown diagramatically in Fig. 5.14.

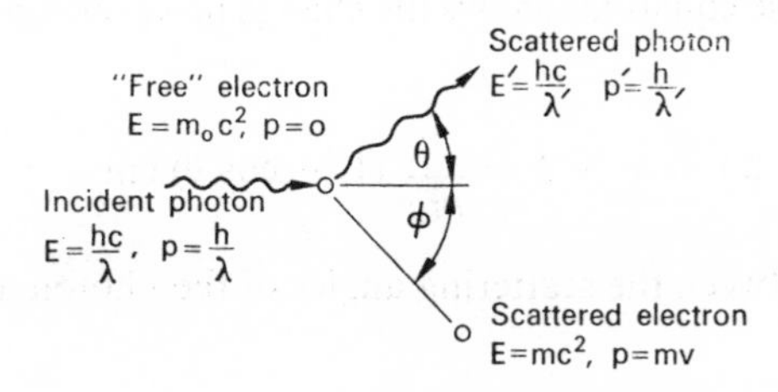

FIG. 5.14. Compton scattering: an elastic collision between a photon and an electron.

In a collision between a photon and a free electron, it is impossible for all the photon's energy to be transferred to the electron if momentum and energy are to be conserved. This can be shown by assuming that such a reaction is possible. If this were true, then, according to the conservation of energy, all the energy of the photon is imparted to the electron, and we have

$$E = mc^2.$$

According to the law of conservation of momentum, all the momentum of the photon, p, must be transferred to the electron if the photon is to disappear:

$$p = \frac{E}{c} = mv.$$

Eliminating m from these two equations and solving for v, we find $v = c$, an impossible solution. The original assumption, that the photon transferred all of its energy to the electron, must therefore be false.

Since all the photon's energy cannot be transferred, the photon must be scattered, and the scattered photon must have less energy—or a longer wavelength—than the incident photon. Only the energy difference between the incident and scattered photons is transferred to the free electron. The amount of energy transferred in any collision can be calculated by applying the laws of conservation of energy and momentum to the situation pictured in Fig. 5.14. To conserve energy, we must have

$$\frac{hc}{\lambda} + m_0c^2 = \frac{hc}{\lambda'} + mc^2, \tag{5.25}$$

and to conserve momentum in the horizontal and vertical directions respectively, we have

$$\frac{h}{\lambda} = \frac{h}{\lambda'} \cos \theta + mv \cos \phi, \tag{5.26}$$

$$0 = \frac{h}{\lambda'} \sin \theta \; mv \sin \phi. \tag{5.27}$$

The solution of these equations shows the change in wavelength of the photon to be

$$\Delta\lambda = \lambda' - \lambda = \frac{h}{m_0c}(1 - \cos \theta) \text{ cm}, \tag{5.28}$$

and the relation between the scattering angles of the photon and the electron to be

$$\cot \frac{\theta}{2} = \left(1 + \frac{h}{\lambda\, m_0c}\right) \tan \phi. \tag{5.29}$$

When the numerical values are substituted for the constants, and centimeters are converted into angstrom units, equation (5.28) reduces to

$$\Delta\lambda = 0.0242\,(1 - \cos\theta)\ \text{Å}. \tag{5.30}$$

Equation (5.29) shows that the electron cannot be scattered through an angle greater than 90°. This scattered electron is of great importance in radiation dosimetry, because it is the vehicle by means of which energy from the scattered photon is transferred to an absorbing medium. The Compton electron dissipates its kinetic energy in the same manner as a beta particle, and is one of the primary ionizing particles produced by gamma radiation. Compton scattering is also important in health physics engineering because of the fact that a high-energy photon loses a greater fraction of its energy when it is scattered than does a low-energy photon. By taking advantage of this fact, required shielding thickness can be reduced and economic savings thereby effected.

Example 5.10

What fractions of their energies do 1 MeV and 0.1 MeV photons lose if they are scattered through an angle of 90°?

Substituting $E = hc/\lambda$ into equation (5.28), and solving for the energy of the scattered photon, we have

$$E' = \frac{E}{1 + (E/m_0c^2)\,(1 - \cos\theta)}, \tag{5.31A}$$

and the fraction of the incident energy carried by the scattered photon is

$$\frac{E'}{E} = \frac{1}{1 + (E/m_0c^2)\,(1 - \cos\theta)}. \tag{5.31B}$$

Substituting the values for the incident photon and the scattering angle into equation (5.31A), we have

$$E' = \frac{1\ \text{MeV}}{1 + (1/0.51)\,(1)} = 0.338\ \text{MeV},$$

and the fractional energy loss is

$$1 - \frac{E'}{E} = \frac{1 - 0.338}{1} = 0.662 = 66.2\%.$$

In the case of the 0.1 MeV gamma-ray, the energy of the scattered photon is, from equation (5.30), 0.0835 MeV, and the fractional energy loss is only 0.165, or 16.5%.

The probability of a Compton interaction decreases with increasing quantum energy and with increasing atomic number of the absorber. In light atomic numbered elements, Compton scattering is the main mechanism of interaction. In Compton scattering every electron acts as a scattering center, and the bulk scattering properties of matter depends mainly on the electronic density per unit mass. Probabilities for Compton scattering are therefore given on a per electron basis. The theoretical cross sections for Compton

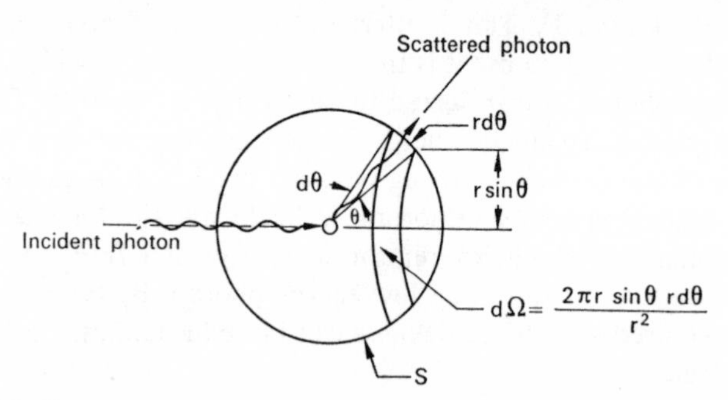

FIG. 5.15. Compton scattering diagram to illustrate differential scattering cross section. S is a sphere of unit radius whose center is the scattering electron.

scattering were derived by Klein and Nishina. For scattering into a differential solid angle d Ω at an angle θ to the direction of the incident photon, Fig. 5.15, they give the differential total scattering coefficient as

$$\frac{d\sigma_t}{d\Omega} = \frac{e^4}{2\, m_0^2 c^4} \left[\frac{1}{1 + \alpha\,(1 - \cos\theta)}\right]^2 \left[\frac{1 + \cos^2\theta + \alpha^2\,(1 - \cos\theta)^2}{1 + \alpha\,(1 - \cos\theta)}\right], \quad (5.32)$$

where e, m_0, and c have the usual meaning, and $\alpha = hf/m_0c^2$. Equation (5.32) and Fig 5.16 give the probability of scattering a photon into a solid angle d Ω through an angle θ. The total probability of scattering, Ω_t, can be obtained by substituting $d\Omega = 2\pi \sin\theta\, d\theta$ and integrating the differential scattering coefficient over the entire sphere. The result of this calculation, for quantum energies up to 10 MeV, is presented graphically in Fig. 5.17.

The cross section for any reaction, when viewed from the point of view of an individual particle or a single photon, is the probability that the photon will undergo that reaction. From the point of view of a beam of radiation, the cross section gives the fraction of the particles in the beam that reacts in a given manner. In Compton scattering, energy is transferred from the photon to the scattered electron, and the scattered photon has less energy than the incident photon. The fraction of the incident energy that is carried by the scattered photon is given by equation (5.31B). The fraction of the energy in a gamma-ray beam that is scattered is given by the product of the fraction of the photons that are scattered and the fractional energy loss per collision:

$$\frac{d\sigma_s}{d\Omega} = \frac{d\sigma_t}{d\Omega} \cdot \frac{E'}{E}. \quad (5.33)$$

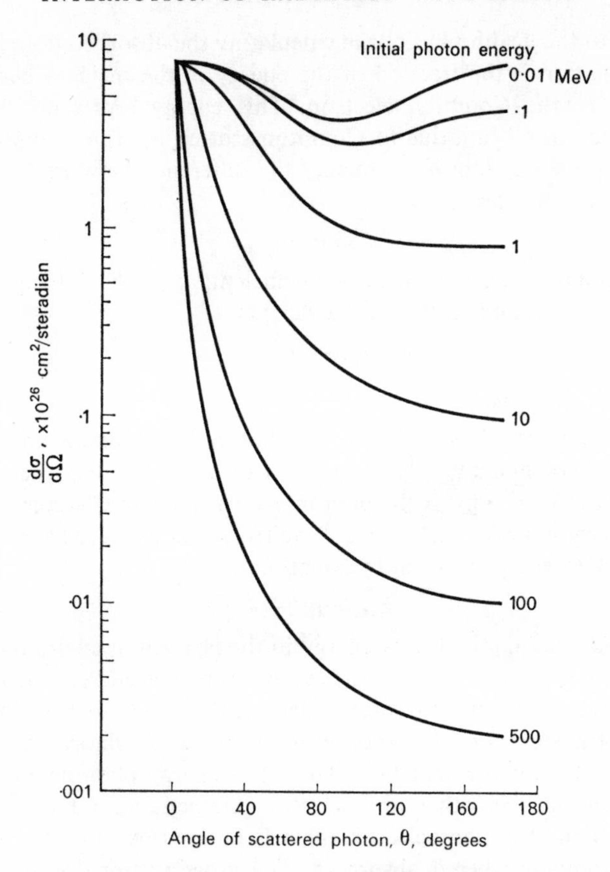

FIG. 5.16. Differential scattering coefficient showing the probable angular distribution of Compton scattered photons.

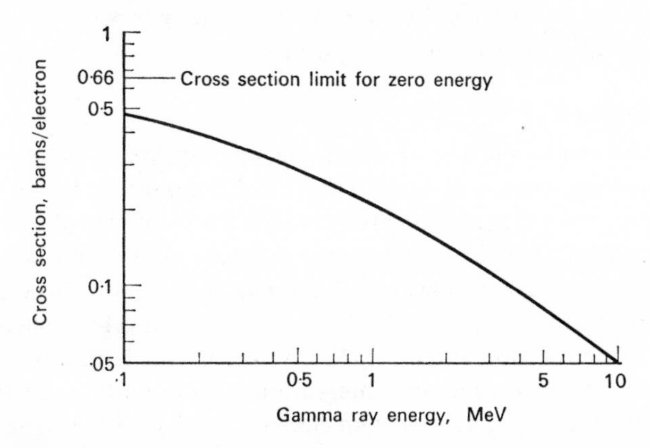

FIG. 5.17. Total Compton cross sections for a free electron.

Of interest to the health physicist in calculating the absorbed dose from X or gamma radiation is the fraction of the energy in the incident beam that is transferred to the Compton electron. This energy represents the energy absorbed from the beam due to Compton scattering. The Compton energy absorption cross section, σ_e, is merely the difference between the total and scattering cross sections

$$\sigma_e = \sigma_t - \sigma_s, \tag{5.34A}$$

and the Compton energy absorption coefficient, μ_{ce}, is the difference between the Compton total and scattering coefficients

$$\mu_{ce} = \mu_{ct} - \mu_{cs}. \tag{5.34B}$$

Photoelectric absorption

The photoelectric effect, in which the photon disappears, is an interaction between a photon and a tightly bound electron whose binding energy is equal to or less than the energy of the photon, as discussed in Chapter 3.

The primary ionizing particle resulting from this interaction is the photoelectron, whose energy is given by equation (3.13),

$$E_{pe} = hf - \phi.$$

The photoèlectron dissipates its energy in the absorbing medium mainly by excitation and ionization. The photoelectric effect is considered true absorption because all the energy of the photon is deposited in the absorber. The binding energy ϕ is transferred to the absorber by means of the fluorescent radiation that follows the initial interaction. These low-energy photons are absorbed by outer electrons, in other photoelectric interactions, not far from their points of origin. The photoelectric effect is favored by low-energy photons and high-atomic-numbered absorbers. The cross section for this reaction varies approximately as $Z^4 \lambda^3$. It is this very strong dependence of photoelectric absorption on the atomic number, Z, that makes lead such a good material for shielding against X-rays. For very low atomic numbered absorbers, the photoelectric effect is relatively unimportant.

Photodisintegration

In photodisintegration, the absorber nucleus captures a gamma-ray and, in most instances, emits a neutron. This is a threshold reaction in which the quantum energy must exceed a certain minimum value that depends on the absorbing nucleus. This is a high-energy reaction, and with few exceptions is not an absorption mechanism for gamma-rays from radioisotopes. An important exception is the case of ^{9}Be, in which the threshold energy is only 1.666 MeV. The reaction ^{9}Be (γ, n) ^{8}Be is useful as a laboratory source of monoenergetic neutrons. Photodisintegration is an important reaction in the case of very high energy photons from electron accelerators such as betatrons and synchrotrons. Here, too, interest is centred on the fact that photodisintegration results in neutron production. Generally, the cross sections for photodisintegration are very much smaller than the total cross section given in equation (5.35). In shielding calculations, therefore, the photodisintegration cross sections are usually considered insignificant, and are neglected.

Photodisintegration is a threshold reaction because the energy added to the absorber nucleus must be at least equal to the binding energy of a nucleon. Furthermore, a neutron is preferentially emitted rather than a proton because it has no coulombic potential barrier to overcome in order to escape from the nucleus, and hence has a lower threshold. The range of energy thresholds for photodisintegration by neutron emission varies from 1.66 MeV for beryllium to about $8\frac{1}{2}$ MeV. For light nuclei, the thresholds fluctuate unsystematically; in the range of atomic mass numbers 20–130, the thresholds increase slowly to about $8\frac{1}{2}$ MeV, and then decreases slowly to about 6 MeV as the atomic mass numbers increase. Quantum energies greater than the threshold appear as kinetic energy of the emitted neutrons or, if great enough, may cause the emission of charged particles from the absorber nucleus.

Combined effects

The absorption coefficients or cross sections give the probabilities of removal of a photon from a beam under conditions of good geometry, where it is assumed that any of the possible interactions removes the photon from the beam. The total absorption coefficient, therefore, is the sum of the coefficients for each of the three reactions discussed above:

$$\mu_t = \mu_{pe} + \mu_{cs} + \mu_{pp}, \tag{5.35}$$

where the three right-hand terms are the absorption coefficients respectively for the photoelectric effect, for Compton scattering, and for pair production. In computing attenuation of radiation for purposes of shielding design, the total absorption coefficient as defined in equation (5.35) is used.

Equation (5.35) gives the fraction of the energy in a beam that is removed by an absorber. The fraction of the beam's energy that is deposited in the absorber considers only the energy transferred to the absorber by the photoelectron, by the Compton electron, and by the electron pair. Energy carried away by the scattered photon in a Compton interaction and the energy carried off by the annihilation radiation after pair production is not included. The *energy absorption coefficient*, which is also called the *true absorption coefficient*, is given by

$$\mu_e = \mu_{pe} + \mu_{ce} + \mu_{pp}\left(\frac{hf - 1.02}{hf}\right), \tag{5.36}$$

and is used in calculation of radiation dose. The total and true absorption coefficients for air are shown in Fig. 5.18.

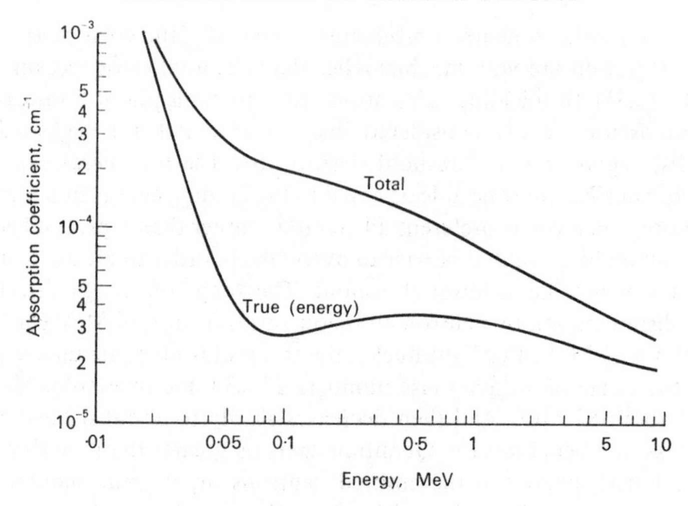

FIG. 5.18. Total absorption (attenuation) coefficient and true (energy) absorption coefficient for gamma-rays of various energy.

Neutrons

Production

Except for several fission fragments of very short half-life, there are no radioisotopes that emit neutrons. All neutron sources therefore must depend on nuclear reactions. The most prolific neutron source is a nuclear reactor. Copious neutron beams may also be produced in accelerators by many different reactions. For example, bombardment of beryllium by high-energy deuterons in a cyclotron produces neutrons according to the reaction

$$^{9}_{4}Be + ^{2}_{1}D \rightarrow (^{11}_{5}B)^* \rightarrow ^{10}_{5}B + ^{1}_{0}n. \tag{5.37}$$

The term in the parenthesis is called a *compound nucleus*, and the asterisk shows that it is in an excited state. The compound nucleus rids itself of its excitation energy instantaneously ($<10^{-8}$ sec) by proceeding to the next step in the reaction. For small laboratory sources of neutrons, the photodisintegration of beryllium may be used. Another commonly used neutron source depends on the bombardment of beryllium with alpha particles. The reaction, in this case, is

$$^{9}_{4}Be + ^{4}_{2}He \rightarrow (^{13}_{6}C)^* \rightarrow ^{12}_{6}C + ^{1}_{0}n. \tag{5.38}$$

For the source of the alpha particles, radium, polonium, and plutonium are used. The alpha emitter, as a powder, is thoroughly mixed with finely powdered beryllium, and the mixture is sealed in a capsule, as shown in Fig. 5.19. The neutrons that are produced are all high energy. In all cases of neutrons based on this reaction, the neutron energy is spread over a broad spectrum, as shown in Fig. 5.20. This spread of energies from a $^{9}Be\ (\alpha, n)^{12}C$ source is in sharp contrast to the monoenergetic neutrons from a photodisintegration source using monoenergetic photons. In the α, n reaction, the energy equivalent of the difference in mass between the reactants and the products plus the kinetic energy of the bombarding particle is divided between the neutron and

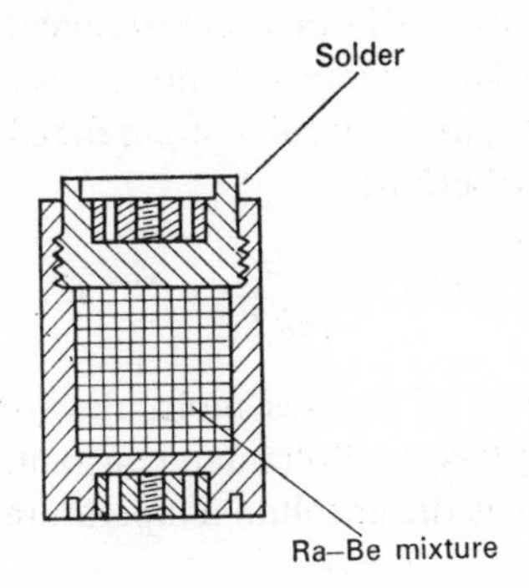

FIG. 5.19. Typical Ra–Be (α, n) neutron source in a sealed container.

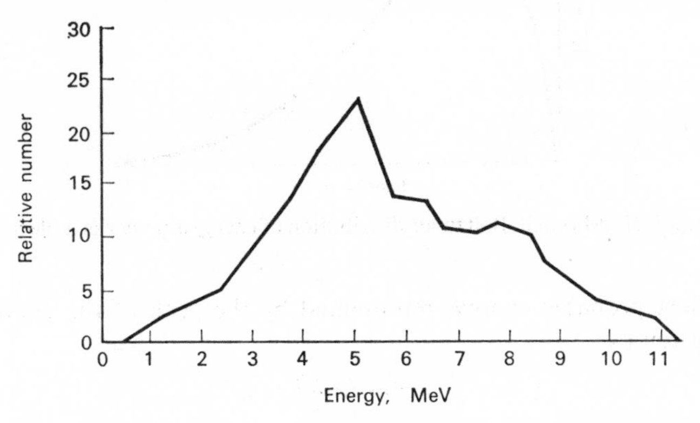

FIG. 5.20. Energy distribution of Po–Be neutrons. (Technical Bulletin NS-1, AECL, Ottawa.)

the recoil nucleus. In practical α, n sources, some of the alpha particle energy is dissipated by self-absorption within the source. As a consequence, the alphas that initiate the reaction have a wide range of energy, thereby contributing to the spectral spread of the neutrons. The neutron yield from an α, n source increases with increasing alpha energy because of the greater ease with which higher-energy alphas penetrate the coulomb barrier at the nucleus. Tables 5.3 and 5.4 list some γ, n and α, n neutron sources respectively.

Classification

Neutrons are classified according to their energy because the type of reaction that a neutron undergoes depends very strongly on its energy. High-energy neutrons, those whose energies exceed about 0.1 MeV, are called *fast neutrons*. *Thermal neutrons*, on the other hand, have the same average kinetic energy as gas molecules in their environment. In this respect, thermal neutrons are indistinguishable from gas molecules at the same temperature. The kinetic energies of gas molecules are related to temperature by the Maxwell–Boltzman distribution:

$$f(E) = \frac{2\pi}{(\pi kT)^{\frac{3}{2}}} e^{-E/kT} E^{\frac{1}{2}}, \tag{5.39}$$

where $f(E)$ is the fraction of the gas molecules (or neutrons) of energy E per unit energy interval; k is the Boltzman constant, 1.38×10^{-16} erg/°K or 8.6×10^{-5} eV/°K; and T is the absolute temperature of the gas, °K.

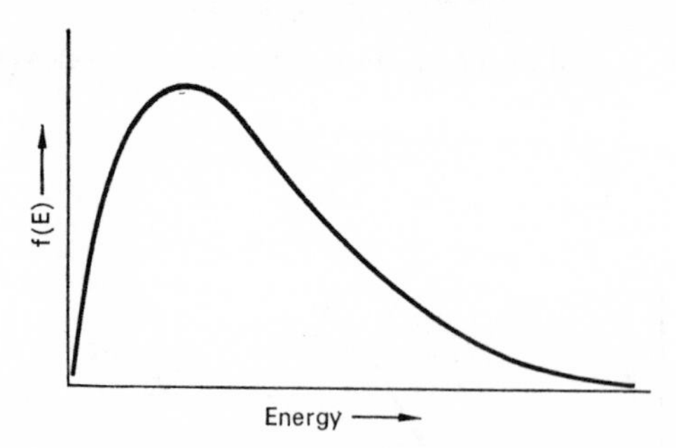

FIG. 5.21. Maxwell–Boltzman distribution of energy among gas molecules.

The most probable energy, represented by the peak of the curve in Fig. 5.21, is given by:

$$E_{mp} = kT, \tag{5.40}$$

while the average energy of gas molecules at any given temperature is:

$$\bar{E} = \tfrac{3}{2}\, kT. \tag{5.41}$$

For neutrons at a temperature of 293°K, the most probable energy is 0.025 eV. This is the energy often implied in the term "thermal" neutrons. The velocity corresponding to this energy, which is given by

$$\tfrac{1}{2}\, mv^2 = kT, \tag{5.42}$$

is 2.2×10^5 cm/sec.

In the region of energy between thermal and fast, neutrons are called by various names, including *intermediate neutrons*, *resonance neutrons*, and *slow neutrons*. All these descriptive adjectives are used loosely, and their exact meaning must be inferred from the context in which they are used.

TABLE 5.3. γ, n PHOTONEUTRON SOURCES

Source	Half-life	Average neutron energy MeV	Yield n/sec/Ci
$^{24}Na + Be$	15 hr	0.83	1.3×10^5
$^{24}Na + D_2O$	15 hr	0.22	2.7×10^5
$^{56}Mn + Be$	2.58 hr	0.1(90%), 0.3(10%)	2.9×10^4
$^{56}Mn + D_2O$	2.58 hr	0.22	3.1×10^3
$^{72}Ga + Be$	14.2 hr	0.78	5×10^4
$^{72}Ga + D_2O$	14.2 hr	0.13	6×10^4
$^{88}Y + Be$	88 d	0.16	1×10^5
$^{88}Y + D$	88 d	0.31	3×10^3
$^{116}In + Be$	54 min	0.30	8.2×10^3
$^{124}Sb + Be$	60 d	0.024	1.9×10^5
$^{140}La + Be$	40 hr	0.62	3×10^3
$^{140}La + D_2O$	40 hr	0.15	8×10^3
$Ra + D_2O$	1600 yr	0.12	1×10^3

TABLE 5.4. α, n NEUTRON SOURCES

Source	Half-life	Average neutron energy MeV	Yield n/sec/Ci
$Ra + Be$	1600 yr	5	1.7×10^7
$Ra + B$	3.8 d	3	6.8×10^6
$^{222}Em + Be$	3.8 d	5	1.5×10^7
$^{210}Po + Be$	138 d	4	3×10^6
$^{210}Po + B$	138 d	2.5	9×10^5
$^{210}Po + F$	138 d	1.4	4×10^5
$^{210}Po + Li$	138 d	0.42	9×10^4
$^{239}Pu + Be$	24,000 yr	4	10^6

Interaction

All neutrons, at the time of their birth, are fast. Generally, fast neutrons lose energy by colliding elastically with atoms in their environment, and then, after being slowed down to thermal or near thermal energies, they are captured by nuclei of the absorbing material. Although a number of possible neutron reaction types exist, for the health physicist the chief reactions are elastic scattering and capture followed by the emission of a photon or another particle from the absorber nucleus.

When absorbers are placed in a collinated beam of neutrons, and the transmitted neutron intensity is measured, as was done for gamma-rays in Fig. 5.10, it is found that neutrons, too, are removed exponentially from the beam. Instead of using linear or mass absorption coefficients to describe the ability of a given absorber material to remove neutrons from the beam, it is customary to designate only the microscopic cross section, σ, for the absorbing material. The product σN, where N is the number of absorber atoms per cm^3, is the macroscopic cross section Σ. The removal of neutrons from the beam is thus given by

$$I = I_0 e^{-\sigma N t}. \tag{5.43}$$

Neutron cross sections are strongly energy dependent. If removal of a neutron from the beam may be effected by more than one mechanism, the total cross section is the sum of the cross sections for the various possible reactions.

Example 5.11

In an experiment designed to measure the total cross section of lead for 10 MeV neutrons, it was found that a 1-cm thick lead absorber attenuated the neutron flux to 84.5% of its initial value. The atomic weight of lead is 207.21, and its specific gravity is 11.3. Calculate the total cross section from these data.

The atomic density of lead is

$$\frac{6.03 \times 10^{23}\ \text{atoms/mole}}{207.21\ \text{g/mole}} \times 11.3\ \text{g/cm}^3 = 3.29 \times 10^{22}\ \text{atoms/cm}^3,$$

$$\frac{I}{I_0} = e^{-\sigma N t},$$

$$0.845 = e^{-\sigma \times 3.29 \times 10^{22} \times 1},$$

$$\ln \frac{1}{0.845} = 3.29 \times 10^{22} \sigma,$$

$$\sigma = \frac{0.168}{3.29 \times 10^{22}} = 5.1 \times 10^{-24}\ \text{cm}^2,$$

$\sigma = 5.1$ barns, and the macroscopic cross section is

$$\Sigma = \sigma N = 5.1 \times 10^{-24}\ \text{cm}^2 \times 3.29 \times 10^{22}\ \text{cm}^{-3} = 0.168\ \text{cm}^{-1}.$$

Scattering

Neutrons may collide with nuclei, and undergo either inelastic or elastic scattering. In the former case, some of the kinetic energy that is transferred to the target nucleus excites the nucleus, and the excitation energy is emitted as a gamma-ray photon. This interaction is best described by the compound nucleus model, in which the neutron is captured, then re-emitted by that target nucleus together with the gamma photon. This is a threshold phenomena; the neutron energy threshold varies from infinity for hydrogen (inelastic scattering cannot occur) to about 6 MeV for oxygen to less than 1 MeV for uranium. Generally, the cross section for inelastic scattering is small, on the order of 1 barn or less, for low energy fast neutrons, but increases with increasing energy, and approaches a value corresponding to the geometrical cross section of the target nucleus.

Elastic scattering is the most likely interaction between fast neutrons and low-atomic-numbered absorbers. This interaction is a "billiard ball" type collision, in which kinetic energy and momentum are conserved. By applying these conservation laws, it can be shown that the energy, E, of the scattered neutron after a head-on collision is

$$E = E_0 \left[\frac{M - m}{M + m}\right]^2, \tag{5.44}$$

where E_0 = energy of the incident neutron,
m = mass of the incident neutron,
M = mass of the scattering nucleus.

The energy transferred to the target nucleus is $E_0 - E$. From equation (5.44), we have

$$E_0 - E = E_0 \left[1 - \left(\frac{M - m}{M + m}\right)^2\right]. \tag{5.45}$$

According to equations (5.44) and (5.45), it is possible, in a head-on collision with a hydrogen nucleus, for a neutron to transfer all its energy to the hydrogen nucleus. With heavier nuclei, all the kinetic energy of the neutron cannot be transferred in a single collision. In the case of oxygen, for example, equation (5.45) shows that the maximum fraction, $(E_0 - E)/E_0$, of the neutron's kinetic energy that can be transferred during a single collision is only 22.2%. This shows that nuclei with small mass numbers are more effective, on a "per collision" basis, than nuclei with high mass numbers for slowing down neutrons.

Equations (5.44) and (5.45) are valid only for head-on collisions. Most collisions are not head-on, and the energy transferred to the target nuclei are consequently less than the maxima given by the two equations above.

In the course of the successive collisions suffered by a fast neutron as it passes through a slowing down medium, the average decrease, per collision, in the logarithm of the neutron energy (which is called the average logarithmic energy decrement) remains constant. It is independent of the neutron energy, and is a function only of the mass of scattering nuclei. The average energy decrement is defined as

$$\xi = \overline{\Delta \ln E} = \overline{\ln E_0 - \ln E} = \overline{\ln \frac{E_0}{E}} = -\overline{\ln \frac{E}{E_0}}, \tag{5.46}$$

and can be shown to be given by

$$\xi = 1 + \frac{\alpha \ln \alpha}{1 - \alpha}, \tag{5.47}$$

where $\alpha = [(M - m)/(M + m)]^2$, as used in equation (5.44). If the slowing-down medium contains n kinds of nuclides, each of microscopic scattering cross section σ_s and average logarithmic energy decrement ξ, then the mean value of ξ for the n species is

$$\xi = \frac{\sum_{i=1}^{n} \sigma_{si} N_i \xi_i}{\sum_{i=1}^{n} \sigma_{si} N_i}. \tag{5.48}$$

Since

$$\overline{\ln \frac{E}{E_0}} = -\xi,$$

$$\frac{E}{E_0} = e^{-\xi},$$

and the median fraction of the incident neutron's energy that is transferred to the target nucleus during a collision is

$$f = 1 - \frac{\bar{E}}{E_0} = 1 - e^{-\xi}. \tag{5.49}$$

Thus for hydrogen, $\xi = 1$, the median energy transfer during a collision with a fast neutron is 63% of the kinetic energy of the neutron. In the case of carbon, $\xi = 0.159$, an average of only 14.7% of the neutron's kinetic energy is absorbed by the struck nucleus during an elastic collision. The struck nucleus, as a result of the kinetic energy imparted to it by the neutron, becomes an ionizing particle, and dissipates its kinetic energy in the absorbing medium by excitation and ionization.

The distance traveled by a fast neutron between its introduction into a slowing down medium and its thermalization depends on the number of collisions made by the neutron and the distance between collisions. Although the actual path of the neutron is tortuous because of deflections due to collisions, the average straight line distance covered by the neutron can be determined; it is called the *fast diffusion length*, or the *slowing-down length*. (The square of the fast diffusion length is called the Fermi age of the neutron.) The distance traveled by the thermalized neutron until it is absorbed is measured by the *thermal diffusion length*. The thermal diffusion length is defined as the thickness of a slowing down medium that attenuates a beam of thermal neutrons by a factor of e. Thus, attenuation of a beam of thermal neutrons by a substance of thickness t cm whose thermal diffusion length is L cm is given by

$$n = n_0 e^{-t/L}. \tag{5.50}$$

(The terms fast diffusion length and thermal diffusion length are applicable only to materials in which the absorption cross section is very small. When this condition is not met, as in the case of boron or cadmium, the attenuation of a beam of thermal neutrons is given by equation (5.43).) Although fast and thermal diffusion lengths may be calculated, the assumptions inherent in the calculations make it preferable to use measured values for these parameters. Values for fast (fission neutrons) and thermal diffusion lengths for certain slowing down media are given in Table 5.5.

TABLE 5.5. FAST AND THERMAL DIFFUSION LENGTHS OF SELECTED MATERIALS

Substance	Fast diffusion length	Thermal diffusion length
H_2O	5.75 cm	2.88 cm
D_2O	11	171
Be	9.9	24
C (graphite)	17.3	50

For the case of a point source of n_0 thermal neutrons per second in a spherically shaped non-multiplying medium (a medium which contains no fissile material) of radius R and thermal diffusion length L, the flux of neutrons escaping from the surface is

$$\phi = \frac{n_0}{4\pi RL \sinh R/L}, \tag{5.51}$$

and, for $L < R$,

$$\phi = \frac{n_0}{2\pi RL} e^{-R/L}. \tag{5.52}$$

Example 5.12

A Pu–Be neutron source that emits 10^6 neutrons per second is in the center of a spherical water shield whose diameter is 50 cm. How many thermal neutrons are escaping per cm²/sec from the surface of the shield?

Since the radius of the water shield is much greater than the fast diffusion length, Table 5.5, we may assume (for the purpose of this calculation) that essentially all the fast neutrons are thermalized, and that the thermal neutrons are diffusing outward from the center. Substituting the appropriate numbers into equation (5.52), we have

$$\phi = \frac{10^6}{2\pi \times 25 \times 2.88} e^{-25/2.88} = 0.375 \frac{\text{neutrons}}{\text{cm}^2/\text{sec}}.$$

Absorption

From the discussion above, it is seen that fast neutrons are rapidly degraded in energy by elastic collisions if they interact with low-atomic-numbered substances. As neutrons reach thermal or near thermal energies, their likelihood of capture by an absorber nucleus increases. The absorption cross

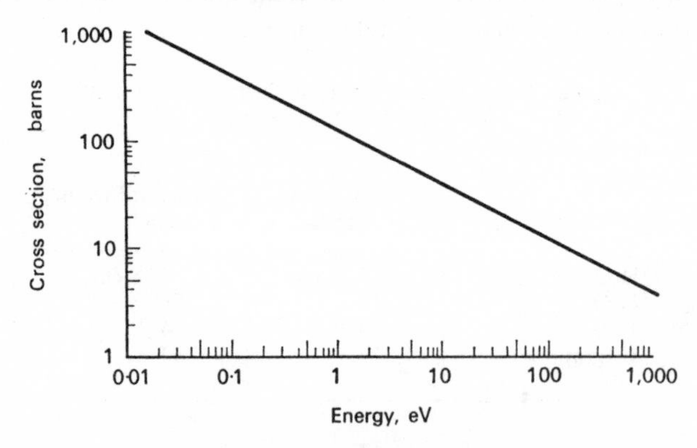

FIG. 5.22. Neutron absorption cross section for boron, showing the validity of the $1/v$ law for neutrons from 0.02 to 1000 eV in energy. The equation of the curve is $\sigma = 116\sqrt{\text{eV}}$ barns.

section of many nuclei, as the neutron energy becomes very small, has been found to be inversely proportional to the square root of its kinetic energy, and thus to vary inversely with its energy:

$$\sigma \propto \frac{1}{\sqrt{E}} \propto \frac{1}{v}. \tag{5.53}$$

Equation (5.53) is called the "one over v law" for slow neutron absorption. For ^{10}B, this relationship is valid for the span of energies from 0.001 to 1000 eV, as shown in Fig. 5.22. Thermal neutron cross sections are usually given for neutrons whose most probable energy is 0.025 eV. If the cross section at energy E_0 is σ_0, then the cross section for any other energy within the range of validity of the $1/v$ law is given by

$$\frac{\sigma}{\sigma_0} = \sqrt{\frac{E_0}{E}}. \tag{5.54}$$

Example 5.13

The cross section of boron for the ^{10}B (n, α) ^{7}Li reaction is 753 barns for 0.025 eV neutrons. What is the boron cross section for 50 eV neutrons? Substituting into equation (5.54) gives

$$\sigma = 753 \sqrt{\left(\frac{0.025}{50}\right)} = 16.8 \text{ barns.}$$

Some capture reactions of practical importance in health physics include the following:

$$^{1}H\ (n, \gamma)\ ^{2}H \qquad \sigma = 0.33 \text{ barns,} \tag{5.55}$$

$$^{14}N\ (n, p)\ ^{14}C \qquad \sigma = 1.70 \text{ barns,} \tag{5.56}$$

$$^{10}B\ (n, \alpha)\ ^{7}Li \qquad \sigma = 4.01 \times 10^{3} \text{ barns,} \tag{5.57}$$

$$^{113}Cd\ (n, \gamma)\ ^{114}Cd \qquad \sigma = 2.1 \times 10^{4} \text{ barns.} \tag{5.58}$$

Equations (5.55) and (5.56) are important in neutron dosimetry, since H and N are major constituents of tissue. Equation (5.57) is important in the design of instruments for measuring neutrons as well as neutron shielding, while the last equation is important mainly in shielding. It should be noted that the neutron reactions with hydrogen and with cadmium results in the emission of high-energy gamma-rays, while the capture of a thermal neutron by ^{10}B releases a low-energy (0.48 MeV) gamma-ray in 93% of the reactions. When a thermal neutron is captured by ^{14}N, a 0.6 MeV proton is emitted.

Neutron activation

Neutron activation is the production of a radioactive isotope by absorption of a neutron, such as the *n, p* reaction of equation (5.56). In that instance, ^{14}C is produced. Activation by neutrons is important to the health physicist for several reasons. First, it means that any substance that was irradiated by neutrons may be radioactive; a radiation hazard may therefore persist after the irradiation by neutrons is terminated. Secondly, it provides a convenient tool for measuring neutron flux. This is done simply by irradiating a known amount of the material to be activated, measuring the induced activity, and then, with a knowledge of the activation cross section, computing the neutron flux. In case of a criticality accident (an accidental attainment of an uncontrolled chain reaction), the measurement of induced radioactivity due to neutron irradiation permits calculation of the neutron dose. This same principle is applied by the chemist in neutron activation analysis. This method, which for many elements is more sensitive than other physical or chemical procedures, involves irradiation of that unknown sample in a neutron field of known intensity, measurement of the induced activity, and then calculation of the amount of the unknown in the sample. Furthermore, by spectroscopic examination of the induced radiation, qualitative analysis of the unknown is also possible.

If a radionuclide is being made by neutron irradiation, and is decaying at the same time, the net number of radioactive atoms present in the sample at any time is the difference between the rate of production and the rate of decay. This may be expressed mathematically by the equation net rate of increase of radioactive atoms = rate of production − rate of decay,

$$\frac{dN}{dt} = \phi\sigma n - \lambda N, \tag{5.59}$$

where
ϕ = flux, neutrons per cm² per sec,
σ = activation cross section, cm²,
λ = decay constant of the induced activity,
N = number of radioactive atoms,
n = number of target atoms.

Equation (5.59) is a linear differential equation which may be integrated to yield

$$\lambda N = \phi\sigma n(1 - e^{-\lambda t}). \tag{5.60}$$

In equation (5.60), $\phi\sigma n$ is sometimes called the saturation activity; for an infinitely long irradiation time, it represents the maximum obtainable activity with any given neutron flux.

Example 5.14

A sample containing an unknown quantity of chromium is irradiated for 1 week in a thermal neutron flux of 10^{11} n/cm²/sec. The resulting ^{51}Cr gamma-rays give a counting rate of 600 counts per minute in a scintillation counter whose overall efficiency is 10%. How many grams of chromium were there in the original sample? The reaction in this case is

$$^{50}\text{Cr} + {}^{1}_{0}n \rightarrow {}^{51}\text{Cr} + \gamma.$$

The thermal neutron activation cross section for ^{50}Cr is 13.5 barns, and ^{50}Cr forms 4.31% by number of the naturally occurring chromium atoms. Chromium-51 decays by orbital electron capture with a half-life of 27.8 days, and emits a 0.323 MeV gamma-ray in 9.8% of the decays. The atomic weight of Cr is 52.01.

The activity is given by λN in equation (5.60). This equation may therefore be solved for n, the number of target atoms. Substituting the numerical values into equation (5.60), we have

$$10\,\frac{\text{counts}}{\text{sec}} \times 10\,\frac{\text{dis}}{\text{count}} = 10^{11}\,\frac{1}{\text{cm}^2\text{sec}} \times 1.35 \times 10^{-23}\,\frac{\text{cm}^2}{\text{atom}} \times$$

$$0.098 \times 0.0431n \text{ atoms } (1 - e^{-0.693/27.8 \times 7}),$$

$$n = \frac{10^2 \times 10^{23}}{10^{11} \times 1.35 \times 1.6 \times 10^{-1} \times 9.8 \times 10^{-2} \times 4.31 \times 10^{-2}}$$

$$= 1.095 \times 10^{18} \text{ atoms Cr.}$$

Since there are 52.01 g Cr/mole, the weight of chromium in the unknown is

$$\frac{1.095 \times 10^{18} \text{ atoms}}{6.027 \times 10^{23} \text{ atoms/mole}} \times 52.01\,\frac{\text{g}}{\text{mole}} = 9.46 \times 10^{-5} \text{ g.}$$

Problems

1. What is the thickness of Cd that will absorb 50% of an incident beam of thermal neutrons? The capture cross section for the *element* Cd is 2550 barns for thermal neutrons; the specific gravity of Cd is 8.65, and its atomic weight is 112.4.

2. Compare the electronic densities of a piece of aluminum 5 mm thick and a piece of iron of the same density thickness.

3. In surveying a laboratory, a health physicist wipes a contaminated surface, and runs an absorption curve using a thin end window counter and aluminum absorbers. The range of the beta-rays (no gammas were found) was found to be 0.08 mm aluminium. What could the contaminant be? What further studies could be done in the smear sample to help verify the identification of the contaminant?

4. A Compton electron that was scattered straight forward ($\phi = 0°$) was completely stopped by an aluminum absorber 460 mg/cm² thick. (a) What was the kinetic energy of the Compton electron? (b) What was the energy of the incident photon?

5. The following gamma-ray absorption data were taken with lead absorbers.

Absorber thickness, mm	0	2	4	6	8	10	15	20	25
Counts per minute	1000	880	770	680	600	530	390	285	210

(a) Determine the linear and mass, and atomic absorption coefficients.
(b) What was the energy of the gamma-ray?

6. The following absorption data were taken with aluminum absorbers:

Absorber thickness, mm	0	0.02	0.04	0.06	0.08	0.1	0.12	0.14	0.16	0.2	0.4	0.8	1.5	2	2.8
Counts per minute	1000	576	348	230	168	134	120	107	96	95	90	82	68	60	50

(a) Plot the data; what types of radiation does the curve suggest?
(b) If a beta particle is present, what is its energy?
(c) If a gamma-ray is present, what is its energy?
(d) What isotope is compatible with the absorption data?
(e) Write the equation that fits the absorption data.

7. A small ^{124}Sb gamma-ray source, whose activity is 1 Ci, is completely surrounded by 25 g beryllium. Calculate the number of neutrons per second from the ^{9}Be (γ, n) ^{8}Be reaction if the cross section is 1 millibarn.

8. Cadmium is used as a thermal neutron shield in an average flux of 10^{12} neutrons cm^2/sec. How long will it take to use up 10% of the ^{113}Cd atoms?

9. The cross section for the ^{32}S (*n*, P) ^{32}P reaction is 300 millibarns for neutron energies greater than 2.5 MeV. How many microcuries of ^{32}P activity can we expect if 100 mgm ^{32}S is irradiated in a fast flux of 10^2 neutrons/cm^2 sec for 1 week?

10. Calculate the energy released by the thermal neutron reaction ^{10}B (n, α) ^7Li.

11. If the absorption coefficient of the high energy component of cosmic radiation is 2.5×10^{-3} per meter water, calculate the reduction in intensity of these cosmic rays at the bottom of the ocean, at the depth of 10,000 m.

12. If deuterium is irradiated with 2.62 MeV gamma-rays from ^{208}Tl (^{11}ThC), the nucleus disintegrates into its component parts of 1 proton and 1 neutron. If the neutron and proton each has 0.225 MeV of kinetic energy, and if the proton has a mass of 1.007593 atomic mass units, calculate the mass of the neutron.

13. Calculate the gamma-ray threshold energy for the reaction ^{11}C (n, γ) ^{12}C.

14. X-rays are generated as bremsstrahlung by causing high-speed electrons to be stopped by a high atomic numbered target, as shown in the figure below. If the electrons are accelerated by a constant high voltage of 250 kV, and if the electron beam current is 10 mA,

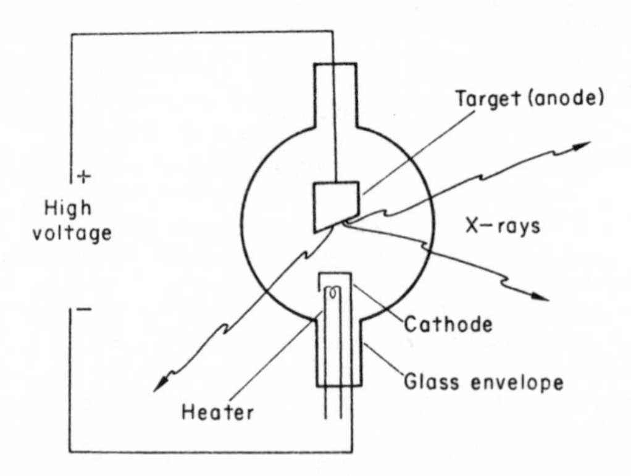

calculate the X-ray energy flux at a distance of 1 m from the tungsten target. Neglect absorption by the glass tube, and assume that the bremsstrahlung are emitted isotropically.

15. If the most energetic photon results from the instantaneous stopping of an electron in a single collision, what voltage must be applied across an X-ray tube in order to generate X-rays whose shortest wave length approaches 0.124 Å?

16. A beta particle whose kinetic energy is 0.159 MeV passes through a 4 mg/cm^2 window into a helium-filled Geiger tube. How many ion pairs will the beta particle produce inside the tube?

17. A beam of fast neutrons includes two energy groups, one group, of 1 MeV neutrons, includes 99% of the total neutron flux. The remaining 1% of the neutrons have an energy of 10 MeV.

(a) What will be the relative proportions of the two groups after passing through 25 cm of water?

(b) What would be the relative proportion of the two groups after passing through a slab of lead of the same density thickness?

The removal cross sections are:

	1 MeV	10 MeV
H	4.2 barns	0.95
O	8	1.5
Pb	5.5	5.1

Suggested References

1. GLASSTONE, S.: *Sourcebook on Atomic Energy*, D. Van Nostrand, Princeton, 1958.
2. LAPP, R. E. and ANDREWS, H. L.: *Nuclear Radiation Physics*, Prentice Hall, Englewood Cliffs, 1963.
3. SEMAT, H.: *Introduction to Atomic and Nuclear Physics*, Rinehart & Co., New York, 1964.
4. EVANS, R. D.: *The Atomic Nucleus*, McGraw-Hill, New York, 1955.
5. KAPLAN, I.: *Nuclear Physics*, Addison-Wesley Pub. Co., Reading, 1962.
6. CORK, J. A.: *Radioactivity and Nuclear Physics*, D. Van Nostrand, Princeton, 1957.
7. BLATZ, H. (Ed.): *Radiation Hygiene Handbook*, McGraw-Hill, New York, 1959.
8. ETHERINGTON, H. (Ed.): *Nuclear Engineering Handbook*, McGraw-Hill, New York, 1958.
9. *Radiation Health Handbook*, U.S. Department of Health, Education, and Welfare, Public Health Service, Washington, 1960.
10. JOHNS, H. E.: *The Physics of Radiology*, Charles C. Thomas, Springfield, 1964.
11. *Physical Aspects of Irradiation*, National Bureau of Standards Handbook 85, Government Printing Office, Washington, 1964.
12. CURTISS, L. F.: *Introduction to Neutron Physics*, D. Van Nostrand, Princeton, 1959.
13. MURRAY, R. L.: *Introduction to Nuclear Engineering*, Prentice-Hall, Englewood Cliffs, 1954.

CHAPTER 6

RADIATION DOSIMETRY

Units

During the early days of radiological experience, there was no precise unit of radiation dose that was suitable either for radiation protection or for radiation therapy. For purposes of radiation protection, a common "dosimeter" was a piece of dental film with a paper clip attached. A daily exposure great enough to just produce a detectable shadow was considered a maximum permissible dose. For greater doses and for therapy purposes, the dose unit was frequently the "skin erythema unit". Because of the great energy dependence of these dose units, as well as other inherent defects, neither of these two units could be biologically meaningful or useful either in the quantitative study of the biological effects of radiation or for radiation protection purposes. Furthermore, since the fraction of the energy in a radiation field that is absorbed by the body is energy dependent, it is necessary to distinguish between radiation *exposure* and radiation *absorbed dose*. Until recently, the several units related to radiation dosimetry were used loosely and ambiguously. To eliminate such ambiguity, the International Commission on Radiological Units (ICRU) explicitly defined the various radiological quantities and units in its report 10a, *Radiation Quantities and Units*, in 1962.

Absorbed Dose: The Rad

Radiation damage depends on the absorption of energy from the radiation, and is proportional to the concentration of absorbed energy in tissue. For this reason, the basic unit of radiation dose is expressed in terms of absorbed energy per unit mass of tissue. This unit is called the *rad* (Radiation Absorbed Dose) and is defined as:

One rad is an absorbed radiation dose of 100 ergs per gram.

The rad is universally applicable to all types of radiation dosimetry—irradiation due to external fields of gamma-rays, neutrons, or charged particles as well as that due to internally deposited radioisotopes.

Exposure: The Roentgen

For external radiation of any given energy flux, the absorbed dose to any point within an organism depends on the type and energy of radiation, the depth within the organism of the point at which the absorbed dose is desired, and elementary constitution of absorbing medium at this point. For example bone, consisting of higher-atomic-numbered elements (calcium and phosphorous) than soft tissue (carbon, oxygen, hydrogen, and nitrogen), absorbs more energy from an X-ray beam, per unit mass of absorber, than soft tissue. For this reason, the X-ray fields to which an organism may be exposed are usually specified in a unit that tells us the amount of energy transferred from the X-ray field to a unit mass of air. This unit of exposure is called *the roentgen* (abbreviated R). It was introduced at the Radiological Congress held in Stockholm in 1928, and has been the most widely used unit of X-ray dose since its introduction. According to the original definition: "The roentgen shall be the quantity of X- or gamma-radiation such that the associated corpuscular emission per 0.001293 g of air produces, in air, ions carrying one electrostatic unit of quantity of electricity of either sign." More recently (1962) the ICRU defined exposure as "the quotient of ΔQ by Δm, where ΔQ is the sum of the electrical charges on all the ions of one sign produced in air when all the electrons (negatrons and positrons), liberated by photons in a volume element of air whose mass is Δm, are completely stopped in air". The special unit of exposure in air is the roentgen.

$$1 \text{ R} = 2.58 \times 10^{-4} \frac{\text{coulombs}}{\text{kg}}. \tag{6.1}$$

Equation (6.1) is mathematically equivalent to the original definition of the roentgen.

It is important to emphasize that the roentgen is operationally defined. Ionization of air is a convenient measure of a radiation exposure because of the relative ease with which radiation induced electrical charge can be measured. However, at energies less than several keV and more than several MeV, it becomes very difficult to fulfil the criteria for measuring the roentgen. Accordingly, the roentgen is not used as a unit of X-ray exposure for quantum energies in excess of 3 MeV. For such high energy, exposure is expressed in units of watt-seconds per cm^2; exposure rate is expressed in watts per cm^2. Although this method for describing radiation dose has strongly influenced the hypotheses for the mechanism of the biological effects of radiation, the implication that ionization is the sole mechanism for the production of biological damage was not intended. The operational definition of the roentgen may be easily converted into the more fundamental units of energy absorbed per unit mass of air by applying the fact that the charge on a single ion is 4.8×10^{-10} sC and that the average energy dissipated in the production of a single ion pair in air is 34 eV. Therefore:

$$\begin{aligned}1 \text{ R} &= \frac{1 \text{ sC}}{0.001293 \text{ g air}} \times \frac{1 \text{ ion}}{4.8 \times 10^{-10} \text{ sC}} \times 34 \text{ eV/ion} \times 1.6 \times 10^{-12} \text{ ergs/eV} \\ &= 87.6 \text{ ergs/g air.}\end{aligned}$$

It should be noted that the roentgen is an integrated measure of exposure, and is independent of the time over which the exposure occurs. The strength of a radiation field is usually given as an exposure rate, such as roentgens per minute or milliroentgens per hour. (A milliroentgen, which is abbreviated "mR", is equal to 0.001 roentgens.) The total exposure, of course, is the product of exposure rate and time.

Exposure Measurement: The Free Air Chamber

The operational definition of the roentgen can be satisfied by the instrument shown in Fig. 6.1. The X-ray beam enters through the portal and interacts with the cylindrical column of air defined by the entry port diaphragm. The ions resulting from interactions between the X-rays and the volume of

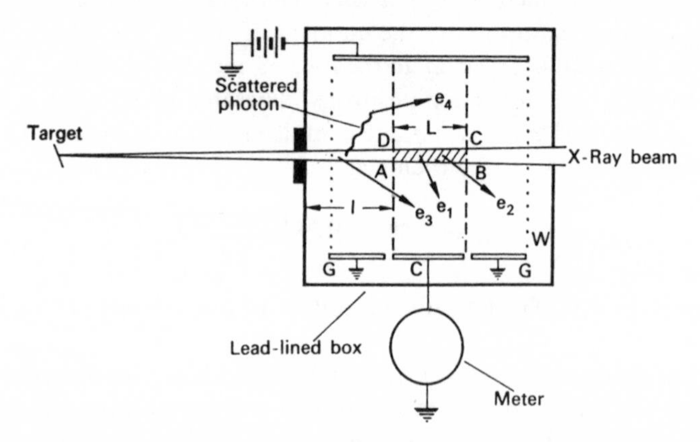

FIG. 6.1. Schematic diagram of a parallel plate free air ionization chamber. (From N.B.S. Handbook 64, *Design of Free Air Ionization Chamber*, 1957.)

air A B C D, which is determined by the intersection of the X-ray beam with the electric lines of force from the edges of the collector plate C, is collected by the plates, causing current to flow in the external circuit. The guard ring, G, and the guard wires, W, help to keep these electric field lines straight and perpendicular to the plates. The electric field intensity between the plates is on the order of 100 V/cm—high enough to collect the ions before they recombine, but not great enough to cause secondary ionization by the electrons released by the primary ionizing particles. The guard wires are connected to a voltage-dividing network to ensure a uniform potential drop across the plates. The number of ions collected because of X-ray interactions in the collecting volume is calculated from the current flow, and the dose rate, in roentgens per unit time, is then computed. For the roentgen to be measured in this way, all the energy of the primary electrons must be dissipated in the air within the meter. This condition can be satisfied by making the air chamber larger than the maximum range of the primary electrons. (For 300 keV X-rays, the spacing between the collector plates is about 30 cm, and the overall box is a cube about 50 cm on edge.) The fact that most of the ions produced as a consequence of X-ray interactions within the sensitive volume are not collected is of no significance if as many electrons from interactions elsewhere in the X-ray beam enter the sensitive volume as leave it. This condition is known as *electronic equilibrium*. When electronic equilibrium is attained, an electron of equal energy enters into the sensitive volume for every electron that leaves. A sufficient thickness of air, dimension 1 in Fig. 6.1, must be allowed between the beam entrance port and the sensitive volume in order to attain electronic equilibrium. For highly filtered 250 kV X-rays, 9 cm air is required, for 500 kV X-rays, the air thickness required for electronic equilibrium in the sensitive volume increases to 40 cm. Under conditions of electronic equilibrium, and assuming negligible attenuation of the X-ray beam by the air in length 1, the ions collected from the sensitive volume result from primary photon interactions at the beam entrance port; and the measured exposure, consequently, is at that point, and not in the sensitive volume. Free air chambers are in use that measure quantity of X-rays whose quantum energies reach as high as 500 keV. Higher-energy radiation necessitates much greater size free air chambers. The technical problems arising from the use of such large chambers makes it impractical to use the free air ionization chamber as a primary measuring device for quantum energies in excess of 500 keV.

The use of the free air ionization chamber to measure X-ray exposure rate in roentgens per unit time may be illustrated by the following example:

Example 6.1

The opening of the diaphragm in the entrance port of a free air ionization chamber is 1 cm in diameter, and the length AB of the sensitive volume is 5 cm. A 200 kV X-ray beam projected into the chamber produces a steady current in the external circuit of 0.01 μA. The temperature at the time of the measurement was 27°C and the pressure was 750 mm Hg. What is the exposure rate from this beam of X-rays?

A current of 0.01 μA corresponds to a flow of electrical charge of 10^{-8} C/sec. Since there are 3×10^9 sC per coulomb, the current flow due to 0.01 μA is

$$10^{-2}\ \mu\text{A} \times 10^{-6}\ \text{C/sec}/\mu\text{A} \times 3 \times 10^9\ \text{sC/C} = 30\ \text{sC/sec}.$$

The sensitive volume in this case is 3.927 cm^3. When the pressure and temperature are corrected to standard conditions, we have

$$\frac{\text{R}}{\text{sec}} = \frac{30\ \text{sC/sec}}{3.927\ \text{cm}^3} \times \frac{293}{273} \times \frac{760}{750} = 8.32\ \text{R/sec}.$$

Exposure Measurement: The Air Wall Chamber

The free air ionization chamber described above is practical only as a primary laboratory standard. For field use, a more portable instrument is required. Such an instrument could be made by compressing the air around the measuring cavity. If this were done, then the conditions for defining the roentgen would continue to be met. In practice, of course, it would be quite difficult to construct an instrument whose walls are made of compressed air. However, it is possible to make an instrument with walls of "air equivalent" material, that is, a wall material whose X-ray absorption properties are very similar to those of air. Such a chamber can be built in the form of an electrical capacitor; its principle of operation can be explained with the aid of the following diagram.

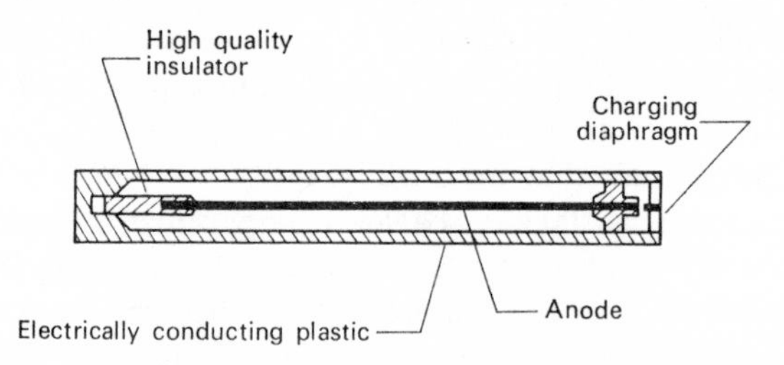

FIG. 6.2. Non self-reading condenser-type pocket ionization chamber.

The instrument consists of an outer cylindrical wall, about $\frac{3}{16}$ in. thick made of electrically conducting plastic. Coaxial with the outer wall, but separated from it by a very high quality insulator, is a center wire. This center wire, or central anode, is positively charged with respect to the wall. When the chamber is exposed to X or to gamma radiation, the ionization produced in the measuring cavity as a result of interactions between photons and the wall discharges the condenser, thereby decreasing the potential of the anode. This decrease in the anode voltage is directly proportional to the ionization produced in the cavity, which in turn is directly proportional to the radiation dose. For example, consider the following instance:

Example 6.2

Chamber volume = 2 cm^3.
Electrical capacity = 5 $\mu\mu$F.
Voltage across chamber before exposure to radiation = 180 V.
Voltage across chamber after exposure to radiation = 160 V.
Exposure time = $\frac{1}{2}$ hr.
Calculate the radiation exposure and the exposure rate.

The exposure in roentgens is calculated as follows:

$$C \times \Delta V = \Delta Q$$

$$5 \times 10^{-12} \text{ farads} \times (180 - 160) \text{ volts} = 1 \times 10^{-10} \text{ coulombs.}$$

Since there are 3×10^9 statcoulombs per coulomb, the charge collected er unit volume is

$$\frac{1 \times 10^{-10}\ \text{C} \times 3 \times 10^9\ \text{sC/C}}{2\ \text{cm}^3} = 0.150\ \text{sC/cm}^3.$$

And since an exposure of 1 roentgen results in 1 statcoulomb per cubic entimeter, the exposure in this case is

$$\frac{0.150\ \text{sC/cm}^3}{(1\ \text{sC/cm}^3)/\text{R}} = 0.150\ \text{R},$$

r 150 milliroentgens (mR). The exposure rate was

$$\frac{150\ \text{mR}}{0.5\ \text{hr}} = 300\ \text{mR/hr}.$$

A chamber built according to this principle is called an "air wall" chamber. Vhen such a chamber is used, care must be taken that the walls are of the roper thickness for the energy of the radiation being measured. If the walls re too thin, an insufficient number of photons will interact to produce rimary electrons; if too thick, the primary radiation will be absorbed to a ignificant degree by the wall, and an attenuated primary electron flux will esult.

The determination of the optimum thickness may be illustrated by an xperiment in which the ionization produced in the cavity of an ionization hamber is measured as the wall thickness is increased from a very thin wall ntil it becomes relatively thick. In performing this experiment, care must be aken to prevent secondary electrons that are formed outside the chamber valls and beta rays from the gamma ray source from reaching the sensitive olume of the chamber. When this is done, and the cavity ionization is plotted gainst the wall thickness, the curve shown in Fig. 6.3. results.

Since the cavity ionization is caused mainly by primary electrons resulting rom gamma-ray interactions with the wall, increasing the wall thickness llows more photons to interact, thereby producing more primary electrons which ionize the gas in the chamber as they traverse the cavity. However, vhen the wall thickness reaches a point where a primary electron produced t the outer surface of the wall is not sufficiently energetic to pass through the vall into the cavity, the ionization in the cavity begins to decrease. The wall hickness at which this just begins is the *equilibrium wall thickness*.

As the wall material departs from air equivalence, the response of the ionization chamber becomes energy dependent. By proper choice of wall material and thickness, the maximum in the curve of Fig. 6.3. can be made quite broad, and the ionization chamber, as a consequence, made relatively energy independent over a wide range of quantum energies. In practice, this approximately flat response spans the energy range from about 200 keV to about 2 MeV. In this range of energies, the Compton effect is the predominant mechanism of energy transfer. For lower energies, the probability of a Compton interaction increases approximately as the wavelength, while the probability of a photoelectric interaction is approximately proportional to the cube of the quantum wavelength. The total number of primary electrons therefore increases, and the sensitivity of the chamber consequently increases; the

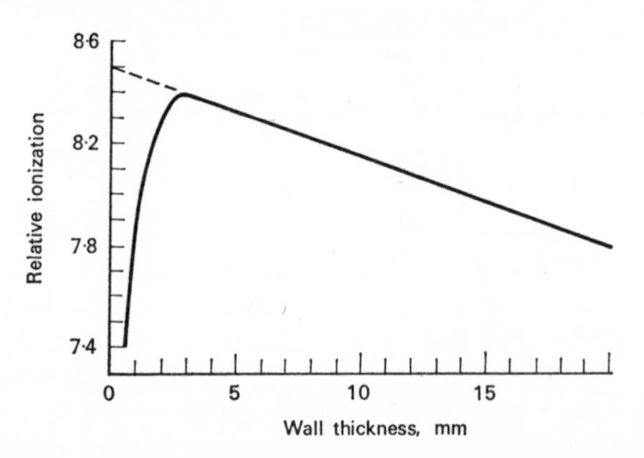

FIG. 6.3. Ion pairs per unit volume as a function of wall thickness. The ionization chamber in this case was made of pure carbon, and was a cylinder 20 mm inside diameter and 20 mm long. (W. V. Mayneord and J. E. Roberts, *British Journal of Radiology* **10**, 365, 1937.)

increased sensitivity, however, reaches a peak as the quantum energy decreases and then, because of the severe attenuation of the incident radiation by the chamber wall, the sensitivity rapidly decreases. These effects are shown in Fig. 6.4, a curve showing the energy correction factor for a pocket dosimeter.

For quantum energies greater than 3 MeV, the roentgen is not used as the unit of measurement of exposure. This is due to the fact that the high energy and consequently the long range of the primary electrons produced in the wall, makes it impossible to build an instrument that meets the criteria for measuring the roentgen. Because of the long range of the primary electrons, very thick walls are necessary. However, when the walls are sufficiently thick, on the basis of the range of the primary electrons, they attenuate the gamma radiation to a significant degree, as shown in Fig. 6.5. Under these

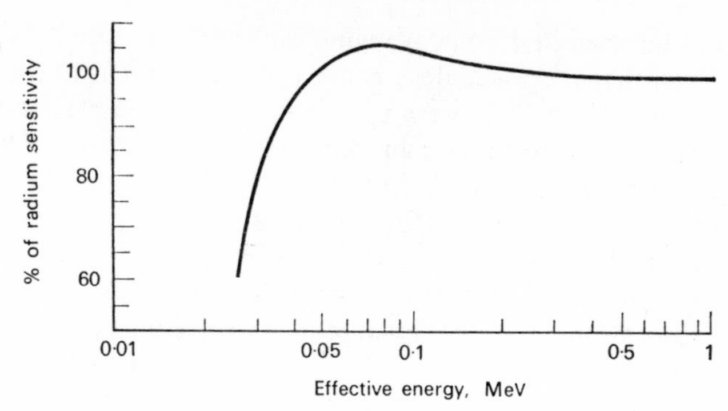

FIG. 6.4. Energy dependence characteristics of the pocket dosimeter shown in Fig. 9.15.

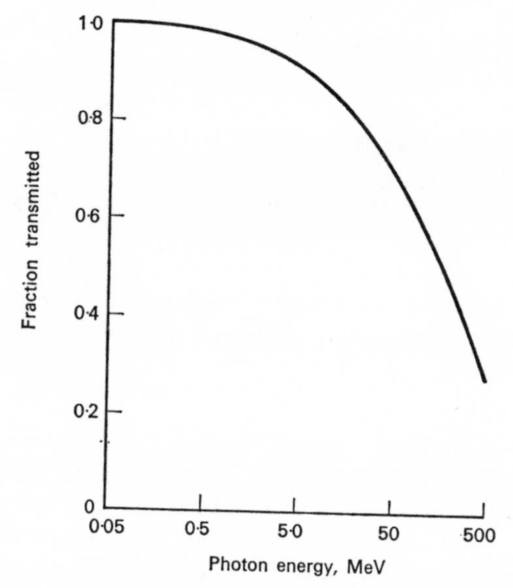

FIG. 6.5. Fractional number of photons transmitted through an air wall of thickness equal to the maximum range of the secondary electrons. (From N.B.S. Handbook 55, 1954.)

conditions, it is not possible to attain electronic equilibrium, since the radiation intensity within the wall is not constant, and the primary electrons, consequently, are not produced uniformly throughout the entire volume of wall from which they may reach the cavity.

Rad–Roentgen Relationship

The air wall chamber, as the name implies, measures the energy absorption in air. In most instances we are interested in the energy absorbed in tissue. Since energy absorption is approximately proportional to the electronic density of the absorber in the energy region where the roentgen is used, it ca easily be shown that the tissue dose is not necessarily equal to the air dos for any given radiation field. For example, if we consider muscle tissue to hav a specific gravity of 1, and to have an elementary composition of 5.98×10^2 hydrogen atoms per gram, 2.75×10^{22} oxygen atoms per gram, 0.172×10^2 nitrogen atoms per gram, and 6.02×10^{21} carbon atoms per gram, then th electronic density is 3.28×10^{23} electrons per gram. For air, whose densit is 1.293×10^{-3} g/cm^3, the electronic density is 3.01×10^{23} electrons per gram The energy absorption, in ergs per gram of tissue, corresponding to 1 roentge in air is, therefore,

$$\frac{3.28}{3.01} \times 87.8 = 95 \text{ ergs per gram tissue.}$$

This value agrees very well with calorimetric measurements of energy absorp tion by soft tissue exposed to an air dose of 1 roentgen. This tissue dose 95 ergs per gram, resulting from an exposure of 1 roentgen, is very close to th tissue dose of 100 ergs per gram which corresponds to 1 rad. For this reason an exposure of 1 roentgen is frequently considered approximately equivalen to an absorbed dose of 1 rad, and the unit "roentgen" is loosely (but incor rectly) used to mean "rad".

The roentgen bears a simple quantitative relationship to the rad that per mits the calculation of absorbed dose in any medium exposed to a given ai dose measured in roentgens. This relationship may be illustrated by the follow ing example.

Example 6.3

Consider a gamma-ray beam of quantum energy 0.3 MeV. If the photo flux is 1000 quanta per cm^2/sec, what is the exposure rate at a point in thi beam and what is the absorbed dose rate for soft tissue at this point?

From Fig. 5.18, the energy absorption coefficient for air, μ_a, at 20°C, fo 300 keV photons is found to be 3.46×10^{-5} cm^{-1}. The exposure rate i roentgens per second is given by

$$D_R = \frac{\phi \text{ photons/cm}^2 \text{ sec} \times E \text{ MeV/photon} \times 1.6 \times 10^{-6} \text{ ergs/MeV} \times \mu_a \text{ cm}^{-1}}{\rho_a \text{ g/cm}^3 \times 87.7 \text{ ergs/g/R}} \quad (6.2$$

Substituting the appropriate numerical values into equation (6.1), we hav

$$D_R = \frac{10^3 \times 0.3 \times 1.6 \times 10^{-6} \times 3.46 \times 10^{-5}}{(1.293 \times 10^{-3} \times 273/293) \times 87.7}$$

$$= 2.06 \times 10^{-7} \text{ R/sec.}$$

he absorbed dose rate, in rad per second, is given by the equation

$$D_{rad} = \frac{\phi \text{ photons/cm}^2 \text{ sec} \times E \text{ MeV/photon} \times 1.6 \times 10^{-6} \text{ ergs/MeV} \times \mu_m \text{ cm}^{-1}}{\rho_m \text{ g/cm}^3 \times 100 \text{ ergs/g/rad}}. \quad (6.3)$$

When the value for the energy absorption coefficient for tissue for 300 keV photons, $\mu_m = 0.0312$ cm^{-1}, and a tissue density of 1 g/cm^3 are substituted into equation (6.3), the absorbed dose rate is found to be 1.5×10^{-7} rad/sec.

The relationship between roentgens and rads is simply obtained from the ratio of equations (6.2) and (6.3):

$$\frac{D_{rad}}{D_R} = \frac{(\phi \times E \times 1.6 \times 10^{-6} \times \mu_m)/\rho_m \times 100}{(\phi \times E \times 1.6 \times 10^{-6} \times \mu_a)/\rho_a \times 87.7}, \quad (6.4)$$

$$D_{rad} = \frac{87.7}{100} \times \frac{\mu_m/\rho_m}{\mu_a/\rho_a} \times D_R. \quad (6.5)$$

Equation (6.5) shows the radiation dose absorbed by any medium exposed to a given air dose to be determined by the ratio of the mass absorption coefficients of the medium to that of air. In the case of tissue, the ratio of rad

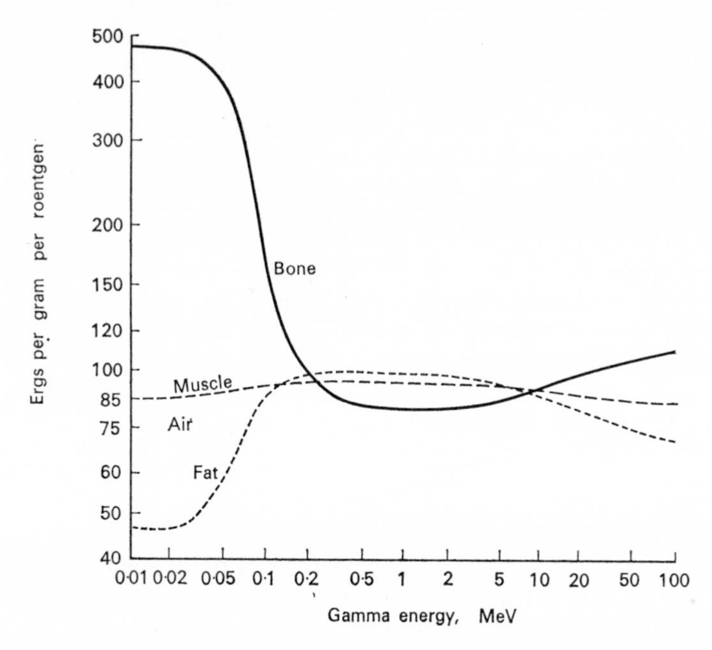

FIG. 6.6. Energy absorption per roentgen of various tissues. (From O. Glasser, *Medical Physics*, vol. II.)

to roentgen remains approximately constant over the quantum energy range of about 0.1–10 MeV because the chief means of interaction between the tissue and the radiation is Compton scattering, and the cross section for Compton scattering depends mainly on electronic density of the absorbing medium. In the case of lower energies, photoelectric absorption becomes important, and the cross section for this mode of interaction increases with atomic number of the absorber. As a consequence of this dependence on atomic number, bone, which contains approximately 10% by weight of calcium absorbs much more energy than does soft tissue from a given air dose of low energy X-rays. This point is illustrated in Fig. 6.6, which shows the number of ergs per gram absorbed per roentgen of exposure for fat, muscle, and bone as a function of quantum energy.

Absorbed Dose Measurement: Bragg–Gray Principle

If a cavity ionization chamber is built with wall material whose radiation absorption property is similar to that of tissue, then, by taking advantage of the *Bragg–Gray* principle, an instrument can be built to measure tissue dose directly. According to the Bragg–Gray principle, the amount of ionization produced in a small gas-filled cavity surrounded by a solid absorbing medium is proportional to the energy absorbed by the solid. Implicit in the practical application of this principle is that the gas cavity be small enough relative to the mass of the solid absorber to leave unchanged the angular and velocity distributions of the primary electrons. This requirement is fulfilled if the primary electrons lose only a very small fraction of their energy in traversing the gas-filled cavity. If the cavity is surrounded by a solid medium of proper thickness to establish electronic equilibrium, then the energy absorbed per unit mass of wall, dE_m/dM_m, is related to the energy absorbed per unit mass of gas in the cavity, dE_g/dM_g, by

$$\frac{dE_m}{dM_m} = \frac{S_m}{S_g} \times \frac{dE_g}{dM_g}, \tag{6.6}$$

where S_m is the mass stopping power of the wall material and S_g is the mass stopping power of the gas. Since the ionization per unit mass of gas is a direct measure of dE_g/dM_g, equation (6.6) can be rewritten as

$$\frac{dE_m}{dM_m} = \rho_m \times w \times J, \tag{6.7}$$

where ρ_m is the ratio of the mass stopping powers of the solid relative to that of the gas, S_m/S_g, w is the mean energy dissipated in the production of an ion pair in the gas, and J is the number of ion pairs per unit mass of gas. Using the appropriate equations for stopping power given in Chapter 5, it becomes a relatively simple matter to compute ρ_m for electrons of any given energy.

For those cases where the gas in the cavity is the same substance as the chamber wall, such as methane and paraffin, ρ_m is equal to unity. Table 6.1 shows stopping power ratios, relative to air, of several substances for monoenergetic electrons. For gamma radiation, however, the problem of evaluating ρ_m is more difficult; the relative fraction of the gamma-rays that will interact by each of the competing mechanisms, as well as the spectral distribution of the primary electrons (Compton, photoelectric, and pair produced electrons)

TABLE 6.1. MEAN MASS STOPPING POWER RATIOS, RELATIVE TO AIR, FOR ELECTRONIC EQUILIBRIUM SPECTRA GENERATED BY INITIALLY MONOENERGETIC ELECTRONS

Initial energy, MeV	Element and state of molecular binding								
	Hydrogen saturated	Hydrogen unsaturated	Carbon saturated	Carbon unsaturated	Carbon highly chlorinated	Nitrogen, amines, nitrates	Nitrogen ring	Oxygen, —O—	Oxygen, O=
0.1	2.52	2.59	1.016	1.021	1.047	0.976	1.018	0.978	0.994
0.2	2.52	2.59	1.015	1.019	1.043	0.978	1.016	0.979	0.995
0.3	2.48	2.55	1.014	1.018	1.040	0.979	1.016	0.981	0.995
0.327	2.48	2.54	1.014	1.018	1.040	0.979	1.015	0.981	0.995
0.4	2.46	2.53	1.014	1.018	1.038	0.980	1.015	0.981	0.996
0.5	2.44	2.51	1.013	1.017	1.037	0.980	1.015	0.982	0.996
0.6	2.44	2.50	1.012	1.016	1.035	0.980	1.013	0.981	0.995
0.654	2.43	2.49	1.011	1.014	1.034	0.979	1.012	0.981	0.994
0.7	2.42	2.48	1.010	1.013	1.033	0.978	1.011	0.980	0.993
0.8	2.40	2.46	1.009	1.012	1.031	0.978	1.010	0.979	0.992
1.0	2.39	2.44	1.004	1.008	1.026	0.975	1.005	0.977	0.988
1.2	2.37	2.42	1.001	1.004	1.022	0.973	1.002	0.974	0.985
1.308	2.36	2.42	0.999	1.002	1.019	0.971	1.000	0.972	0.983
1.5	2.35	2.39	0.995	0.998	1.015	0.967	0.996	0.969	0.980

From N.B.S. Handbook 85, *Physical Aspects of Irradiation*, 1964.

must be considered, and a mean value for relative stopping power must be determined. For the equilibrium electron spectra generated by gamma-rays from ^{198}Au, ^{137}Cs, and ^{60}Co, the values for the mean mass relative stopping powers are given in Table 6.2. For air, w, the mean energy loss for the production of an ion pair in air, has a value of 34 eV. To determine the radiation absorbed dose, it is necessary only to measure the ionization per unit mass of gas, J.

TABLE 6.2. MEAN MASS STOPPING POWER RATIOS, S_m/S_{air} FOR EQUILIBRIUM ELECTRON SPECTRA GENERATED BY ^{198}Au, ^{137}Cs, AND ^{60}Co, WITH THE ASSUMPTION THAT THE ELECTRONS SLOW DOWN IN A CONTINUOUS MANNER

Energy, MeV		Medium		
		Graphite	Water	Tissue
0.411	(^{198}Au)	1.032		
0.670	(^{137}Cs)	1.027	1.162	1.145
1.25	(^{60}Co)	1.017	1.155	1.137

From N.B.S. Handbook 78, *Report of the International Commission on Radiological Units and Measurements*, 1959.

Example 6.4

Calculate the absorbed dose rate from the following data on a tissue equivalent chamber with walls of equilibrium thickness embedded within a phantom and exposed to ^{60}Co gamma-rays for 10 min. The volume of the air cavity in the chamber is 1 cm^3, the capacitance is 5 $\mu\mu$F, and the gamma-ray exposure results in a decrease of 72 V across the chamber. The charge collected by the chamber is

$$Q = C\,\Delta V$$
$$= 5 \times 10^{-12}\text{ F} \times 72\text{ V}$$
$$= 3.6 \times 10^{-10}\text{ C}.$$

The number of electrons collected, which corresponds to the number of ion pairs formed in the air cavity is

$$\frac{3.6 \times 10^{-10}\text{ C}}{1.6 \times 10^{-19}\text{ C/electron}} = 2.25 \times 10^{9}\text{ electrons.}$$

Since 34 electron volts are expended per ion pair formed in air, and since the stopping power of tissue relative to air is 1.137, we have from the Bragg–Gray relationship of equation (6.7)

$$\frac{dE_m}{dM_m} = \rho_m \times w \times J$$

$$= \frac{1.137 \times 34\text{ eV/ip} \times 2.25 \times 10^{9}\text{ ip/cm}^3 \times 1.6 \times 10^{-12}\text{ ergs/eV}}{1.293 \times 10^{-3}\text{ g/cm}^3 \times 100\text{ ergs/g/rad}}$$

$$= 1.08\text{ rads.}$$

The exposure time was 10 min, and the dose rate therefore is 0.108 rad/min.

Source Strength: Specific Gamma-ray Emission

The radiation intensity from any given gamma-ray source is used as a measure of the strength of the source. The gamma radiation exposure rate from a point source of unit activity at unit distance is called the specific gamma-ray emission, and frequently is given either in units of roentgens per hour at 1 meter from a 1 curie point source (rhm) or roentgens per hour at 1 foot from a 1 point curie source (rhf). The source strength may be easily computed if the decay scheme of the isotope is known. In the case of ^{131}I, for example, whose gamma rays are shown in Fig. 4.7, and whose corresponding true absorption coefficients are found in Fig. 5.18, we have the following:

Quantum energy, MeV	Photons / Disintegration	Energy absorption coefficient for air
0.080	0.06	2.9×10^{-5} cm^{-1}
0.284	0.06	3.4×10^{-5}
0.364	0.79	3.5×10^{-5}
0.638	0.15	3.5×10^{-5}

The gamma-radiation exposure level is calculated by considering the energy absorbed per unit mass of air at the specified distance from the one curie point source due to the photon flux at that distance, as shown in equation (6.8):

$$D = \frac{f \text{ phot/dis} \times E \text{ MeV/phot} \times 1.6 \times 10^{-6} \text{ erg/MeV} \times 3.7 \times 10^{10} \text{ dis/sec/Ci} \times \times 3.6 \times 10^{3} \text{ sec/hr} \times \mu \text{ cm}^{-1}}{4\pi (d \text{ cm})^2 \times \rho \text{ g/cm}^3 \times 87.8 \text{ erg/g/R}}, \tag{6.8}$$

D = exposure rate, roentgens per hr per curie,
f = fraction of disintegrations that result in a photon of the quantum energy under consideration,
E = quantum energy, MeV,
μ = linear energy absorption coefficient,
d = distance from the gamma-ray source, cm,
ρ = density of air, g/cm^3.

This calculation is made for each different quantum energy, and the results are summed to obtain the source strength. For the 0.080 MeV gamma-ray, we have

$$D = \frac{6 \times 10^{-2} \times 8 \times 10^{-2} \times 1.6 \times 10^{-6} \times 3.7 \times 10^{10} \times 3.6 \times 10^{3} \times 2.9 \times 10^{-5}}{4\pi(100)^2 \times 1.293 \times 10^{-3} \times 87.8}$$

$$= 0.0021 \text{ R/hr.}$$

The dose rate for each of the other three quanta emitted by ^{131}I is calculated in a similar manner, except that the corresponding decay frequency and absorption coefficient are used for each of the quanta of different energy. The results of this calculation are tabulated below:

Quantum energy, MeV	R/hr at 1 m
0.080	0.002
0.284	0.009
0.364	0.150
0.638	0.050
Total =	0.211 R/hr

Equation (6.8) contains several constants: 3.7×10^{10} disintegrations/sec/Ci, 3.6×10^3 sec/hr, 1.6×10^{-6} erg/MeV, $4\pi(100)^2$, 0.001293 g/cm^3, and 87.8 erg/g/R. If all of these constants are combined, the source strength, Γ, in R/hr/Ci at 1 m is given by

$$\Gamma = 1.49 \times 10^4 \sum_i f_i \times E_i \times \mu_i \frac{\text{R m}^2}{\text{Ci hr}}, \tag{6.9}$$

where f_i is the photons per disintegration of the ith photon, E_i is the energy, in MeV, of the ith photon, and μ_i is the linear energy absorption coefficient in air of the ith photon. For many practical purposes, equation (6.9) may be simplified. For quantum energies from 60 keV to 2 MeV, the linear absorption coefficient varies little with energy; over this range μ is about 3.2×10^{-5} cm^{-1}. Using this value, equation (6.9) may be approximated as

$$\Gamma = 0.48 \sum_i f_i \times E_i \frac{\text{R m}^2}{\text{Ci hr}}. \tag{6.10}$$

The source strength in R per hour per curie at 1 ft is calculated by appropriately changing the constant to give

$$\Gamma = 6 \sum_i f_i \times E_i \frac{\text{R ft}^2}{\text{Ci hr}}. \tag{6.11}$$

Internally Deposited Radioisotopes

The calculation of the absorbed dose from internally deposited radioisotopes follows directly from the definition of the rad. For an infinitely large medium containing a uniformly distributed radioisotope, the concentration of absorbed energy must be equal to the concentration of energy emitted by the isotope. For practical health physics purposes, "infinitely large" may be approximated by a tissue mass whose dimensions exceed the range of the radiation from the distributed isotope. For the case of alpha and most beta emitters, this condition is easily met in practice; for gamma-rays, however, it is more difficult to approximate an infinite mass, since the photon may travel great distances, indeed it may leave the tissue completely, before it interacts.

Beta Emitters

The computation of the radiation absorbed dose due to a uniformly distributed beta emitter within a tissue may be illustrated with the following example:

Example 6.5

Iodine tends to concentrate in the thyroid gland—about 60% of the total iodine in the body is in the thyroid, while most of the remainder circulates in the blood as protein-bound iodine. Compute the beta-ray dose rate to a thyroid gland that weighs 25 g and has 2 mCi ^{131}I uniformly distributed throughout the organ.

Iodine-131 emits two beta-rays: 85% of the disintegrations are accomplished by the emission of a 0.608-MeV beta, while the remaining 15% involves a 0.315-MeV beta. The average beta-ray energy therefore is

$$\bar{E} = 1/3\,(0.85 \times 0.608 + 0.15 \times 0.315) = 0.187 \text{ MeV/dis.}$$

The range in tissue of the high-energy beta is only about 2 mm. The thyroid may therefore be considered infinitely large, and it may be assumed that all the beta ray energy is absorbed by the tissue. This assumption leads to a small overestimate of the dose, since the dose rate at the surface is only one-half the value at a depth of 2 mm. The beta-ray dose rate from Q mCi dispersed in W g tissue is

$$\text{Dose rate} = \frac{Q \text{ mCi} \times 3.7 \times 10^{7} \text{ dps/mCi} \times \bar{E} \text{ MeV/d} \times 1.6 \times 10^{-6} \text{ ergs/MeV} \times 3.6 \times 10^{3} \text{ sec/hr}}{W \text{ g} \times 100 \text{ ergs/g/rad}} \qquad (6.12)$$

$$= \frac{2 \times 3.7 \times 10^{7} \times 1.87 \times 10^{-1} \times 1.6 \times 10^{-6} \times 3.6 \times 10^{3}}{25 \times 100}$$

$$= 31.9 \frac{\text{rads}}{\text{hr}}.$$

Effective Half-life

The total dose absorbed during any given time interval after the uptake of the iodine in the thyroid may be calculated by integrating the dose rate over the required time interval. In making this calculation, two factors must be considered, viz.

1. *In situ* radioactive decay of the isotope.
2. Biological elimination of the isotope.

In most instances, biological elimination follows first-order kinetics. In this case, the equation for the quantity of radioisotope within an organ at any time t after uptake of a quantity Q_0 is given by

$$Q = (Q_0 \, e^{-\lambda_R t}) \, (e^{-\lambda_B t}), \tag{6.13}$$

where λ_R is the radioactive decay constant, and λ_B is the biological elimination constant. The two exponentials in equation (6.13) may be combined

$$Q = Q_0 \, e^{-(\lambda_R + \lambda_B) t}, \tag{6.14}$$

and, if $\lambda_E = \lambda_R + \lambda_B$, we have (6.15)

$$Q = Q_0 \, e^{-\lambda_E t}, \tag{6.16}$$

where λ_E is called the *effective* elimination constant. The effective half-life then is

$$T_E = \frac{0.693}{\lambda_E}. \tag{6.17}$$

From the relationship among λ_E, λ_R, and λ_B, we have

$$\frac{1}{T_E} = \frac{1}{T_R} + \frac{1}{T_B}, \tag{6.18}$$

or

$$T_E = \frac{T_R \times T_B}{T_R + T_B}. \tag{6.19}$$

For ^{131}I, $T_R = 8$ days and T_B, the biological half-life in the thyroid, is 180 days. The effective half-life, therefore, is

$$T_E = \frac{8 \times 180}{8 + 180} = 7.7 \text{ days},$$

and the effective elimination constant is

$$\lambda_E = \frac{0.693}{7.7} = 0.09 \text{ day}^{-1}.$$

Dose Due to Total Decay

The dose, dD, during an infinitesimally small time period, dt, at a time interval t after an initial dose rate D_0 is

$$\begin{aligned} dD &= \text{instantaneous dose rate} \times dt \\ &= D_0\, e^{-\lambda_E t}\, dt, \end{aligned} \tag{6.20}$$

where D_0 is the dose rate at time $t = 0$, the instant when the isotope was taken up by the tissue; and t is the elapsed time after uptake. The total dose during the time interval T after uptake of the isotope is

$$D = D_0 \int_D^T e^{-\lambda_E t}\, dt, \tag{6.21}$$

which, when integrated, yields

$$D = \frac{D_0}{\lambda_E}(1 - e^{-\lambda_E t}). \tag{6.22}$$

For an infinitely long time, that is, when the isotope is completely gone, equation (6.22) reduces to

$$D = \frac{D_0}{\lambda_E}. \tag{6.23}$$

It should be noted that the dose due to total decay is merely equal to the product of the initial dose rate, D_0, and the average life of the radioisotope within the organ, $1/\lambda_E$. For the case in Example 6.5, the total absorbed dose during the first 5 days after deposition of the radioiodine in the thyroid is, according to equation (6.22), and after substituting $\lambda_E = 0.693/T_E$,

$$\begin{aligned} D &= \frac{31.9 \text{ rad/hr} \times 24 \text{ hr/day} \times 7.7 \text{ day}}{0.693}\left(1 - \exp - \frac{0.693}{7.7} \times 5\right) \\ &= 3090 \text{ rads}, \end{aligned}$$

and the dose due to complete decay is, from equation (6.22),

$$D = \frac{31.9 \text{ rad/hr} \times 24 \text{ hr/day} \times 7.7 \text{ day}}{0.693} = 8520 \text{ rads}.$$

Gamma Emitters

For a uniformly distributed gamma emitting isotope, the dose rate at any point P due to the isotope in the infinitesimal volume dV at any other point at a distance r from point p, as shown in Fig. 6.7, is

$$dD = C\,\Gamma \frac{e^{-\mu r}}{r^2}\, dV \text{ rad/hr}, \tag{6.24}$$

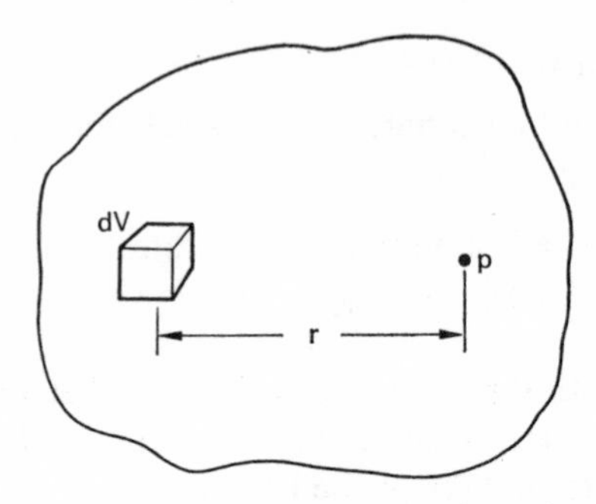

FIG. 6.7. Diagram for calculating dose at point p from the gamma-rays emitted from the volume element dV in a tissue mass containing a uniformly distributed isotope.

where C is the concentration of the isotope, in millicuries per cm³, Γ is the source strength, in rads per hour per millicurie at 1 cm, and μ is the linear energy absorption coefficient per centimeter tissue. The dose at point p due to all the isotope in the tissue is computed by the contributions from all the infinitesimal volume elements

$$D = C\,\Gamma \int_0^V \frac{e^{-\mu r}}{r^2}\, dV \text{ rad/hr.} \tag{6.25}$$

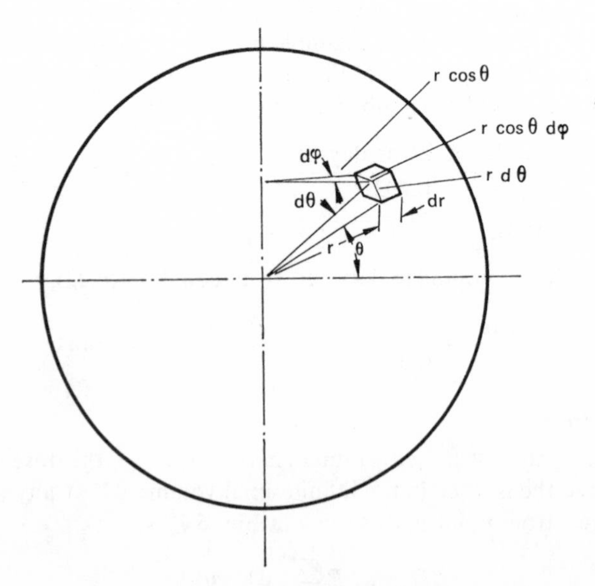

FIG. 6.8. Geometry for evaluating equation (6.25) for the center of a sphere.

For the case of a sphere, the dose rate at the center is

$$D = 4C\,\Gamma \int_{r=0}^{r=R} \int_{\theta=0}^{\theta=\pi/2} \int_{\varphi=0}^{\varphi=\pi} \frac{e^{-\mu r}}{r^2} \cdot r\,d\theta \cdot r\cos\theta\,d\varphi \cdot dr. \qquad (6.26)$$

Integrating with respect to each of the variables, we have, for the dose rate at the center of the sphere,

$$D = C\,\Gamma \cdot \frac{4\pi}{\mu}(1 - e^{-\mu R}) \text{ rad/hr.} \qquad (6.27)$$

From an examination of equations (6.24), (6.25), and (6.26), it is seen that the factor that multiplies $C\Gamma$ depends only on the geometry of the tissue mass, and hence is called the geometry factor. The geometry factor, g, is defined by

$$g = \int_0^V \frac{e^{-\mu r}}{r^2}\,dV \text{ cm.} \qquad (6.28)$$

Equation (6.25) may therefore be rewritten as

$$D = C\Gamma g \text{ rad/hr.} \qquad (6.29)$$

The definition of g in equation (6.28) applies to a given point within a volume of tissue. In most health physics instances, we are interested in the average dose rate rather than the dose rate at a specific point. For this purpose, we may define an average geometry factor

$$\bar{g} = \frac{1}{V}\int g\,dV. \qquad (6.30)$$

For a sphere,

$$\bar{g} = \tfrac{3}{4}\,(g)_{\text{center}}. \qquad (6.31)$$

At any other point in the sphere at a distance d from the center, the geometrical factor is given by

$$g_p = (g)_{\text{center}} \left[0.5 + \frac{1 - (d^2/R)}{4\,(d/R)} \ln \frac{1 + d/R}{|1 - d/R|}\right]. \qquad (6.32)$$

For a cylinder, the average geometry factor depends on the radius and height. Table 6.3 gives the numerical values of $\bar{g}$ for cylinders of various heights and radii.

Because of the complex shapes of the various organs and tissues, it is extremely difficult to compute a geometry factor for the total body. However, if the organs and tissues are approximated by a combination of spheres and cylinders, the geometry factors given in Table 6.4 may be calculated for gamma emitters that are uniformly distributed throughout the body of people of various heights and weights.

TABLE 6.3. AVERAGE GEOMETRY FACTORS FOR CYLINDERS CONTAINING A UNIFORMLY DISTRIBUTED GAMMA EMITTER

Cylinder height, cm	Radius of cylinder, cm							
	3	5	10	15	20	25	30	35
2	17.5	22.1	30.3	34.0	36.2	37.5	38.6	39.3
5	22.3	31.8	47.7	56.4	61.6	65.2	67.9	70.5
10	25.1	38.1	61.3	76.1	86.5	93.4	98.4	103
20	25.7	40.5	68.9	89.8	105	117	126	133
30	25.9	41.0	71.3	94.6	112	126	137	146
40	25.9	41.3	72.4	96.5	116	131	143	153
60	26.0	41.6	73.0	97.8	118	134	148	159
80	26.0	41.6	73.3	98.4	119	135	150	161
100	26.0	41.6	73.3	98.5	119	136	150	162

From HINE and BROWNELL, *Radiation Dosimetry*, Academic Press.

TABLE 6.4. AVERAGE GEOMETRY FACTORS FOR A GAMMA EMITTING ISOTOPE UNIFORMLY DISTRIBUTED THROUGHOUT THE BODY

Weight, kg	Height, cm						
	140	150	160	170	180	190	200
40	110	109	108	106	105	104	102
50	122	119	117	116	114	113	112
60	128	125	122	120	119	118	117
70	135	131	129	126	125	124	123
80	141	139	136	134	131	130	129
90	148	146	143	140	138	136	134
100	154	150	147	145	142	139	138

From HINE and BROWNELL, *Radiation Dosimetry*, Academic Press.

Example 6.6

Compute the gamma-ray dose rate to the thyroid gland of Example 6.5, in which 2 mCi ^{131}I are uniformly distributed, and which weighs 25 g.

The thyroid gland may be assumed to be a sphere of unit density. Its radius therefore may be calculated from

$$4/3\ \pi\, r^3\ \text{cm}^3 \times 1\ \text{g/cm}^3 = 25\ \text{g},$$

$$r = 1.81\ \text{cm}.$$

From equations (6.27) and (6.31), the average dose rate to the thyroid is

$$D = \frac{3\pi C}{\mu}\,\Gamma(1 - e^{-\mu r})\ \text{rad/hr}. \tag{6.33}$$

For values of $\mu r \ll 1$ (since μ for tissue is on the order of 0.03 cm^{-1}, μr may be considered much less than 1 for values of r less than 10 cm), the exponential term in equation (6.33) may be expanded

$$e^{-\mu r} = 1 - \mu r + \frac{\mu^2 r^2}{2!} - \frac{\mu^3 r^3}{3!} + \ldots \tag{6.34}$$

and terms higher than the second disregarded. Substituting $1 - \mu r$ for $e^{-\mu r}$ in equation (6.33) leads to

$$D = 3\pi\, C\Gamma r \text{ rad/hr}, \tag{6.35}$$

from which we see that, for a sphere of radius less than 10 cm,

$$g = 3\pi r. \tag{6.36}$$

The source strength Γ rads per hour per millicurie at 1 cm may, for purposes of dose calculation in soft tissue, be taken as 10 times the source strength in roentgens per hour per curie at 1 m. Substituting $\Gamma = 2.11$ for ^{131}I, $r = 1.81$ cm, and $C = 2$ mCi per 25 cm^3 = 0.08 mCi/cm^3, we have

$$D = 3\pi \times 0.08 \times 2.11 \times 1.81 = 2.88 \text{ rads/hr}.$$

The total dose rate to the thyroid gland in this illustrative example is the sum of the beta and gamma dose rates, or 31.9 + 2.9 = 34.8 rads/hr. In this case, the gamma-rays contribute only about 8.3% of the total radiation dose. The relative fraction of the total dose due to gamma radiation when a beta-gamma emitter is uniformly distributed throughout a tissue mass increases as the size of the tissue mass increases. When the isotope is distributed throughout the body, the gamma-rays may be responsible for a large fraction of the radiation absorbed dose.

Example 6.7

What is the dose rate to a 70-kg man, who is 160 cm tall, following an intravenous injection of 100 μCi ^{24}Na?

Sodium-24 decays to ^{24}Mg by emitting a beta-ray whose maximum energy is 1.39 MeV and whose average energy is 0.55 MeV; each beta particle is accompanied by two gamma-photons in cascade of 2.75 MeV and 1.37 MeV respectively. The gamma-ray source strength is 15.4 rads per hr per millicurie at 1 cm. According to equation (6.29), and using a geometry factor of 129 from Table 6.4, the gamma-ray dose rate is

$$\begin{aligned} D_\gamma &= 10^{-1} \text{ mCi}/7 \times 10^4 \text{ cm}^3 \times 15.4 \text{ rad cm}^2/\text{mCi hr} \times 129 \text{ cm} \\ &= 2.84 \times 10^{-3} \text{ rad/hr}. \end{aligned}$$

The beta ray dose rate is

$$D_\beta = \frac{10^2\ \mu\text{Ci} \times 3.7 \times 10^4\ \text{dps}/\mu\text{Ci} \times 0.55\ \text{MeV/d} \times 1.6 \times 10^{-6}\ \text{ergs/MeV} \times \times 3.6 \times 10^3\ \text{sec/hr}}{7 \times 10^4\ \text{g} \times 100\ \text{ergs/g/rad}}$$

$$= 16.7 \times 10^{-3}\ \text{rad/hr}.$$

The total dose rate, $D_\gamma + D_\beta$, is 4.51×10^{-3} rad/hr. In this example, the gamma-rays contribute about 63% of the total radiation absorbed dose.

Neutrons

The absorbed dose from a beam of neutrons may be computed by considering the energy absorbed by each of the tissue elements that react with the neutrons. The type of reaction, of course, depends on the neutron energy. For fast neutrons, up to about 20 MeV, the main mechanism of energy transfer is elastic collision, while thermal neutrons may be captured and initiate nuclear reactions. In cases of elastic scattering, the scattered nuclei dissipate their energy in the immediate vicinity of the primary neutron interaction. The radiation dose absorbed locally in this way is called the first collision dose, and is determined entirely by the primary neutron flux; the scattered neutron is not considered after this primary interaction. For fast neutrons, the first collision dose rate is given by

$$D_n(E) = \frac{\phi E \sum_i N_i \sigma_i f_i}{100\ \text{ergs/g/rad}}, \tag{6.37}$$

where ϕ = neutron flux, neutrons/cm²/sec,
E = neutron energy, ergs,
N_i = atoms per gram of the ith element,
σ_i = scattering cross section of the ith element for neutrons of energy E, barns $\times 10^{-24}\text{cm}^2$,
f = mean fractional energy transfer to scattered atom of mass M atomic mass units during collision with neutron of m atomic mass units. For isotropic scattering, the average fraction of the energy lost in an elastic collision with a nucleus of atomic mass number M is

$$f = \frac{2M}{(M+1)^2}. \tag{6.38}$$

The composition of soft tissue may, for the purpose of dose calculation, be assumed to be given by the formula $C_{0.5}\ H_8\ O_{3.5}\ N_{0.14}$, and to have a formula weight of 72. According to this formula, the atomic density, in atoms per gram, of each of the tissue elements is as shown in Table 6.5.

TABLE 6.5. SYNTHETIC TISSUE COMPOSITION

Element	N, atoms/g	f
H	6.70×10^{22}	0.500
C	4.18×10^{21}	0.142
N	1.17×10^{21}	0.124
O	2.93×10^{22}	0.111

Table 6.5 also gives the average fraction of the neutron energy transferred to each of the tissue elements.

Example 6.8

What is the absorbed dose rate to soft tissue in a beam of 5 MeV neutrons whose intensity is 2000 neutrons per square centimeter per sec?

The scattering cross sections of each of the tissue elements for 5 MeV neutrons are listed below:

Element	σ, cm^2
H	1.50×10^{-24}
C	1.65×10^{-24}
N	1.00×10^{-24}
O	1.55×10^{-24}

Substituting the values for hydrogen into equation (6.37), we have

$$D_{nH} = \frac{2000 \text{ neut/cm}^2 \text{ sec} \times 5 \text{ MeV/neut} \times 1.6 \times 10^{-6} \text{ erg/MeV} \times \times 6.70 \times 10^{22} \text{ atoms/g} \times 1.5 \times 10^{-24} \text{ cm}^2\text{/atom} \times 0.5}{100 \text{ ergs/rad/g}}$$

$$= 8.05 \times 10^{-6} \text{ rad/sec,}$$

or

$$8.05 \times 10^{-6} \text{ rad/sec} \times 10^3 \text{ mrad/rad} \times 3.6 \times 10^3 \text{ sec/hr}$$
$$= 28.8 \text{ mrad/hr.}$$

By using the appropriate values for the cross-section and atomic concentration for each of the other tissue elements, the doses due to neutron absorption by these elements are calculated as in the example for hydrogen. The results are: carbon, 0.56; nitrogen, 0.08; and oxygen, 2.9 mrad/hr. The total dose rate is the sum of these contributions, or 32.34 mrad/hr.

In the example above, the neutron beam was monoenergetic, and only one neutron energy was considered. If the beam contains neutrons of several energies, then the calculation must be carried out separately for each of the energy groups.

For thermal neutrons, two reactions are considered, viz, the $^{14}N\ (n, p)\ ^{14}C$ reaction and the $^{1}H\ (n, \gamma)\ ^{2}H$ reaction. For the former reaction, the dose rate may be calculated from the equation

$$D_{n,p} = \frac{\phi\, N\, \sigma\, Q \times 1.6 \times 10^{-6}\ \text{erg/MeV}}{100\ \text{ergs/g/rad}} \tag{6.39}$$

where ϕ = thermal flux, neutrons/cm²/sec,
N = number of nitrogen atoms per gram tissue, 1.17×10^{21},
σ = absorption cross section for nitrogen, 1.75×10^{-24} cm²,
Q = energy released by the reaction = 0.63 MeV.

The latter reaction, $^{1}H(n, \gamma)\ ^{2}H$ is equivalent to having a uniformly distributed gamma-emitting isotope throughout the body, and results in an autointegral gamma-ray dose. The "specific activity" of this distributed gamma emitter, the number of reactions per second per gram, is governed by the neutron flux and is given by the equation

$$A = \phi\, N\, \sigma, \tag{6.40}$$

where ϕ = thermal flux, neutrons/cm²/sec,
N = number of hydrogen atoms per gram tissue = 6.7×10^{22},
σ = absorption cross section for hydrogen = 0.33×10^{-24} cm².

Example 6.9

What is the absorbed dose rate to a man weighing 70 kg and whose height is 160 cm from an average total body exposure of 10,000 thermal neutrons/cm²/sec?

The dose due to the n, p reaction is, from equation (6.39)

$$D_{n,p} = \frac{10^{4}\ n/\text{cm}^2\ \text{sec} \times 1.17 \times 10^{21}\ \text{atom/g} \times 1.75 \times 10^{-24}\ \text{cm}^2/\text{atom} \times \times 0.63\ \text{MeV}/n \times 1.6 \times 10^{-6}\ \text{erg/MeV}}{100\ \text{erg/g/rad}}$$

$$= 2.06 \times 10^{-7}\ \text{rad/sec, or } 0.74\ \text{mrad/hr.}$$

The autointegral gamma-ray dose may be calculated with the aid of equations (6.40) and (6.29), using the appropriate value from Table 6.4 for the geometric factor. The gamma-ray activity, from equation (6.40), is

$$A = 10^{4}\ \text{cm}^{-2}\ \text{sec}^{-1} \times 6.7 \times 10^{22}\ \text{atoms/g} \times 3.3 \times 10^{-25}\ \text{cm}^2/\text{atom}$$

$$= 221\ \frac{\text{phot}}{\text{sec}} \Big/ \text{g},$$

or

$$\frac{2.21\ \text{phot/sec/g}}{3.7 \times 10^{7}\ \text{phot/sec/“mCi”}} \times 1\ \text{g/cm}^3$$

$$= 6 \times 10^{-6}\ \text{“mCi”/cm}^3.$$

The source strength for 2.23 MeV photons is 10 rads per hr per millicurie at 1 cm, and the geometrical factor is 129. The autointegral dose rate, therefore, is

$$D = 6 \times 10^{-6} \times 10 \times 129$$
$$= 7.7 \times 10^{-3} \text{ rad/hr}$$
$$= 7.7 \text{ mrad/hr.}$$

We cannot, in this case, add the autointegral gamma-ray dose to that due to the *n*, *p* reaction because an absorbed dose of 1 rad of gamma-radiation is not biologically equivalent to 1 rad of protons. This point about the relative biological effectiveness of various radiations is discussed in a later chapter.

Problems

1. A 200-mR pocket dosimeter with air equivalent walls has a sensitive volume whose dimensions are 0.5 in. diameter and 2.5 in. long; the volume is filled with air at atmospheric pressure. The capacitance of the dosimeter is 10 pfd. If 200 V are required to charge the chamber, what is the voltage across the chamber when it reads 200 mR?

2. A beam of 1 MeV gamma-rays and another of 0.1 MeV gamma-rays each produce the same ionization density in air. What is the ratio of 1–0.1 MeV photon flux?

3. Assuming a specific heat of the body of 1 calorie/g, what is the temperature rise due to a total body dose of 500 rad?

4. Compute the exposure rate, in mR/hr, at a distance of 50 cm from a small vial containing 10 ml of an aqueous solution of (a) 50 mCi ^{51}Cr, (b) 50 mCi ^{22}Na, based on the decay schemes shown below:

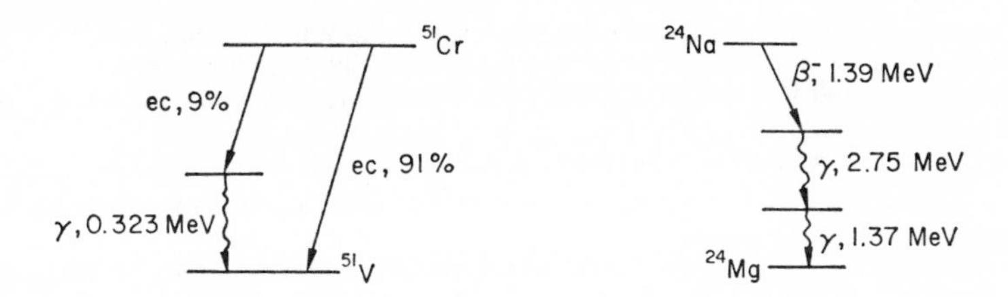

5. In an experiment, a 250-g rat is injected with 10 μCi ^{203}Hg in the form of $Hg(NO_3)_2$. The rate was counted daily in a total body counter, and the following equation was fitted to the whole body-counting data

$$Y = 0.55e^{-0.346t} + 0.45e^{0.0346t},$$

where Y is the fraction of the injected dose retained t days after injection. If the long-lived component of the curve represents clearance from the kidneys, while the short-lived component represents clearance from the rest of the body, calculate the radiation absorbed dose to the whole body and to the kidneys, if each kidney weighs 0.7 g. Assume the mercury to be uniformly distributed in the whole body and in the kidneys. Base the calculation on the decay scheme given in Fig. 4.5.

6. A patient with cancer of the thyroid has been found to have a thyroid iodine uptake of 50%. How much ^{131}I must be injected to deliver a dose to the thyroid, which weighs 30 g, of 1500 rad in 3 days?

7. Calculate the annual radiation dose to a man from the ^{40}K and from the ^{14}C deposited in his body. The specific activity of carbon is 6.9 pCi/g. Assume, in both instances, that the radioisotopes are uniformly distributed throughout the body.

8. Calculate the gamma-ray dose rate to a man in the center of a very large crowd due to the ^{40}K in his own body and in the bodies of the other people.

9. A thin-walled carbon-wall ionization chamber, whose volume is 2 cm^3, is placed inside a tank of water to make a depth-dose measurement. A 24-MeV betatron beam produces a current of 0.02 μA in the chamber. What was the absorbed dose rate?

10. An aqueous suspension of virus is irradiated by X-rays whose half-value layer is 2 mm Cu. If the exposure was 1.3×10^6 R, and if the depth of the suspension is 5 mm, what was the absorbed dose, and what was the mean ionization density?

11. A child drinks 1 liter of milk per day containing ^{131}I at a mean concentration of 900 pCi/l over a period of 30 days. Assuming that the child has no other intake of ^{131}I, calculate the dose to the thyroid at the end of the 30-day ingestion period, and the dose for the 90-day period following the start of the ^{131}I intake.

12. A patient who weighs 50 kg is given a tracer dose of an organic compound containing 100 μCi ^{14}C. On the basis of excretion measurements, the following quantities of the radioisotope were found in the patient's body:

Day	0	1	2	3	4	5	6	8	10	12	14
μCi	100	73.5	58.0	47.4	40.4	34.9	30.0	23.5	19.0	14.9	11.9

Assuming the ^{14}C to be uniformly distributed, calculate the absorbed dose to the patient 7 days and 14 days after exposure, and the dose due to the total decay of the ^{14}C.

13. A 2-MeV electron beam is used to irradiate a sample of plastic whose thickness is 0.5 g/cm^2. If a 250-μ amp beam passes through a port 1 cm in dia to strike the plastic, calculate the absorbed dose rate.

Suggested References

1. HINE, G. J. and BROWNELL, G. L. (Ed.): *Radiation Dosimetry*, Academic Press, New York, 1956.
2. WHYTE, G. N.: *Principles of Radiation Dosimetry*, John Wiley & Sons, New York, 1959.
3. JOHNS, H. E.: *The Physics of Radiology*, Charles C. Thomas, Springfield, 1964.
4. ATTIX, F. H. and ROESCH, W. C. (Ed.): *Radiation Dosimetry* Vol. II, Academic Press, New York, 1966.
5. *Selected Topics in Radiation Dosimetry*, I.A.E.A., Vienna, 1961.

The following Handbooks from the National Bureau of Standards, Govt. Printing Office, Washington:

(a) No. 55, *Protection Against Betatron—Synchrotran Radiations up to 100 Million Electron Volts*, 1954.
(b) No. 57, *Photographic Dosimetry of X and Gamma-Rays*, 1954.
(c) No. 64, *Design of Free Air Ionization Chambers*, 1957.
(d) No. 72, *Measurement of Neutron Flux and Spectra for Physical and Biological applications*, 1960.
(e) No. 75, *Measurement of Absorbed Dose of Neutrons, and of Mixtures of Neutrons and Gamma-Rays*, 1961.
(f) No. 78, *Report of the International Commission on Radiological Units and Measurements*, 1959, 1961.
(g) No. 84, *Radiation Quantities and Units*, 1964.
(h) No. 85, *Physical Aspects of Irradiation*, 1964.
(i) No. 87, *Clinical Dosimetry*, 1963.
(j) No. 88, *Radiobiological Dosimetry*, 1963.

CHAPTER 7

BIOLOGICAL EFFECTS OF RADIATION

RADIATION ranks among the most thoroughly investigated etiologic agents associated with disease. Although much still remains to be learned about the interaction between ionizing radiation and living matter, more is known about the mechanism of radiation damage on the molecular, cellular, and organ system levels than is known for most other environmental stressing agents. Indeed, it is precisely this vast accumulation of quantitative dose-response data that is available to the health physicist that enables him to specify environmental radiation levels for occupational exposure, thus permitting the continuing industrial, scientific, and medical exploitation of nuclear energy in safety.

Dose-response Characteristics

Observed radiation effects (or effects of other types of noxious agents) may be broadly classified into two groups, viz. threshold and non-threshold effects. Most phenomena fall into the category of threshold effects; that is, a certain minimum dose must be exceeded before the particular effect under investigation is observed. For such effects, when the magnitude of the effect is plotted as a function of dose in order to obtain a *quantitative relationship* between dose and effect, the dose-response curve *A* shown in Fig. 7.1 is obtained.

When experimentally determining a dose-response curve, the 50% dose, or the dose to which 50% of the exposed animals respond, is found to be statistically the most reliable. For this reason, the 50% dose is most frequently used as an index of relative effectiveness of a given agent in eliciting a particular response. When death of the experimental animal is the biological endpoint, the 50% dose is called the LD-50 dose. The time required for the toxic substance to act is also important, and is always specified with the dose. Thus, if 50% of the animals die within 30 days, we refer to the LD-50/30 day dose. This index, the LD-50/30 day dose, is widely used in toxicology to designate the relative toxicity of a substance.

Non-threshold effects, as the name suggests, are those for which no threshold has been observed, regardless of how small a dose is applied to the experimental animal. For such non-threshold effects, a dose-response curve *B* as shown in Fig. 7.1 is obtained.

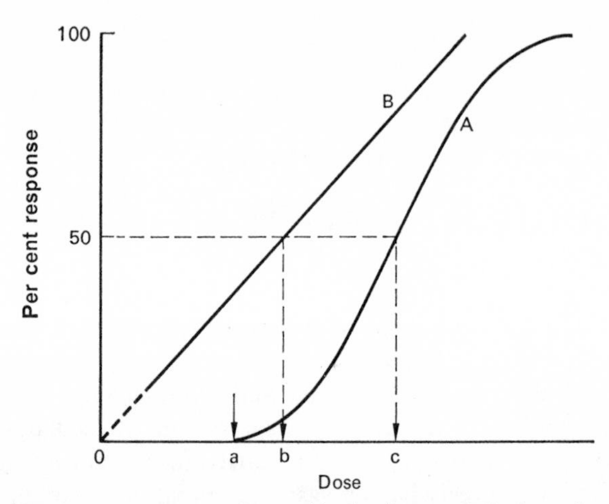

FIG. 7.1. Dose-response curves. Curve *A* is the characteristic shape for a biological effect that exhibits a threshold dose—point *a*. The spread of the curve, from the threshold at *a* until the 100% response is thought to be due to "biological variability" around the mean dose, point *c*, which is called the 50% dose. Curve *B* represents a non-threshold, or linear response; point *b* represents the 50% dose for the non-threshold biological effect.

When biological experiments are performed to obtain dose-response data, control animals are used. The reason for this is simply that various diseases and other extrinsic factors may produce the same effects as those due to the experimental agent, and the control animals are necessary to allow for the spontaneous incidence among an animal population of the biological end point used in the experiment. In "non-threshold" phenomena it is usually extremely difficult to obtain data for very low doses, since the number of animals necessary to obtain a statistically significant difference between the control animals and the experimental animals is enormous. In practical cases, therefore, an effect is said to be of the non-threshold variety if the smallest dose that produces a measurable effect lies on a curve that can be extrapolated to the origin of the coordinate axes.

Direct Action

The gross biological effects resulting from overexposure to radiation are the sequellae to a long and complex series of events that are initiated by ionization or excitation of relatively few molecules in the organism. For example, the LD-50/30 day dose for man of gamma-rays is about 400 rad. Since 1 rad corresponds to an energy absorption of 100 ergs/g, or 6.25×10^{13} eV/g, and since about 34 eV are expended in producing a single ionization, the lethal dose produces, in tissue,

$$\frac{400 \text{ rad} \times 6.25 \times 10^{13} \text{ eV/g/rad}}{34 \text{ eV/ion}} = 7.65 \times 10^{14}$$

ionized atoms per gram tissue. If we estimate that about nine other atoms are excited for each one ionized, we find that about 7.65×10^{15} atoms/g of tissue are directly affected by a lethal radiation dose. In soft tissue, there are about 8×10^{22} atoms/g. The fraction of directly affected atoms, therefore, is

$$\frac{7.65 \times 10^{15}}{8 \times 10^{22}} \approx 1 \times 10^{-7},$$

or about 1 atom in 10 million.

Effects of radiation in which no threshold dose has been observed are thought to be the result of a direct insult to a molecule by ionization and excitation and the consequent dissociation of the molecule. Point mutations, in which there is a change in a single gene locus, is an example of such an effect. The dissociation, due to ionization or excitation, of an atom on the DNA molecule prevents the information originally contained in the gene from being transmitted to the next generation. Such point mutations may occur in the germinal cells, in which case the point mutation is passed on to the next individual; or it may occur in somatic cells, which results in a point mutation in the daughter cell. Since these point mutations are thereafter transmitted to succeeding generations of cells (except for the highly improbable instance where one mutated gene may suffer another mutation), it is clear that for those biological effects of radiation that depend on point mutations, the radiation dose is cumulative; every little dose may result in a change in the gene burden which is then continuously transmitted. When dealing quantitatively with such phenomena, however, we must consider the probability of observing a genetic change among the offspring of an irradiated individual. For radiation doses down to about 25 rad, the magnitude of the effect, as measured by frequency of gene mutations, is proportional to the dose. Below doses of about 25 rad, the mutation probability is so low that enormous numbers of animals must be used in order to detect a mutation that could be ascribed to the radiation. For this reason, no reliable experimental data are available for genetic changes in the range of 0 to about 25 rad.

Indirect Action

Direct effects of radiation, ionization, and excitation are non-specific, and may occur anywhere in the body. When the directly affected atom is in a protein molecule, or in a molecule of nucleic acid, then certain specific effects due to the damaged molecule may ensue. However, most of the body is water, and most of the direct action of radiation therefore is on water. The result of this energy absorption by water is the production, in the water, of highly reactive free radicals that are chemically toxic (a free radical is a fragment of a compound or an element that contains an unpaired electron) and which may exert their toxicity on other molecules. When pure water is irradiated we have

$$H_2O \rightarrow H_2O^+ + e^-, \tag{7.1}$$

and the positive ion dissociates immediately according to the equation

$$H_2O^+ \rightarrow H^+ + OH, \tag{7.2}$$

while the electron is picked up by a neutral water molecule

$$H_2O + e^- \rightarrow H_2O^-, \tag{7.3}$$

which dissociates immediately

$$H_2O^- \rightarrow H + OH^- \tag{7.4}$$

The ions H^+ and OH^- are of no consequence, since all body fluids already contain significant concentrations of both these ions. The free radicals H and OH may combine with like radicals, or they react with other molecules in solution. Their most probable fate is determined chiefly by the LET of the radiation. In the case of a high rate of linear energy transfer, such as results from passage of an alpha particle or other particle of high specific ionization, the free OH radicals are formed close enough together to enable them to combine with each other before they can recombine with free H radicals, which leads to the production of hydrogen peroxide

$$OH + OH \rightarrow H_2O_2, \tag{7.5}$$

while the free H radicals combine to form gaseous hydrogen. Whereas the products of the primary reactions of equations (7.1) through (7.4) have very short lifetimes, on the order of a microsecond, the hydrogen peroxide, being a relatively stable compound, persists long enough to diffuse to points quite remote from their point of origin. The hydrogen peroxide, which is a very powerful oxidizing agent, can thus affect molecules or cells that did not suffer radiation damage directly. If the irradiated water contains dissolved oxygen, the free hydrogen radical may combine with oxygen to form the hydroperoxyl radical,

$$H + O_2 \rightarrow HO_2, \tag{7.6}$$

which is not as reactive, and therefore has a longer lifetime, than the free OH radical. This greater stability allows the hydroperoxyl radical to combine with a free hydrogen radical to form hydrogen peroxide, thereby further enhancing the toxicity of the radiation.

Radiation is thus seen to produce biological effects by two mechanisms, viz. directly by dissociating molecules following their excitation and ionization; and indirectly by the production of free radicals and hydrogen peroxide in the water of the body fluids. The greatest gap in our knowledge of radiobiology is the sequence of events between the primary effects described above, and the gross biological effects that may be observed long after irradiation.

Radiation Effects

In health physics, as in other areas of environmental control of harmful agents, we are concerned with two types of exposure: (1) a single accidental exposure to a high dose of radiation during a short period of time, which is commonly called *acute* exposure, and which may produce biological effects within a short time after exposure; (2) long-term, low level overexposure, commonly called *continuous* or *chronic* exposure, where the results of the overexposure may not be apparent for years, and which is likely to be the result of improper or inadequate protective measures.

Acute Effects

Acute radiation overexposure affects all the organs and systems of the body. However, since not all organs and organ systems are equally sensitive to radiation, the pattern of response, or disease syndrome, in an overexposed individual depends on the magnitude of the dose. To simplify classification, the acute radiation syndrome is subdivided into three classes; in order of increasing severity, these are: (1) the hemopoietic syndrome, (2) the gastro-intestinal syndrome, and (3) the central nervous system syndrome. Certain effects are common to all categories; these include:

(a) nausea and vomiting,
(b) malaise and fatigue,
(c) increased temperature,
(d) blood changes.

In addition to these effects, numerous other changes are seen.

Blood changes

Of the four common effects listed above, changes in the peripheral blood count are the most sensitive biological indicators of acute overexposure. These changes are seen even in cases of mild overexposure, which results in none of the three syndromes. Although blood changes have been seen in individuals with exposure doses as low as 14 R, they usually do not appear until doses of 25–50 R are experienced. Beyond 50 R, blood changes are almost certain to appear.

The blood consists of about 55% (by volume) fluid, called the blood plasma, and about 45% of formed elements, including white blood cells, called leucocytes, red blood cells, called erythrocytes, and platelets, or thrombocytes. The white blood cells, which number about 7000/mm^3 of blood in the average adult, function in the body as a major line of defense against bacterial invasion. An infection anywhere in the body stimulates the production of leucocytes in order to combat the infecting organisms. Several major types of leucocytes are found: the granulocytes and the lymphocytes, each with certain specialized functions to aid in the fight against infection. Under normal conditions, the relative proportions of each of these remain approximately constant—the granulocytes form about 70–75% of the white blood cells while the lymphocytes account for about 25–30%. The granulocytes are produced in the bone marrow, and circulate for about 3 days before death and destruction, while the lymphocytes are produced in the lymph nodes and spleen and remain alive in the blood for about 24 hr. The red blood cells are the most abundant of the formed elements; their concentration in the blood is about 5 million per cubic millimeter. The main function of the red blood cells is to transport oxygen from the lungs to the body cells, and to carry the carbon dioxide waste from the cells to the lung. The erythrocytes are formed in the bone marrow, and survive in the circulating blood for about 90–120 days. The platelets, or thrombocytes, which number about 200,000–400,000/mm³ blood, are concerned with the clotting of the blood. They are manufactured in the marrow, and have a useful lifetime of about 8–12 days.

Following an acute radiation exposure in the sub-lethal range, there is a transitory sharp increase in the number of granulocytes followed within a day

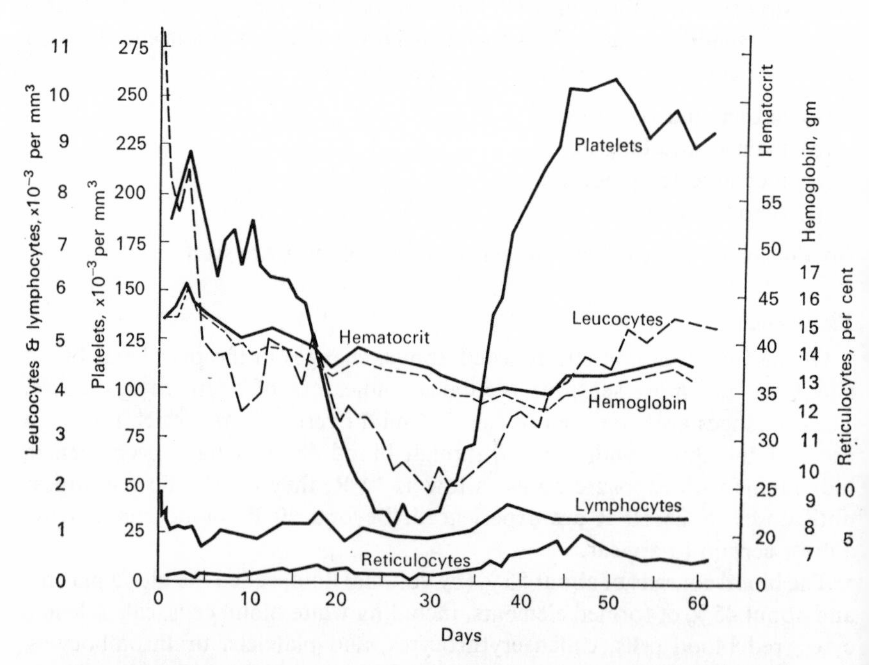

FIG. 7.2. Hematologic effect of radiation overexposure. Average values for five patients who were exposed to 236–365 rad (estimated) during a criticality accident at the Y-12 plant in Oak Ridge on 16 June 1958. (G. H. Andrews, B. W. Sitterson, A. L. Kretchman and M. Brucer, Criticality accident at the Y-12 plant, *Diagnosis and Treatment of Acute Radiation Injury*, pp. 27–48, World Health Organization, Geneva, 1961.)

by a decrease which reaches a minimum several weeks after exposure, and then returns to normal after a period of several weeks to several months. The lymphocytes drop sharply after exposure, and remain depressed for a period of several months. In constrast to the very rapid response of the white cells to radiation overexposure, the red blood count does not reflect an overexposure until about a week after exposure. Depression in the erythrocyte count continues until a minimum is reached between 1 to 2 months after exposure, followed by a slow recovery over a period of weeks. The platelet count falls steadily until a minimum is reached about a month after exposure; recovery is very slow, and may take several months. In all cases, the degree of change in the blood, as well as the rate of change, is a function of the radiation dose. Figure 7.2 shows graphically the trends in rate and degree of blood changes for several different exposure doses.

Hemopoietic syndrome

The hemopoietic syndrome appears after an exposure dose of about 200 R. This disease is characterized by depression or ablation of the bone marrow, and the physiological consequences of this damage. The onset of the disease is rather sudden, and is heralded by nausea and vomiting within several hours after the overexposure occurred. Malaise and fatigue are felt by the victim, but the degree of malaise does not seem to be correlated with the size of the dose. Epilation, which is almost always seen, appears between the second and third week post exposure. Death may occur within 1 to 2 months after exposure. The chief effects to be noted, of course, are in the bone marrow and in the blood. Marrow depression is seen at 200 R; at about 400–600 R, complete ablation of the marrow occurs. In this case, however, spontaneous regrowth of the marrow is possible if the victim survives the physiological effects of the denuding of his marrow. An exposure of about 700 R or greater leads to irreversible ablation of the bone marrow. The blood picture reflects the damage to the marrow; the changes noted above are seen, and the magnitude of the change is roughly correlated with the dose. A very low lymphocyte count, 500/mm^3 or less within the first day or two after exposure suggests that death will probably ensue.

Gastrointestinal syndrome

This disease follows a total body exposure of about 1000 R or greater, and is a consequence of the desquamation of the intestinal epithelium. All the signs and symptoms of the hemopoietic syndrome are seen—with the addition of severe nausea, vomiting, and diarrhea which begins very soon or immediately after exposure. Death within 1 to 2 weeks after exposure is the most likely outcome.

Central nervous system syndrome

A total body exposure in excess of about 2000 R damages the central nervous system as well as all the other organ systems in the body. Unconsciousness follows within minutes after exposure, and death in a matter of hours to several days; the rapidity of onset of unconsciousness is directly related to dose. In one instance in which a 200 μsec burst of mixed neutrons and gamma-rays delivered a mean total body dose of about 4400 rad, the victim was ataxic and disoriented within 30 sec. In 10 min he was unconscious and in shock. Thirty-five minutes after the accident, analysis of faecal fluid from an explosive watery diarrhea showed a copius passage of fluids into the gastrointestinal tract. Vigorous symptomatic treatment kept the patient alive for $34\frac{1}{2}$ hr after the accident.

Other acute effects

Several other immediate effects of acute overexposure should be noted. Because of its physical location, the skin is subject to more radiation exposure, especially in the case of low energy X-rays and beta-rays, than most other tissues. An exposure of about 300 R of low energy (in the diagnostic range) X-rays results in erythema; higher doses may cause changes in pigmentation, epilation, blistering, necrosis, and ulceration. Radiation dermatitis of the hands and face was a relatively common occupational disease among radiologists who practiced during the early years of the twentieth century.

The gonads are particularly radiosensitive. A single exposure of only 30R to the testes results in temporary sterility among men; for women, an exposure of 300 R produces temporary sterility. Higher doses increase the period of temporary sterility; one man, whose exposure to the gonads was less than 440 R, was aspermatic for a period of several years. In women, temporary sterility is evidenced by a cessation of menstruation for a period of 1 month or more, depending on the dose. Irregularities in the menstrual cycle, which suggest functional changes in the gonads, may result from local irradiation of the ovaries with doses smaller than that required for temporary sterilization.

The eyes, too, are relatively radiosensitive. A local exposure of several hundred roentgens can result in acute conjunctivitis and keratitis.

Delayed Effects

The delayed effects of radiation may be due either to a single large overexposure or continuing low-level overexposure.

Continuing overexposure can be due to exposure to external radiation fields, or can result from inhalation or ingestion of a radioisotope which then becomes fixed in the body through chemical reaction with the tissue protein or, because of the chemical similarity of the radioisotope with normal metabolites, may be systemically absorbed within certain organs and tissues. In ither case, the internally deposited radioisotope may continue to irradiate he tissue for a long time. In this connection, it should be pointed out that the djectives "acute" and "chronic", as ordinarily used by toxicologists to escribe single and continuous exposures respectively, are not directly appli-able to inhaled or ingested radioisotopes, since a single, or "acute" exposure nay lead to continuous, or "chronic" irradiation of the tissue in which the adioactive material is located. In the case of either a single massive over-xposure or continuous low level overexposure, the end results are the same. Among the delayed consequences of overexposure which are of greatest con-ern are cancer, shortening of life span, and cataracts.

Cancer

Although any organ or tissue may develop neoplasia following over-xposure to radiation, certain organs and tissues seem to be more sensitive n this respect than others. Radiation-induced cancer is observed most requently in the skin, the hemopoietic system, the bone, and the thyroid land. In all these cases, the tumor induction time in man is relatively long—n the order of 5 to 20 years after exposure. Carcinoma of the skin was the irst type of malignancy that was associated with exposure to X-rays. Early X-ray workers, including physicists and physicians, had a much higher inci-ence of skin cancer than could be expected from random occurrence of this lisease. Well over 100 cases of radiation-induced skin cancer are documented n the literature. As early as 1900, a physician who had been using X-rays in nis practice described in writing the irritating effects of X-rays. Erythema and tching progressed to hyper-pigmentation, ulceration, neoplasia, and finally leath from metastatic carcinoma. The entire disease process spanned a period of 9 years.

An occupational disease among dentists was cancer of those fingers that were used to hold dental X-ray films in the mouths of patients while X-raying heir teeth.

Leukemia

Leukemia is among the most likely forms of malignancy resulting from overexposure to total body radiation. Radiologists and other physicians who use radiation in their practice show a significantly higher rate of leukemia han do their colleagues who do not use radiation. Higher leukemia incidence among children who had been irradiated *in utero* or during infancy than unirradiated controls has been observed. A similar increase in leukemia ncidence has been observed among patients who were treated by X-rays for ankylosing spondilitis. Among the survivors of the nuclear bombings of Japan, there is a significantly greater incidence of leukemia among those people who were within 1500 m of the hypocenter, and presumably received greater radiation doses, than those who were more than 1500 m fro ground zero. All the available data, including results of studies on anima as well as on men, show quite clearly that radiation is a leukemogenic agen The question regarding the existence of a threshold dose, however, is not y resolved; it is not possible at this time (1968) to say anything more than th the threshold for leukemia, if it exists, probably lies between zero an 460 rems.

Bone cancer

Historically, occupational radiation-induced cancer of the bone and of th lung are important in emphasizing the carcinogenicity of internally deposite radioisotopes. Although the cancer producing properties of radium wer known within several years after its discovery by Pierre and Marie Curie i 1902, little or no effort was made to protect people from the harmful effect of radium. One of the first industrial uses to which radium was put was i the manufacture of luminous paint. When powdered radium is mixed wit ZnS crystals, the crystals glow due to absorption of energy from the alph particles emitted by the radium. Luminous paint made in this way soon wa used to paint instrument and clock dials. This application of radium receive a great impetus when World War I began. In order to paint fine lines, th girls who were employed as dial painters pointed their brushes between thei lips. Minute amounts of radium and mesothorium were swallowed each tim that a brush was pointed. In the early 1920's several girls who had worked a dial painters died from anemia and with degeneration of the jaw bones. A dentist who was treating another dial painter suspected the occupationa etiology of the disease. Further investigation revealed more cases of bon damage, including osteogenic sarcoma, and definitely established radium a the etiologic agent. Follow-up studies on radium dial painters and on patient who had received radium injections therapeutically confirmed this finding. I also showed that radium is tenaciously retained in the bones. Significan deposits of radium were found 25–35 years after exposure. Further experi mental work with laboratory animals revealed a number of "bone-seeking" radioisotopes which produced the same type of damage as radium. Radio strontium and barium, as well as radium, are chemically similar to calcium and are therefore incorporated into the mineral structure of the bone an into the epiphysis. The bone-seeking rare earth fission products such a ^{144}Ce–^{144}Pr also tend to accumulate in the mineral structure. Plutonium, ar extremely toxic element, is found to accumulate in the periosteum, endosteum, and trabeculae of the bone. All bone seekers are considered very toxic because they can damage the radiosensitive hemopoietic tissue in the bone marrow; all the bone seekers produce bone cancer when they are injected into labora- tory animals in sufficient quantity.

ung cancer

The susceptibility of the lung to radiation-induced carcinoma has been :nown for a long time. The mines in Joashimsthal and Schneeberg are rich n pitchblende, a radium-bearing ore (as well as in other minerals which are ılso suspected of being carcinogens). As a consequence of the radium in the ground, radon gas, the radioactive daughter of radium, is produced and difuses out of the ground into the air in the mine shafts. Since radon is itself adioactive and gives rise to several radioactive descendants (Table 4.3) a ;ood deal of radioactivity due to the radon daughter products is also present. The atmospheric concentration of radon in the two European mining centers vas about 3×10^{-6} μCi/ml air. To this activity must be added the daughter ctivities, which are chiefly RaA and RaC′. A very high percentage of workers xposed to this atmosphere developed bronchogenic carcinoma within 15 ears after beginning to work in the mines. The carcinogenicity of inhaled adon was confirmed in the laboratory using mice who were exposed daily to tmospheric concentrations of 10^{-3} μCi/ml. In other laboratory studies, nvolving the deposition of radioactive materials in the lung by intratracheal njection, by inhalation, by trans-pleural injection of ^{90}Sr-loaded glass beads nd by surgical implantation in a bronchus of a small cylinder of ^{106}Ru, adiation was found to be carcinogenic to the lung. From these studies it vas found that long term continuous radiologic insult from radioactive naterials residing in the lung was required in order to produce lung cancer. On the order of 100 or more days elapsed, in most instances, between exposure and the observation of lung cancer. However, in one instance, a squamous cell carcinoma was observed only 48 days after intratracheal injection of 15 μCi ^{144}Ce as CeF_3 particulates. Irradiation of the lungs with X-rays has been found to produce fibrosis and pneumonitis, but no neoplasia.

Hazard and toxicity

The experimental work referred to above, in which lung tumors resulted from radioactivity in the lung that was deposited unphysiologically by injection or surgical implantation, clearly cannot serve as a measure of the *hazard* from radioactive dusts. They can serve only to indicate the *toxicity* of radioactive material after the radioactivity is located at the site of its toxic action. The hazard from inhaled radioactive dusts—or indeed the hazard from any toxic material—must include a consideration of the liklihood that the toxic substance will reach the site of its toxic action. In the case of inhaled dusts, this site is assumed to be the bronchial epithelium, the alveolar epithelium, and the pulmonary lymph nodes. The two main factors that influence the degree of hazard from toxic airborne dusts are: (1) the deposition in the lung of the dust particle, and (2) the retention of the particle within the lung.

The deposition of particles within the lung depends mainly on the particle size of the dust, while the retention in the lung depends on the physical an chemical properties of the dust as well as on the physiological status of th lung. Dusts generated by almost any process are found to be randoml distributed in size, or "diameter", around a mean value. This size distributio is found to be "log-normal" that is, the logarithm of the particle size is foun to be normally distributed, rather than the size directly. In this case, the mea size, which is called the geometric mean, is defined by

$$m_g = \text{antilog}\left[\frac{\Sigma \log d_n}{n}\right], \qquad (7.7)$$

where d_n is the particle diameter, which most often is expressed in micron ($1\ \mu = 10^{-4}$ cm), of each of the different particles that is measured, and n is th number of particles that are sized. The standard deviation of the size dis tribution is defined as

$$\sigma = \sqrt{\left(\frac{\Sigma (\log d_n)^2}{n} - (m_g)^2\right)}. \qquad (7.8)$$

It corresponds to the inflection point of the Gaussian distribution curve, a shown in Fig. 7.3(a). In a normal distribution, 68.2% of the population fall within the limits bounded by plus and minus one standard deviation from the mean; 95.4% of the population is included within plus and minus twc

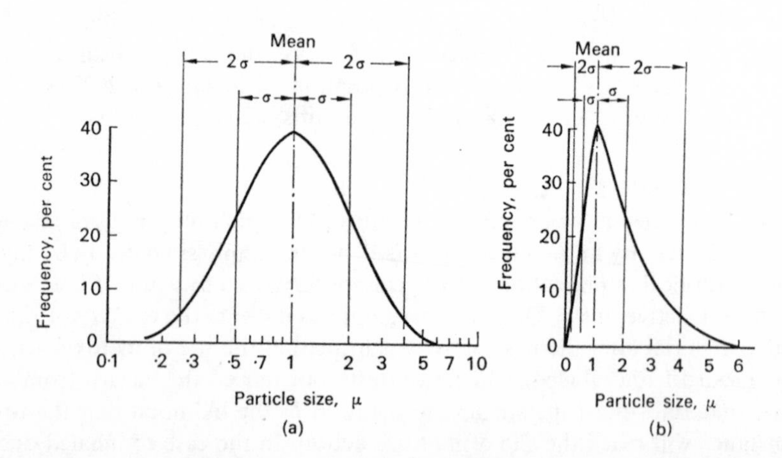

FIG. 7.3. (a). Log-normal distribution of dust particle sizes. The logarithm of the diameter is normally distributed, thus giving a Gaussian distribution when the logarithm of the particle size is plotted against frequency. In this curve, the mean size is 1 μ, and the geometric standard deviation is 2. (b) Size distribution curve for the same data as shown in (a). In this curve, the diameter is plotted against frequency, thus yielding a skewed distribution curve.

andard deviations of the mean. Because of the logarithmic nature of the istribution, the size distribution must be given as

$$m_g \overset{\times}{\div} \sigma_g, \tag{7.9}$$

ather than, as in the case of an arithmetic mean,

$$m_a \pm \sigma_a. \tag{7.10}$$

or example, if the log-normal size distribution of a dust is given as $1.4 \overset{\times}{\div} 2\ \mu$, ais means that 68% of all the particles lie between 0.7 (or $1.4 \div 2$) and 2.8 r 1.4×2) μ, and that about 96% lie between 0.35 and 5.6 μ.

The respiratory tree may be roughly divided into two categories on the asis of physiologic function: the upper respiratory tract and the deep spiratory tract. The functional part of the lung, where gaseous exchange tween the inspired air and the dissolved gas in the blood occurs, is e deep respiratory tract. The rest of the respiratory tree, or upper spiratory tract, including the naso-pharyngeal passages, the trachea, the onchi, and the terminal bronchioles, are only avenues that permit gas to ass from the atmosphere to and from the alveoli, which form the deep resratory tract. The volume of the upper respiratory tract, which is really a dead ace as far as gas exchange is concerned, is about 150 ml in an average adult; e tidal volume, that is, the volume of air inspired or expired during a single spiratory excursion, is about 500 ml. The functional residual capacity, hich is the volume of air remaining in the lung after an ordinary expiration, about 2.4 liters. It is thus seen that a relatively small fraction, only about %, of the air normally present in the lung is exchanged during a single spiratory cycle. The nasal passages are lined with hair, which acts to filter at coarse particles, while the upper respiratory tract from the trachea down the terminal bronchi are lined with cells containing very fine, hair-like aips, called cilia, which beat rhythmically and thereby carry upward towards e throat those dust particles which are on their surface. The depth of penеation of airborne particulates into the respiratory tract depends on the size the air-borne particulate. Large dust particles, in excess of about 5 μ, are kely to be filtered out by the nasal hair or to impact on the naso-pharyngeal rface. It should be pointed out here that particles must possess a great deal kinetic energy in order to have motion, except for gravitational settling, dependent of the motion of the air. The effect of gravitational settling comes less pronounced as the particle size decreases. For example, a 20-μ ameter particle of unit density has a settling velocity of 1 cm/sec while a μ particle of this material settles at a rate of only 0.0035 cm/sec. For practil purposes, therefore, small particles may be regarded as remaining susnded in the atmosphere, and all but very large particles may be considered be carried by moving air. In the respiratory tree, because of the relatively aall cross-sectional areas of many of the air passages, the inspired air may attain relatively high velocities. Large particles that escape the hair-filte in the nose therefore have high kinetic energies as they pass through the ai passages. As a consequence of the momentum of such a heavy particle, cannot follow the inspired air around sharp curves, and is impacted on to th walls of the upper respiratory tract. As the particle size decreases below 5 μ this inertial impaction decreases, and an increasing number of particles i carried down into the lung. The air in the alveoli is relatively still—since onl a small fraction of the air there is exchanged with incoming air during respiratory excursion. Particles that are carried into the deep respirator tract, therefore, have the opportunity to settle out under the force of gravity Gravitational settling, however, decreases with decreasing particle size, an vanishes when the particle size is about $\frac{1}{2}\,\mu$. As the particle size decrease

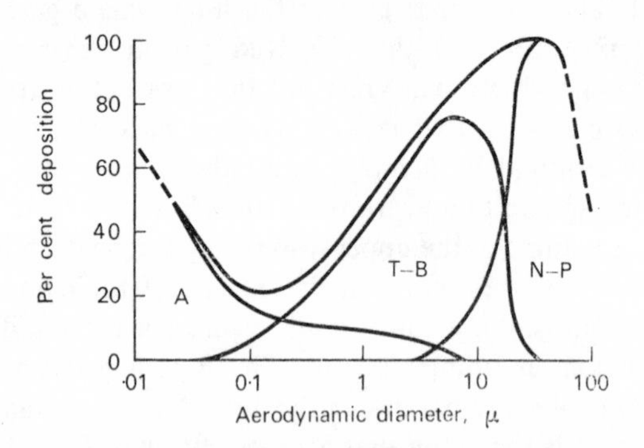

FIG. 7.4. Deposition of dust in the respiratory tract. Region *A* represents principally alveolar deposition, region *T–B* is mainly tracheobronchial deposition, and region *N–P* represents naso-pharyngeal deposition. Total deposition is the sum of these three regions. (From P. E. Morrow, Evaluation of inhalation hazards based upon the respirable dust concept and the philosophy and application of selective sampling, *A.I.H.A. Journ.* **25**, 213–36, 1964.)

below about 0.1 μ, the effect of Brownian motion becomes significant. As th particles move randomly about, they may strike the alveolar wall—and b trapped on its moist surface. The combination of these three effects, inertia impaction, gravitational settling, and Brownian motion, leads to a maximum likelihood of deposition in the deep respiratory tract for particles in the 1–2 size range, and a minimum deposition for particles between 0.1 and 0.5 μ as shown above in Fig. 7.4.

The retention of particles in the lung depends on the area within the respiratory tract where the particles were deposited, on the physical and chemica properties of the particles, and on the physiologic properties of the lung Retention of the inhaled particles, or its inverse—pulmonary clearance—i important in determining the degree of hazard because of its role in tissu exposure time and total dose. Studies of pulmonary retention of various dust show that the curve of dust remaining in the lung after cessation of exposure is fitted by a complex exponential curve that includes at least two components; one of half retention time on the order of several hours, and the other on the order of days. Very often, the long-lived component is also complex, and may be resolved into two or three components. Algebraically, this curve is given by the equation

$$D = D_1 e^{-k_1 t} + D_2 e^{-k_2 t} + \dots \quad (7.11)$$

Figure 7.5 illustrates a typical two-component retention curve, which is described by the first two terms of equation (7.11).

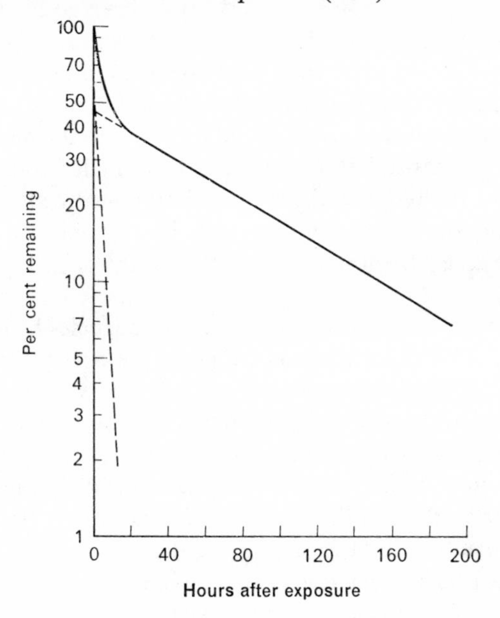

FIG. 7.5. Pulmonary retention curve showing amount of $BaSO_4$ particulate remaining in the lung as a function of time after exposure. According to this curve, 53% of the dust was deposited in the upper respiratory tract; the clearance rate from the URT is 27% per hour. Forty-seven per cent of the dust was deposited in the deep respiratory tract, and was cleared out at a rate of 1% per hour.

In this curve, the first component represents the dust in the upper respira-ɔry tract; D_1 is the amount of dust deposited there, while k_1 gives the rate at hich it is cleared from the upper respiratory tract. D_2 represents the dust eposited initially in the deep respiratory tract, and k_2 is the deep respiratory-act clearance rate.

At least three distinct mechanisms are thought to operate simultaneously remove foreign particulates from the lung. The first of these mechanisms, liary clearance, can act only in the upper respiratory tract. The rhythmic beating of the cilia propel particles upward into the throat—from whence they are swallowed—at very high speeds. Particle velocities ranging from about 2 mm/min in the bronchi to about 3 cm/min in the trachea have been observed. The other two clearance mechanisms deal mainly with particulates in the deep respiratory tract. They include: solubility and absorption into the capillary bed across the alveolar membrane, and removal by phagocytosis. It should be pointed out that the solubility in water of any given substance may not necessarily be a good index of solubility in the lung. For example, mercury from ^{203}HgS, one of the most insoluble compounds known, was found in significant quantities in the kidneys and in the urine of rats exposed to HgS particles of about 1 μ; the tagged mercury could have gotten there only after solution of the particles. On the other hand, ^{144}Ce intratracheally injected as $CeCl_3$ solution was found to be tenaciously retained in the lung for very long periods of time. In this case, the Ce in solution was bound to the tissue protein in the lung; very little of the cerium found its way into the blood. The case of cerium, however, seems to be exceptional. Most inhaled soluble particulates are absorbed into the blood and their chemical constituents translocated to other organs and tissues, where they may be systemically absorbed. The lung can thus be an excellent portal of entry into the body for many different radionuclides.

Phagocytosis, which is the engulfing of foreign particles by alveolar macrophages and their subsequent removal either up the ciliary "escalator" or by entrance into the lymphatic system, is a major pulmonary clearance mechanism. Phagocytes loaded with radioactive particles may be trapped in the sinuses of the tracheobronchial lymph nodes, and may remain there for long periods of time. This accumulation of radioactive dust in the lymph nodes may result in a higher radiation dose to the lymph nodes than to the lungs. The rate of phagocytosis depends to a large degree on the nature of the dust particle. Different particles have been found to be phagocytized at different rates. Furthermore, it has been found that radioactive particles are phagocytized more slowly than non-radioactive particles of the same chemical composition, physical form, and size distribution.

It is clear, from the multiplicity of factors that play a role in determining the biological effects of inhaled radioactive materials, that no simple quantitative relationship between gross atmospheric concentration of a radioisotope and lung effects can be assumed. Very conservative criteria, therefore, must be applied to environmental control of radioactive dust.

Life shortening

Cancer resulting from overexposure to radiation naturally shortens the life span of persons thus overexposed. In addition to this mode of accelerating death, radiation shortens the life span by increasing the rate of physiological aging. Such a life shortening effect has been observed among groups of experimental animals. The exact cause of death among these animals cannot be uniquely attributed to radiation; the causes of death are those that are expected among an animal population. Only the age-specific death rate is different from the controls; death occurs earlier among the irradiated animals. Although this life-shortening effect has been clearly demonstrated in animals, no similar clean-cut effect has been seen among man. Comparison of the duration of life among medical specialists who used X-rays in their work with other medical practitioners who did not use radiation in their work gives conflicting implications. While some of the data suggest an increased death rate from non-specific causes among users of X-rays, other data show no statistically significant difference in life expectancy between these two population groups. Quite clearly, therefore, radiation exposure at the levels encountered by radiologists and other X-ray users among physicians is not high enough to accelerate the aging process to the degree that will cause a statistically significant shortening of life span.

Cataracts

Much higher incidences of cataracts among physicists in cyclotron laboratories whose eyes had been exposed intermittently for long periods of time to relatively low radiation fields, and among atomic-bomb survivors whose eyes had been exposed to a single high radiation dose, showed that both chronic and acute overexposure of the eyes could lead to cataracts. Radiation may injure the cornea, conjunctiva, iris, and the lens of the eye. In the case of the lens, the principal site of damage is the proliferating cells of the anterior epithelium. This results in abnormal lens fibers, which eventually disintegrate to form an opaque area, or cataract, that prevents light from reaching the retina. The cataractogenic dose to the lens is on the order of 500 rad of beta or gamma radiation. Fast neutrons are more effective in producing cataracts than X- or beta-rays; cataracts have been reported after a dose of 200 rad from mixed gamma and neutron irradiation of the lens.

Not all radiations are equally effective in producing cataracts; neutrons are much more efficient than the other radiations. The cataractogenic dose has been found, in laboratory experiments with animals, to be a function of age; young animals are more sensitive than old animals. On the basis of occupational exposure data, it is estimated that the threshold dose for cataracts lies between about 15 and 45 rad of neutrons. No radiogenic cataracts resulting from occupational exposure to X-rays have been reported. From patients who suffered irradiation of the eye in the course of X-ray therapy and developed cataracts as a consequence, the cataractogenic threshold is estimated at about 200 rad. In cases either of occupationally or therapeutically induced radiation cataracts, a long latent period, on the order of several years, usually elapsed between exposure and the appearance of the lens opacity.

Relative Biological Effectiveness (RBE) and Quality Factor (QF)

Not only have neutrons been found more effective than X-rays in producing cataracts, but alpha radiation too has been found to be more toxic than beta or gamma radiation. When comparing the relative toxicity, or damage producing potential of various radiations, it is assumed, of course, that the comparison is on the basis of equal amounts of energy absorption. Generally, the higher the rate of linear energy transfer (LET) of the radiation, the more effective it is in damaging an organism. The ratio of the amount of energy of 200 keV X-rays required to produce a given effect to the energy required of any radiation to produce the same effect is called the *relative biological effectiveness* (abbreviated RBE) of that radiation. The RBE of a specific radiation depends on the exact biological effect on a given species of organism under a given set of experimental conditions. The term RBE is thus restricted in application to radiation biology. For health physics purposes, a conserva-

TABLE 7.1. RELATIONSHIP BETWEEN QUALITY FACTOR AND LINEAR ENERGY TRANSFER

LET keV per micron in water	QF
3.5 or less	1
3.5–7.0	1–2
7.0–23	2–5
23–53	5–10
53–175	10–20

TABLE 7.2. QUALITY FACTOR VALUES FOR VARIOUS RADIATIONS

Radiation	QF
Gamma-rays from radium in equilibrium with its decay products (filtered by 0.5 mm platinum)	1
X-rays	1
Beta-rays and electrons of energy >0.03 MeV	1
Beta rays and electrons of energy <0.03 MeV	1.7
Thermal neutrons	3
Fast neutrons	10
Protons	10
Alpha-rays	10
Heavy ions	20

tive upper limit of the RBE for the most important effect due to a radiation other than the reference radiation (200 keV X-rays) is used as a normalizing factor in adding doses from different radiations. This normalizing factor, which is called the *quality factor* (abbreviated QF), is related to LET as shown in Table 7.1. For convenience in making health-physics measurements and for administrative purposes, the various radiations most frequently encountered by health physicists are assumed to have the quality factors listed in Table 7.2.

Dose Equivalent: The Rem

The rem is the unit of radiation dose equivalent (DE) that is used for radiation safety purposes and for administrative purposes. The dose equivalent, expressed in rems, considers the QF of radiation as well as the absorbed dose, plus other factors, such as non-uniform distribution of an internally deposited radioisotope, DF, that may influence the biologic effect of a given absorbed dose. (The distribution factor is discussed in Chapter 8, in connection with the concentration of radioisotopes in drinking water based on comparison with radium.) The dose equivalent is defined by

$$\text{DE, rems} = \text{dose, rads} \times \text{QF} \times \text{DF}. \tag{7.12}$$

According to equation (7.12), an absorbed dose of 1 mrad of X- or beta-rays is equal to 1 mrem, while a fast neutron dose of 1 mrad corresponds to 10 mrem, and an absorbed heavy ion dose of 1 mrad is 20 mrem. The rad is based on only physical factors, while the rem considers both physical and biological factors. Maximum allowable radiation dose for the various radiations is given in units of rems or millirems. The use of the rem unit in routine radiation surveying is illustrated by the following example:

> The dose rate outside the shielding of a cyclotron is found to be 0.5 mR/hr gamma, 0.2 mrad/hr thermal neutrons, and 0.1 mrad/hr fast neutrons. What is the total dose rate of the combined radiations?

$$\begin{aligned} \text{gamma-rays: } 0.5 \text{ mR/hr} \times 1 &= 0.5 \text{ mrem/hr} \\ \text{thermal neutrons: } 0.2 \text{ mrad/hr} \times 3 &= 0.6 \text{ mrem/hr} \\ \text{fast neutrons: } 0.1 \text{ mrad/hr} \times 10 &= \underline{1.0 \text{ mrem/hr}} \\ \text{mixed radiation dose rate} &= \underline{2.1 \text{ mrem/hr}} \end{aligned}$$

High-energy Radiation

High-energy radiation, such as that found around a synchrocyclotron, which produces particles of hundreds of MeV in energy, or around a proton-synchrotron, which accelerates particles to the GeV range, poses special problems. The rad dose can be measured with relative ease using a tissue-equivalent ionization chamber. The spectral distribution of the radiation is more difficult to measure, but it can be estimated by film-track technics, by pulse height analysis, by threshold detectors, or by absorption methods. However, the biological effects of these high-energy radiations are not sufficiently well known to permit the determination of numerical values for the quality factor. In this regard, the extremely high rate of LET by very high energy heavy ions is of special interest.

Suggested References

1. CLAUS, W. D. (Ed.): *Radiation Biology and Medicine*, Addison-Wesley, Reading, 1958.
2. LEA, D. E.: *Actions of Radiations on Living Cells*, Cambridge University Press, Cambridge, 1955.
3. SPEAR, F. G.: *Radiations and Living Cells*, John Wiley & Sons, New York, 1953.
4. ALEXANDER, P.: *Atomic Radiation and Life*, Penguin Books, Harmondsworth, 1957.
5. BACQ, Z. M. and ALEXANDER, P.: *Fundamentals of Radiobiology*, Pergamon Press, Oxford, 1961.
6. *Report of the United Nations Scientific Committee on the Effects of Atomic Radiation*, United Nations, New York, 1962.
7. HOLLAENDER, A. (Ed.): *Radiation Biology*, vols. 1 and 2, McGraw-Hill, New York, 1955.
8. SZIRMAI, E. (Ed.): *Nuclear Hematology*, Academic Press, New York, 1965.
9. WOLF, G.: *Isotopes in Biology*, Academic Press, New York, 1964.
10. *Diagnosis and Treatment of Acute Radiation Injury*, World Health Organization, Geneva, 1961.
11. MCLEAN, F. C. and BUDY, A. M.: *Radiation, Isotopes, and Bone*, Academic Press, New York, 1964.

CHAPTER 8

RADIATION PROTECTION GUIDES

Organizations that Set Standards

International Commission on Radiological Protection

The basic responsibility for providing guidance in matters of radiation safety has been assumed by the International Commission on Radiological Protection (ICRP). This organization was established in 1928 by the Second International Congress of Radiology as the International X-ray and Radium Protection Commission. At that time and for many years afterward, its main concern was with the safety aspects of medical radiology. Its interests in radiation protection expanded with the widespread use of radiation outside the sphere of medicine, and in 1950 its name was changed to the ICRP in order to more accurately describe its area of interest. In describing its operating philosophy, the ICRP says that "The policy adopted by the Commission in preparing recommendations is to deal with the basic principles of radiation protection and to leave to the various national protection committees the responsibility of introducing the detailed technical regulations, recommendations, or codes of practice best suited to the needs of their individual countries" (ICRP Publication 6, p. 1, Pergamon Press, Oxford, 1964). The ICRP publishes its reports in the appropriate journals. A listing of such reports is given in the bibliography at the end of this chapter.

International Atomic Energy Agency

The International Atomic Energy Agency (IAEA), a specialized agency of the United Nations that was organized in 1956 in order to promote the peaceful uses of nuclear energy, is concerned with the practical application of the ICRP recommendations. "Under its Statute the International Atomic Energy Agency is empowered to provide for the application of standards of safety for protection against radiation to its own operations and to operations making use of assistance provided by it or with which it is otherwise directly associated. To this end authorities receiving such assistance are required to observe relevant health and safety measures prescribed by the Agency". (From *Safe Handling of Radioisotopes*, Safety Series No. 1, IAEA, Vienna, 1962.) The health and safety measures prescribed by the Agency are published according to subject in the Agency's *Safety Series*.

International Labour Organization

The International Labour Organization (ILO), which was founded in 1919 and then became part of the League of Nations, survived the demise of the League to become the first of the specialized agencies of the United Nations; its concern generally is with the social problems of labor. Included in its work is the specification of international labor standards which deal with the health and safety of workers. These specifications are set forth in the *Model Code of Safety Regulations for Industrial Establishments for the Guidance of Governments and Industries*, in the recommendations of expert committees, and in technical manuals. In regard to radiation, the model code has been amended to incorporate those recommendations of the ICRP that are pertinent to control of occupational radiation hazards, and several manuals dealing with protection of workers against radiation hazards have been published.

International Commission on Radiological Units and Measurements

The International Commission on Radiological Units and Measurements (ICRU), which works closely with the ICRP, has had, since its inception in 1925, as its principle objective the development of internationally acceptable recommendations regarding:

(1) Quantities and units of radiation and radioactivity.
(2) Procedures suitable for the measurement and application of these quantities in clinical radiology and radiobiology.
(3) Physical data needed in the application of these procedures, the use of which tends to assure uniformity in reporting.

In its operating policy, "The ICRU feels it is the responsibility of national organizations to introduce their own detailed technical procedures for the development and maintenance of standards. However, it urges that all countries adhere as closely as possible to the internationally recommended basic concepts of radiation quantities and units." (ICRU Report 10a, 1962, N.B.S. Handbook 84, 1962.)

National Committee on Radiation Protection and Measurements

In accordance with the policy laid down by the ICRP, its recommendations are adapted to the needs and conditions in the various countries by national bodies. In the United States, two such bodies exist. The older one is called the National Committee on Radiation Protection and Measurements (NCRP). This committee, which was originally known as the Advisory Committee on X-ray and Radium Protection (founded in 1929), consists of a group of technical experts who are specialists in radiation protection and scientists who are experts in the disciplines which form the basis for radiation protection.

The concern of the NCRP is only with the scientific and technical aspects of radiation protection. To accomplish its objectives, the NCRP is organized into a Main Committee and twenty sub-committees. Each of the sub-committees is responsible for preparing specific recommendations in its field of competence. The recommendations of the sub-committees require approval of the Main Committee before publication. Finally, the approved recommendations are published by the National Bureau of Standards as Handbooks, such as Handbook No. 93, *Safety Standards for Non-medical X-ray and Sealed Gamma-ray Sources.* It should be emphasized that the NCRP is not an official government agency, despite the fact that its recommendations are published by the U.S. government. However, its recommendations are very often adopted by Federal, state, and local government agencies that are concerned with the regulation of radiation hazards.

Federal Radiation Council

The second national body that deals with radiation-protection guides is the Federal Radiation Council (FRC). An official United States government body, it was formed in 1959 to guide the government in the formulation of Federal policy on human radiation exposure. One of its major functions is to "advise the President with respect to radiation matters directly or indirectly affecting health, including guidance for all Federal agencies in the formulation of radiation standards and in the establishment and execution of programs of cooperation with the states" (Public Law 86-373). The FRC relies heavily on the technical recommendations of the NCRP. However—and this is the point of difference between the NCRP and FRC—the Federal Radiation Council weighs non-scientific aspects of radiation exposure, such as political, military, economic, and social considerations as well as the strictly technical and scientific considerations in making risk versus benefit evaluations. Its recommendations to the President are based on what it considers best in the overall interests of the people of the United States.

Philosophy of Radiation Protection

Engineering control of the environment by industrial hygienists and by public health personnel is very often based on the concept of a threshold dose. If the threshold dose of a toxic substance is not exceeded, then it is assumed that the normally operating physiologic mechanisms can cope with the biological insult from that substance. This threshold is usually determined from a combination of experimental animal data and clinical data; then it is reduced by an appropriate factor of safety from which we derive a maximum allowable concentration (MAC) for the toxic substance. The MAC is then used as the criterion of safety in environmental control. The maximum allowable concentration was defined by the International Association on Occupational Health in 1959 as "The term maximum allowable concentration for any substance shall mean that average concentration in air which causes no signs or symptoms of illness or physical impairment in all but hypersensitive workers during their working day on a continuing basis, as judged by the most sensitive internationally accepted tests."

An apparently different approach is used in the control of environmental radiations. For the purpose of setting maximum allowable exposure limits, the most sensitive radiation induced change is assumed to be genetic damage. It is further assumed that there is no threshold for this effect, that the dose-response curve is linear down to zero dose, and that the effect is independent of the dose rate, that is, that only the total dose is of biological significance. Since this means that every increment of dose, no matter how slight, increases the likelihood of an adverse effect, and because the benefits to man of the use of nuclear energy and radiation are manifold, the problem in environmental control is to limit exposure to a level that results in an acceptable risk to the individual and to the population. On the basis of the "acceptable risk con-concept", the maximum permissible dose for an individual was defined in 1959 by the International Commission on Radiological Protection as: "That dose, accumulated over a long period of time or resulting from a single exposure which, in the light of present knowledge, carries a negligible probability of severe somatic or genetic injuries; furthermore, it is such a dose that any effects that ensue more frequently are limited to those of a minor nature that would not be considered unacceptable by the exposed individual and by competent medical authorities."

The distinction between non-radioactive and radioactive agents based on the existence or absence of a threshold is not as clear cut as may first appear. For those substances where a threshold has indeed been established, the threshold is for an individual. Different people have different thresholds. Thus, although the average threshold value for blood changes due to radiation is taken as 25 R, changes have been observed in persons exposed to as little as 14 R, while others whose exposure doses lay between 30 and 40 R showed no blood changes. Obviously, if large numbers of exposed people were to be examined blood changes would be very likely among a very small fraction of those whose dose was even less than 14 R.

It is not unreasonable to expect a distribution of sensitivity somewhat like that shown in Fig. 8.1, in which the sensitivity distribution curve is skewed to the right. The curve on the high-dose end of the distribution should actually intersect the abscissa, since we are reasonably certain that there exists some dose that will affect everyone. On the other hand, it is known that there are "hypersensitive" individuals who respond to extremely low doses which would not affect most people. On this basis, it may be assumed that the distribution curve to the left of the mode passes through the origin of the coordinate axes.

In effect, the distribution of susceptibility among the individuals of a population means that the concept of a threshold dose cannot be applied to a very large population. In setting an MAC for a large population group, therefore, a value judgement must be exercised. Someone must decide on what is an acceptable fraction of the population that may be adversely affected by the harmful substance for which the MAC is being set in return for the benefits to be derived from the use of that substance. The MAC is usually set so conservatively that an extremely large number of people would have to be exposed at that level before the hypersensitive person is found.

This same type of reasoning prevails among those who are concerned with recommending maximum radiation exposures. For occupational exposure, the question of recommending maximum exposures as a guide to radiation

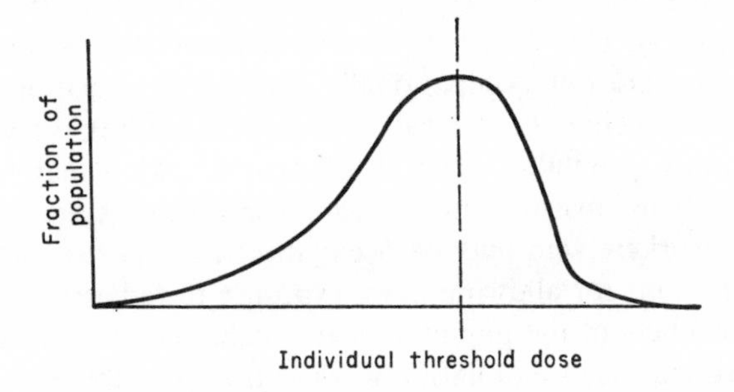

FIG. 8.1. Distribution of individual thresholds among a population.

protection is relatively simple. A vast amount of human experience was gained from the promiscuous exposures to radium and to X-rays during the first quarter of the twentieth century, and a much larger amount of data were gathered from laboratory experiments with animals. On the basis of this information, maximum-exposure limits can be set which, if applied to the relatively small number of people who may be exposed to radiation during the course of their work, would assure for practical purposes that no worker would suffer any observable damage. He may, however, suffer unobservable damage, such as more rapid physiological aging or increased susceptibility to leukemia or other neoplastic disease. It should be emphasized that there is no evidence whatsoever that continuous exposure at the recommended maximum limits does, in fact, lead to these effects. The possibility of the existence of these consequences of low-level exposure is suggested only by extrapolation of experimental results of high-level irradiation. Because of this possibility, and because of the possible genetic effects of radiation, which may not manifest itself until one or more generations later in the descendants of the exposed person, more stringent criteria are applied. Possible genetic damage, which is considered the most conservative criterion of safety, is the principal effect on which recommended maximum exposure is based. Genetic defects are observed in about 4% of all live births in the United States. These defects may be due to a number of chemical mutagens and physical causes, or to radiation. Geneticists estimate that between about one-twentieth and one-eighth of the spontaneously occurring defects are due to natural radiation. On this basis, they believe the "doubling dose", that is, the increased radiation to the entire population that has not yet passed the child-bearing age that would double the present rate of spontaneous mutations, to lie between 30 rad and 80 rad. If every one in this population group received a doubling dose, then the mutation rate would increase slowly over several centuries to a plateau twice as high as the present one. Such an increase in mutation rate is not considered acceptable. Accordingly, the recommended maximum average exposure to the *total gene pool* is 5 rem per generation (a generation is considered 30 years). This dose is to be averaged over the entire population. Since people who are not occupationally exposed to radiation receive, on the average (and including medical radiation exposure), very much less radiation than this acceptable mean, it follows that others may be permitted to receive more than this maximum mean. The higher acceptable exposure is reserved for workers who may be occupationally exposed. In addition to the statistical reasons for allowing more exposure to radiation workers than to the other members of the population, the higher acceptable occupational exposure is justified because radiation workers are specially selected, work in a controlled environment, are continuously monitored, and are under continuing medical surveillance. Although upper exposure limits are recommended, the basic philosophy of radiation protection is to keep exposure as low as practicable.

Basic Radiation Safety Criteria

For purposes of radiation safety, the International Commission on Radiological Protection (ICRP) recognizes two categories of exposure:

1. Occupational exposure—to adults who are exposed to ionizing radiation in the course of their work. Persons in this exposure category may be called radiation workers. This category contains two sub-groups: (a) women of reproductive capacity and (b) all other radiation workers.
2. Exposure of the general public.

For occupational exposure, the basic recommendations for maximum exposure limits in the case of routine, long-term (lifetime), continuous exposure are:

1. Whole body, gonads, lenses of the eye, or hemopoietic system: the total accumulated lifetime dose may not exceed that given by the formula

$$D = 5\,(N - 18)\ \text{rem}, \tag{8.1}$$

where N is the worker's age in years. The only restriction to this formula is that the dose not exceed 3 rem for any consecutive 13-week period for workers who are not women of reproductive capacity. For women of reproductive capacity, occupational exposure is limited to 1.3 rem per consecutive 13-week period. Under this restriction, the dose to an embryo during the first 2 months of pregnancy would normally be less than 1 rem. If the woman is pregnant the total accumulated dose during the remaining 7 months is restricted to 1 rem.

2. Organs other than those listed above: a worker shall not receive, during a whole year and during any consecutive 13-week period, more than the following:

Organ or tissue	$\frac{\text{Rem}}{\text{year}}$	$\frac{\text{Rem}}{\text{13 weeks}}$
Bone	30	8
Thyroid	30	8
Skin of the whole body	30	8
Hands, forearms, feet, and ankles	75	20
Any other single organ	15	4

3. Planned emergency exposure: a total of 12 rem total body dose per year. Following such a planned exposure, the worker's exposure is restricted during the following 5 years, so that his total accumulated dose at the end of the 5-year period conforms to the limit of 5 (N-18) rem.

Example. Assume a 25-year-old worker whose total accumulated dose is $5(25 - 18) = 35$ rem is given a planned emergency dose of 12 rem. His total accumulated dose is then $35 + 12 = 47$ rem. After the next 5 years, when he will be 30 years old, his maximum accumulated dose should not exceed $5(30 - 18) = 60$ rem. Since he now has 47 rem, he may receive during the next 5 years a total dose of $60 - 47 = 13$ rem, or 13 rem/5 years = 2.6 rem per year.

4. Accidental overexposure: up to 25 rem total body dose is compensated for in the same manner as the planned emergency exposure. If the overexposure exceeds 25 rem, then the worker should be placed under competent medical surveillance, and the advice of the physician with respect to restriction of exposure should be followed.

The recommendation that the total accumulated dose due to occupational exposure shall not exceed $5(N-18)$ rem implies an average exposure rate of 5 rem per year for a radiation worker who begins to work at age 18. For radiation control purposes, this 5000 mrem per year is averaged over 50 weeks to yield a mean weekly dose rate of 100 mrem. This value, 100 mrem per week, is frequently, but incorrectly, referred to as the maximum permissible dose rate.

The maximum acceptable dose is the sum of doses due to all types of occupational exposure—it includes external irradiation as well as radiation due to internally deposited radioisotopes. For example, if a radiation worker has deposited in his body a sufficient amount of a radioisotope to irradiate his hemopoietic system at a rate of 25 mrem per week, then his dose due to external radiation is designed to be less than 75 mrem per week.

Internally Deposited Radioisotopes

For protection against external radiation, the recommended maximum exposure limits are biologically meaningful and directly applicable, since it is a simple matter to determine the rad dose due to exposure for any given time in a radiation field. For protection against internal radiation due to inhaled or ingested radioactive materials, two different criteria are used, depending on the type of isotope under consideration. For bone-seeking radioisotopes, such as ^{90}Sr, ^{227}Ac, ^{232}Th, ^{231}Pa, ^{237}Np, ^{239}Pu, etc., which emit alpha or beta radiation, the maximum permissible body burden is based on a comparison with ^{226}Ra and its daughters. For all other radioisotopes the matter is, in principle, as clear cut as in the case of external radiation. The amount of a radioisotope is computed which, if deposited in the body, will deliver a dose not greater than 100 mrem per week if the total body, or gonads, or hemopoietic system is the *critical organ* (the critical organ is defined as that part of the body that is most susceptible to radiation damage under the specific conditions considered), or not greater than 300 mrem per week for any other critical organ. Having determined this quantity of radioisotope, which is called the maximum permissible body burden, then, if biological information regarding the metabolism of the radioisotope is known, it is possible to compute concentrations of the isotope in air and water which would result in the accumulation in the body after continuous exposure of no more than the maximum permissible body burden. Maximum allowable concentrations for all known radioisotopes, to which lifetime exposure is assumed, have been calculated and published in a report by the ICRP. Included in this report are the detailed methods of calculation, together with the biological and physical factors necessary for the calculations. Several of the maximum body burdens and allowable concentrations are given below in Table 8.1. A short term or accidental exposure to environmental concentrations greater than the recommended maxima does not therefore necessarily mean that the person is overexposed.

Calculation of Permissible Concentration in Drinking Water Based on Dose to Critical Organ

The principles involved in the calculation of a value for the maximum permissible concentration in drinking water may be demonstrated with a

TABLE 8.1. MAXIMUM PERMISSIBLE BODY BURDENS AND MAXIMUM PERMISSIBLE CONCENTRATIONS OF RADIONUCLIDES IN AIR AND IN WATER FOR OCCUPATIONAL EXPOSURE

Radionuclide and type of decay		Organ of reference[a] (critical organ in **bold** type)	Maximum permissible burden in total body $q(\mu c)$	Maximum permissible concentrations			
				For 40-hr week		For 168-hr week[b]	
				$(MPC)_w$ $\mu Ci/cm^3$	$(MPC)_a$ $\mu Ci/cm^3$	$(MPC)_w$ $\mu Ci/cm^3$	$(MPC)_a$ $\mu Ci/cm^3$
${}^3_1H_2(H^3O)$ (β^-)	(Sol)	**Body tissue**	10^3	0.1	2×10^{-5}	0.03	5×10^{-6}
		Total body	2×10^3	0.2	2×10^{-5}	0.05	7×10^{-6}
(H^3_2)	(Immersion)	**Skin**			2×10^{-3}		4×10^{-4}
${}^{14}_6C(CO_2)$ (β^-)	(Sol)	**Fat**	300	0.02	4×10^{-6}	8×10^{-3}	10^{-6}
		Total body	400	0.03	5×10^{-6}	0.01	2×10^{-6}
		Bone	400	0.04	6×10^{-6}	0.01	2×10^{-6}
	(Immersion)	**Total body**			5×10^{-5}		10^{-3}
${}^{32}_{15}P$ (β^-)	(Sol)	**Bone**	6	5×10^{-4}	7×10^{-8}	2×10^{-4}	2×10^{-8}
		Total body	30	3×10^{-3}	4×10^{-7}	9×10^{-4}	10^{-7}
		GI (LLI)		3×10^{-3}	6×10^{-7}	9×10^{-4}	2×10^{-7}
		Liver	50	5×10^{-3}	6×10^{-7}	2×10^{-3}	2×10^{-7}
		Brain	300	0.02	3×10^{-8}	8×10^{-3}	10^{-6}
	(Insol)	**Lung**			8×10^{-8}		3×10^{-8}
		GI (LLI)		7×10^{-4}	10^{-7}	2×10^{-4}	4×10^{-8}
${}^{41}_{20}Ca$ (β^-)	(Sol)	**Bone**	30	3×10^{-4}	3×10^{-8}	9×10^{-5}	10^{-3}
		Total body	200	2×10^{-3}	3×10^{-7}	7×10^{-4}	9×10^{-5}
		GI (LLI)		0.01	3×10^{-8}	4×10^{-3}	10^{-6}
	(Insol)	**Lung**			10^{-7}		4×10^{-8}
		GI (LLI)		5×10^{-3}	9×10^{-7}	2×10^{-3}	3×10^{-7}

Notes for this table are given on page 213.

TABLE 8.1 (*cont.*)

Radionuclide and type of decay	Organ of reference[a] (critical organ in **bold** type)	Maximum permissible burden in total body $q(\mu c)$	Maximum permissible concentrations: For 40-hr week		For 168-hr week[b]	
			$(MPC)_w$ μCi/cm^3	$(MPC)_a$ μCi/cm^3	$(MPC)_w$ μCi/cm^3	$(MPC)_a$ μCi/cm^3
$^{51}_{24}Cr$ (ec, γ) (Sol)	GI (LLI)		0.05	10^{-5}	0.02	4×10^{-6}
	Total body	800	0.6	10^{-5}	0.2	4×10^{-5}
	Lung	10^3	1	2×10^{-5}	0.4	8×10^{-6}
	Prostate	2×10^2	2	3×10^{-5}	0.5	10^{-5}
	Thyroid	4×10^2	3	6×10^{-5}	1	2×10^{-5}
	Kidney	8×10^2	6	10^{-4}	2	4×10^{-5}
(Insol)	**Lung**			2×10^{-6}		8×10^{-5}
	GI (LLI)		0.05	8×10^{-6}	0.02	3×10^{-5}
$^{62}_{27}Co$ (β^-, γ) (Sol)	**GI (LLI)**		10^{-3}	3×10^{-7}	5×10^{-4}	10^{-7}
	Total body	10	4×10^{-3}	4×10^{-7}	10^{-3}	10^{-7}
	Pancreas	70	0.02	2×10^{-5}	7×10^{-3}	6×10^{-7}
	Liver	90	0.03	10^{-6}	9×10^{-3}	5×10^{-7}
	Spleen	200	0.05	4×10^{-6}	0.02	2×10^{-6}
	Kidney	200	0.07	6×10^{-6}	0.03	2×10^{-6}
(Insol)	**Lung**			9×10^{-9}		3×10^{-9}
	GI (LLI)		10^{-3}	2×10^{-7}	3×10^{-4}	6×10^{-9}

TABLE 8.1 (*cont.*)

Radionuclide and type of decay		Organ of reference[a] (critical organ in **bold** type)	Maximum permissible burden in total body $q(\mu c)$	Maximum permissible concentrations			
				For 40-hr week		For 168-hr week[b]	
				$(MPC)_w$ $\mu Ci/cm^3$	$(MPC)_a$ $\mu Ci/cm^3$	$(MPC)_w$ $\mu Ci/cm^3$	$(MPC)_a$ $\mu Ci/cm^3$
$^{65}_{30}Zn$ (β^+, ϵ, γ)	Sol)	**Total body**	60	3×10^{-3}	10^{-7}	10^{-3}	4×10^{-8}
		Prostate	70	4×10^{-3}	10^{-7}	10^{-3}	4×10^{-8}
		Liver	80	4×10^{-3}	10^{-7}	10^{-3}	5×10^{-8}
		Kidney	100	6×10^{-3}	2×10^{-7}	2×10^{-3}	7×10^{-8}
		GI (LLI)		6×10^{-3}	10^{-6}	2×10^{-3}	4×10^{-7}
		Pancreas	200	7×10^{-3}	3×10^{-7}	3×10^{-3}	9×10^{-8}
		Muscle	200	0.01	4×10^{-7}	4×10^{-3}	10^{-7}
		Ovary	300	0.01	5×10^{-7}	4×10^{-3}	2×10^{-7}
		Testis	400	0.02	6×10^{-7}	6×10^{-3}	2×10^{-7}
		Bone	700	0.04	10^{-4}	0.01	4×10^{-7}
	(Insol)	**Lung**			6×10^{-8}		2×10^{-8}
		GI (LLI)		5×10^{-3}	9×10^{-7}	2×10^{-3}	3×10^{-7}
$^{78}_{33}As$ (β^-, γ)	(Sol)	**GI (LLI)**		6×10^{-4}	10^{-7}	2×10^{-4}	4×10^{-8}
		Total body	20	0.4	5×10^{-6}	0.1	2×10^{-8}
		Kidney	20	0.6	8×10^{-9}	0.2	3×10^{-4}
		Liver	40	1	10^{-5}	0.4	5×10^{-6}
	(Insol)	**GI (LLI)**		6×10^{-4}	10^{-7}	2×10^{-4}	3×10^{-6}
		Lung			6×10^{-7}		2×10^{-7}
$^{84}_{38}Sr$ (β^-)	(Sol)	**Bone**	4	3×10^{-4}	3×10^{-8}	10^{-4}	10^{-8}
		GI (LLI)		10^{-3}	3×10^{-7}	4×10^{-4}	9×10^{-8}
		Total body	40	2×10^{-3}	2×10^{-7}	7×10^{-4}	6×10^{-8}

TABLE 8.1 (*cont.*)

Radionuclide and type of decay		Organ of reference[a] (critical organ in **bold** type)	Maximum permissible burden in total body $q(\mu c)$	Maximum permissible concentrations			
				For 40-hr week		For 168-hr week[b]	
				$(MPC)_w$ $\mu Ci/cm^3$	$(MPC)_a$ $\mu Ci/cm^3$	$(MPC)_w$ $\mu Ci/cm^3$	$(MPC)_a$ $\mu Ci/cm^3$
	(Insol)	**Lung**			4×10^{-8}		10^{-8}
		GI (LLI)		8×10^{-4}	10^{-7}	3×10^{-4}	5×10^{-8}
${}^{90}_{38}Sr$ (β^-)	(Sol)	**Bone**	2	4×10^{-6}	3×10^{-10}	10^{-6}	10^{-10}
		Total body	20	10^{-5}	9×10^{-10}	4×10^{-6}	3×10^{-10}
		GI (LLI)		10^{-3}	3×10^{-7}	5×10^{-4}	10^{-7}
	(Insol)	**Lung**			5×10^{-9}		2×10^{-9}
		GI (LLI)		10^{-3}	2×10^{-7}	4×10^{-4}	6×10^{-8}
${}^{95}_{40}Zr$ (β^-, γ, e^-).	(Sol)	**GI (LLI)**		2×10^{-3}	4×10^{-7}	6×10^{-4}	10^{-7}
		Total body	20	3	10^{-3}	1	4×10^{-8}
		Bone	30	4	2×10^{-7}	2	6×10^{-8}
		Kidney	30	4	2×10^{-7}	2	6×10^{-8}
		Liver	40	6	3×10^{-7}	2	9×10^{-8}
		Spleen	40	7	3×10^{-7}	2	10^{-7}
	(Insol)	**Lung**			3×10^{-8}		10^{-8}
		GI (LLI)		2×10^{-3}	3×10^{-7}	6×10^{-4}	10^{-7}
${}^{95}_{41}Nb$ (β^-, γ)	(Sol)	**GI (LLI)**		3×10^{-3}	6×10^{-7}	10^{-3}	2×10^{-7}
		Total body	40	10	5×10^{-7}	4	2×10^{-7}
		Liver	60	20	7×10^{-7}	6	3×10^{-7}
		Kidney	60	20	8×10^{-7}	6	3×10^{-7}
		Bone	80	20	9×10^{-7}	7	3×10^{-7}
		Spleen	80	20	10^{-6}	7	3×10^{-7}

TABLE 8.1 (*cont.*)

Radionuclide and type of decay		Organ of reference[a] (critical organ in **bold** type)	Maximum permissible burden in total body $q(\mu c)$	Maximum permissible concentrations			
				For 40-hr week		For 168-hr week[b]	
				$(MPC)_w$ $\mu Ci/cm^3$	$(MPC)_a$ $\mu Ci/cm^3$	$(MPC)_w$ $\mu Ci/cm^3$	$(MPC)_a$ $\mu Ci/cm^3$
	(Insol)	**Lung**			10^{-7}		3×10^{-8}
		GI (LLI)		3×10^{-3}	5×10^{-7}	10^{-3}	2×10^{-7}
${}^{106}_{44}Ru$ (β^-, γ)	(Sol)	**GI (LLI)**		4×10^{-4}	8×10^{-8}	10^{-4}	3×10^{-8}
		Kidney	3	0.01	10^{-7}	4×10^{-3}	5×10^{-8}
		Bone	10	0.04	5×10^{-7}	0.01	2×10^{-7}
		Total body	10	0.06	7×10^{-7}	0.02	3×10^{-7}
	(Insol)	**Lung**			6×10^{-9}		2×10^{-9}
		GI (LLI)		3×10^{-4}	6×10^{-8}	10^{-4}	2×10^{-8}
${}^{131}_{53}I$ (β^-, γ, e^-)	(Sol)	**Thyroid**	0.7	6×10^{-5}	9×10^{-8}	2×10^{-5}	3×10^{-9}
		Total body	50	5×10^{-3}	8×10^{-7}	2×10^{-3}	3×10^{-7}
		GI (LLI)		0.03	7×10^{-6}	0.01	2×10^{-6}
	(Insol)	**GI (LLI)**		2×10^{-3}	3×10^{-7}	6×10^{-4}	10^{-7}
		Lung			3×10^{-7}		10^{-7}
${}^{137}_{55}Cs$ (β, γ, e^-)	(Sol)	**Total body**	30	4×10^{-4}	6×10^{-8}	2×10^{-4}	2×10^{-8}
		Liver	40	5×10^{-4}	8×10^{-8}	2×10^{-4}	3×10^{-8}
		Spleen	50	6×10^{-4}	9×10^{-8}	2×10^{-4}	3×10^{-8}
		Muscle	50	7×10^{-4}	10^{-7}	2×10^{-4}	4×10^{-8}
		Bone	100	10^{-3}	2×10^{-7}	5×10^{-4}	7×10^{-8}
		Kidney	100	10^{-3}	2×10^{-7}	5×10^{-4}	8×10^{-8}
		Lung	300	5×10^{-3}	6×10^{-7}	2×10^{-3}	2×10^{-7}
		GI (SI)		0.02	5×10^{-6}	8×10^{-3}	2×10^{-6}

TABLE 8.1 (*cont.*)

Radionuclide and type of decay		Organ of reference[a] (critical organ in **bold** type)	Maximum permissible burden in total body $q(\mu c)$	Maximum permissible concentrations			
				For 40-hr week		For 168-hr week[b]	
				$(MPC)_w$ $\mu Ci/cm^3$	$(MPC)_a$ $\mu Ci/cm^3$	$(MPC)_w$ $\mu Ci/cm^3$	$(MPC)_a$ $\mu Ci/cm^3$
$^{144}_{58}$Ce (α, β^-, γ)	(Insol)	**Lung**			10^{-8}		5×10^{-9}
		GI (LLI)		10^{-3}	2×10^{-7}	4×10^{-4}	8×10^{-8}
	(Sol)	**GI (LLI)**		3×10^{-4}	8×10^{-8}	10^{-4}	3×10^{-8}
		Bone	5	0.2	10^{-8}	0.08	3×10^{-9}
		Liver	6	0.3	10^{-8}	0.1	4×10^{-9}
		Kidney	10	0.5	2×10^{-8}	0.2	7×10^{-9}
		Total body	20	0.7	3×10^{-8}	0.3	10^{-8}
	(Insol)	**Lung**			6×10^{-9}		2×10^{-9}
		GI (LLI)		3×10^{-4}	6×10^{-8}	10^{-4}	2×10^{-8}
$^{147}_{61}$Pm (α, β^-)	(Sol)	**GI (LLI)**		6×10^{-3}	10^{-6}	2×10^{-3}	5×10^{-7}
		Bone	60	1	6×10^{-8}	0.5	2×10^{-8}
		Kidney	200	4	2×10^{-7}	2	7×10^{-8}
		Total body	300	7	3×10^{-7}	2	10^{-7}
		Liver	300	8	4×10^{-7}	3	10^{-7}
	(Insol)	**Lung**			10^{-7}		3×10^{-8}
		GI (LLI)		6×10^{-3}	10^{-8}	2×10^{-3}	4×10^{-7}
$^{182}_{73}$Ta (β^-, γ)	(Sol)	**GI (LLI)**		10^{-3}	3×10^{-7}	4×10^{-4}	9×10^{-8}
		Liver	7	0.9	4×10^{-3}	0.3	10^{-8}
		Kidney	20	2	8×10^{-8}	0.7	3×10^{-8}
		Total body	20	2	9×10^{-8}	0.7	3×10^{-8}
		Spleen	30	4	10^{-7}	1	5×10^{-8}
		Bone	50	6	3×10^{-7}	2	9×10^{-8}

TABLE 8.1 (*cont.*)

Radionuclide and type of decay		Organ of reference[a] (critical organ in **bold** type)	Maximum permissible burden in total body $q(\mu c)$	Maximum permissible concentrations			
				For 40-hr week		For 168-hr week[b]	
				$(MPC)_w$ $\mu Ci/cm^3$	$(MPC)_a$ $\mu Ci/cm^3$	$(MPC)_w$ $\mu Ci/cm^3$	$(MPC)_a$ $\mu Ci/cm^3$
	(Insol)	**Lung**			2×10^{-8}		7×10^{-9}
		GI (LLI)		10^{-3}	2×10^{-7}	4×10^{-4}	7×10^{-8}
${}^{192}_{77}Ir$ (β^-, γ)	(Sol)	**GI (LLI)**		10^{-3}	3×10^{-7}	4×10^{-4}	9×10^{-8}
		Kidney	6	4×10^{-3}	10^{-7}	10^{-3}	4×10^{-8}
		Spleen	7	4×10^{-3}	10^{-7}	10^{-3}	5×10^{-8}
		Liver	8	5×10^{-3}	2×10^{-7}	2×10^{-3}	6×10^{-8}
		Total body	20	0.01	4×10^{-7}	4×10^{-3}	10^{-7}
	(Insol)	**Lung**			3×10^{-8}		9×10^{-9}
		GI (LLI)		10^{-3}	2×10^{-7}	4×10^{-4}	6×10^{-8}
${}^{198}_{79}Au$ (β^-, γ)	(Sol)	**GI (LLI)**		2×10^{-3}	3×10^{-7}	5×10^{-4}	10^{-7}
		Kidney	20	0.07	3×10^{-6}	0.02	9×10^{-7}
		Total body	30	0.1	4×10^{-6}	0.04	2×10^{-6}
		Spleen	60	0.2	8×10^{-6}	0.07	3×10^{-6}
		Liver	80	0.3	10^{-5}	0.1	4×10^{-8}
	(Insol)	**GI (LLI)**		10^{-3}	2×10^{-7}	5×10^{-4}	8×10^{-8}
		Lung			6×10^{-7}		2×10^{-7}
${}^{222}_{86}Rn$[c] (α, β, γ)		**Lung**			3×10^{-8}		10^{-8}
${}^{226}_{88}Ra$ (α, β^-, γ)	(Sol)	**Bone**	0.1	4×10^{-7}	3×10^{-11}	10^{-7}	10^{-11}
		Total body	0.2	6×10^{-7}	5×10^{-11}	2×10^{-7}	2×10^{-11}
		GI (LLI)		10^{-3}	3×10^{-7}	5×10^{-4}	10^{-7}
	(Insol)	**GI (LLI)**		9×10^{-4}	2×10^{-7}	3×10^{-4}	6×10^{-8}

TABLE 8.1 (*cont.*)

Radionuclide and type of decay		Organ of reference[a] (critical organ in **bold** type)	Maximum permissible burden in total body $q(\mu c)$	Maximum permissible concentrations			
				For 40-hr week		For 168-hr week[b]	
				$(MPC)_w$ $\mu Ci/cm^3$	$(MPC)_a$ $\mu Ci/cm^3$	$(MPC)_w$ $\mu Ci/cm^3$	$(MPC)_a$ $\mu Ci/cm^3$
$^{235}_{92}U$ $(\alpha, \beta^-, \gamma)$	(Sol)	**GI (LLI)**		8×10^{-4}	2×10^{-7}	3×10^{-4}	6×10^{-8}
		Kidney	0.03	0.01	5×10^{-10}	4×10^{-3}	2×10^{-10}
		Bone	0.06	0.01	6×10^{-10}	5×10^{-3}	2×10^{-10}
		Total body	0.4	0.04	2×10^{-9}	0.01	6×10^{-10}
	(Insol)	**Lung**			10^{-10}		4×10^{-11}
		GI (LLI)		8×10^{-4}	10^{-7}	3×10^{-4}	5×10^{-8}
$^{238}_{92}U$ (α, γ, e^-)	(Sol)	**GI (LLI)**		10^{-3}	2×10^{-7}	4×10^{-4}	8×10^{-8}
		Kidney	5×10^{-3}	2×10^{-3}	7×10^{-11}	6×10^{-4}	3×10^{-11}
		Bone	0.06	0.01	6×10^{-10}	5×10^{-3}	2×10^{-10}
		Total body	0.5	0.04	2×10^{-9}	0.01	6×10^{-10}
	(Insol)	**Lung**			10^{-10}		5×10^{-11}
		GI (LLI)		10^{-3}	2×10^{-7}	4×10^{-4}	6×10^{-8}
$^{239}_{94}Pu$ (α, γ)	(Sol)	**Bone**	0.04	10^{-4}	2×10^{-12}	5×10^{-5}	6×10^{-13}
		Liver	0.4	5×10^{-4}	7×10^{-12}	2×10^{-4}	2×10^{-12}
		Kidney	0.5	7×10^{-4}	9×10^{-12}	2×10^{-4}	3×10^{-12}
		GI (LLI)		8×10^{-4}	2×10^{-7}	3×10^{-4}	6×10^{-8}
		Total body	0.4	10^{-3}	10^{-11}	3×10^{-4}	5×10^{-12}
	(Insol)	**Lung**			4×10^{-11}		10^{-11}
		GI (LLI)		8×10^{-4}	2×10^{-7}	3×10^{-4}	5×10^{-8}

TABLE 8.1 (*cont.*)

[a] The abbreviations GI, S, SI, ULI, and LLI refer to gastrointestinal tract, stomach, small intestines, upper large intestine, and lower large intestine, respectively.

[b] It will be noted that the MPC values for the 168-hr week are not always precisely the same multiples of the MPC for the 40-hr week. Part of this is caused by rounding off the calculated values to one digit, but in some instances it is due to technical differences discussed in the ICRP report. Because of the uncertainties present in much of the biological data and because of individual variations, the differences are not considered significant. The MPC values for the 40-hr week are to be considered as basic for occupational exposure, and the value for the 168-hr week are basic for continuous exposure as in the case of the population at large.

[c] The daughter isotopes of ^{220}Rn and ^{222}Rn are assumed present to the extent they occur in unfiltered air. For all other isotopes the daughter elements are not considered as part of the intake and if present must be considered on the basis of the rules for mixtures.

simplified example. In making this calculation, it is assumed that the radioisotope is stored in one or more "compartments" or organs in the body. In many cases, the uptake by the organ in which it is stored may be considered instantaneous with respect to the time that it will be retained by the organ. After uptake, it will be cleared from the organ at a rate determined by its effective half-life. If its passage through subsequent organs or tissues (kidney, liver, etc.) on its way out of the body is rapid compared with the rate of clearance from the storage organ, then this latter rate is the dominant one, and the elimination from the body will appear to follow first-order kinetics. In this simplest case, the elimination from the body is described by

$$q = q_0 e^{-\lambda_e t}, \tag{8.2}$$

where q is the activity remaining at time t after deposition of quantity q_0, λ_e is the effective elimination constant.

In order to calculate the maximum permissible concentration of a radioisotope in drinking water, certain physical and biological data must be available. These include:

A. *Physical*
   1. Type of radiation (α, β, γ).
   2. Energy of the radiation.
   3. Physical half-life.

B. *Biological*
   1. Identification of the critical organ.
   2. Biological clearance rate from critical organ.
   3. Fraction of the ingested isotope that is systemically deposited.
   4. Ratio of amount of the isotope in the critical organ to that in the total body.

For the case of ^{14}C, we have

Average energy = 0.049 MeV beta.
Physical half-life = 5700 years.
Critical organ = fatty tissue, mass = 10,000 g.
Biological half-life = 12 days.
Effective elimination constant = 0.693/12 = 0.058 day^{-1}.
Fraction absorbed into critical organ from GI tract = 0.5.
Fraction of total body carbon in the critical organ = 0.6.
Daily water intake = 2200 ml.

The quantity of a pure beta-emitting radioisotope that is uniformly distributed throughout an organ that will deliver the maximum recommended dose rate to that organ is given by

$$D_{M\beta} = \frac{q\ \mu\text{Ci} \times 3.7\times 10^4\ \text{dps}/\mu\text{Ci} \times 8.64\times 10^4\ \text{sec/day} \times \bar{E}\ \text{MeV/d} \times 1.6\times 10^{-6}\ \text{erg/MeV}}{W\ \text{g} \times 100\ \text{ergs/g/rad}} \tag{8.3}$$

where $D_{M\beta}$ = maximum recommended dose rate to the critical organ. $D_{m\beta} = 0.014$ rad/day (100 mrad/week) if the gonads or hemopoietic tissue is the critical organ, 0.043 rad per day (300 mrad/week) for other cases

q = activity in critical organ

W = weight of critical organ

$\bar{E}$ = average beta-ray energy

For the case of ^{14}C, $D_{M\beta} = 0.043$ rad/day.

We now calculate the quantity of ^{14}C uniformly distributed in the fatty tissue that will irradiate the fatty tissue at a rate of 43 mrad per day:

$$0.043\ \text{rad/day} = \frac{q\ \mu\text{Ci} \times 3.7 \times 10^4\ \text{dps}/\mu\text{Ci} \times 8.64 \times 10^4\ \text{sec/day} \times 4.9 \times 10^{-2}\ \text{MeV/d} \times \times 1.6\times 10^{-6}\ \text{erg/MeV}}{10^4\ \text{g} \times 100\ \text{ergs/g/rad}}$$

$$= 172\ \mu\text{Ci}.$$

Since 60% of the total body carbon is in the fat, the total body burden is

$$Q = \frac{172}{0.6} = 287\ \mu\text{Ci}.$$

To calculate the maximum concentration in drinking water that would load the adipose tissue with a maximum activity of 172 μCi, we equate the activity ingested with the activity eliminated.

$$\text{Activity ingested} = \text{Activity eliminated}, \tag{8.4A}$$

$$C\,\frac{\mu\text{Ci}}{\text{ml}} \times 2.2 \times 10^3\,\frac{\text{ml}}{\text{day}} \times f = \lambda\ \text{day}^{-1} \times q\ \mu\text{Ci}, \tag{8.4B}$$

$$C = \frac{.058\ \text{day}^{-1} \times 172\ \mu\text{Ci}}{2.2 \times 10^3\ \text{ml/day} \times 0.5} = 8 \times 10^{-3}\,\frac{\mu\text{Ci}}{\text{ml}},$$

where $f = 0.5$ = fraction absorbed into critical organ from G.I. tract.

If the radioisotope is continuously ingested in the maximum permissible amounts, then the accumulation of the radioisotope in the critical organ is given by the equation

$$q(t) = q_m (1 - e^{-\lambda_e t}), \tag{8.5}$$

where q_m is the maximum allowable burden in the critical organ, $q(t)$ is the quantity of the radioisotope at any time t after beginning continuous ingestion, and λ_e is the effective half-life of the radioisotope in the critical organ. According to equation (8.5), an infinitely long time is required for the maximum permissible body burden to be attained. However, for practical purposes, the maximum may be assumed to have been reached when 99.5% is attained. This occurs when

$$0.995 = \frac{q(t)}{q_m} = 1 - e^{-\lambda t},$$

$$\lambda t = 5.1,$$

or

$$\frac{0.693\, t}{T_{\frac{1}{2}}} = 5.1,$$

$$t = 7.35\, T_{\frac{1}{2}}.$$

In the case of ^{14}C, whose effective half-life is 12 days, 99.5% of the maximum is attained in

$$7.35 \times T_{\frac{1}{2}} = 7.35 \times 12 = 88.20 \text{ days.}$$

The calculation above for the maximum permissible concentration of a radioisotope in drinking water is based on a compartmental model that is cleared exponentially. In the case of certain bone seekers—Sr, Ra, U, and Pu—clearance data for man are fitted about as well by a power function as by an exponential equation.

$$R(t) = A\, t^{-n}, \tag{8.6}$$

where $R(t)$ is the fraction of the deposited isotope left at time t after deposition, A is the fraction remaining after unit time, and n is a constant. Clearance according to a power function means that the rate of clearance is continuously decreasing. Maximum permissible concentrations calculated with the power function model are greater than those calculated with the exponential model. Accordingly, in the interests of consistency as well as conservatism, recommended maxima are based on the exponential model.

The case of ^{14}C is particularly simple, since the radioisotope is a pure beta emitter. In the case of a beta-gamma-emitting isotope, the dose due to the gamma-rays must be added to that due to the betas.

In arriving at the beta-ray dose, it was assumed that all the energy from the eta-rays was completely absorbed in the critical organ. Because of the high enetrating power of gamma-rays, they may give up only a small fraction of leir energy in passing out of the critical organ. An important factor, there-ore, in determining the absorbed energy from a uniformly distributed gamma mitter is the size and shape of the critical organ. By approximating the shape f organs as spheres and cylinders, and using the appropriate geometry actors given in Tables 6.3 and 6.4, the gamma-ray dose rate may be computed rom equation (6.29).

$$D_\gamma = C\Gamma g. \tag{6.29}$$

C in equation (6.29) may be rewritten as

$$C = \frac{q}{W} \times \rho,$$

which gives

$$D_\gamma = \frac{q\rho}{W}\Gamma g, \tag{8.7}$$

where q is the activity in the critical organ, W is the weight of the critical organ, and ρ is the density of the critical organ. The total beta–gamma dose rate may be computed by combining equations (8.3) and (8.7)

$$D_{M(\beta+\gamma)} = \frac{q\ \mu\text{Ci}}{W\text{g}} \left[\left(\frac{3.7\times 10^4\ \text{dps}/\mu\text{Ci} \times 8.64\times 10^4\ \text{sec/day} \times \bar{E}\ \text{MeV/d} \times 1.6\times 10^{-6}\ \text{erg/MeV}}{100\ \text{ergs/gm/rad}}\right) + \left(\rho\frac{\text{g}}{\text{cm}^3} \times \Gamma\frac{\text{rad cm}^2}{\mu\text{Ci day}} \times \bar{g}\ \text{cm}\right)\right]. \tag{8.8}$$

The use of equation (8.8) may be illustrated by computing the maximum permissible body burden for ^{60}Co. For this isotope, we have:

average beta-ray energy = 0.1 MeV per disintegration,
gamma-ray photons, one each per disintegration; 1.17 and 1.31 MeV,
physical half-life = 1.9×10^3 days,
biological half-life = 9.5 days,
effective half-life = 9.5 days,
effective elimination constant = 0.073 day^{-1},
$\Gamma = 0.29$ rad cm^2/μCi day,
$g = 129$ for a 70-kg man who is 160 cm tall,
ρ = average density of man = 1 g/cm^3,
critical organ = total body, $W = 7 \times 10^4$ g,
fraction of ingested cobalt that is absorbed = 0.3,
$D_{M(\beta+\gamma)} = 0.014$ rad/day (100 mrad/week).

Substituting the appropriate values into equation (8.8) and solving for
we find

$$q = 23\ \mu\text{Ci}.$$

Using this value in equation (8.4B), we have

$$C\,\frac{\mu\text{Ci}}{\text{ml}} = 2.2 \times 10^{3}\ \text{ml/day} \times 0.3 = 7.3 \times 10^{-2}\ \text{day}^{-1} \times 23\ \mu\text{Ci},$$

$$C = 2.3 \times 10^{-3}\,\frac{\mu\text{Ci}}{\text{ml}}.$$

The equations above are based on absorption and storage of the radioisc
tope in a critical organ. Implicit in the equations is the fact that the isotope i
in some form that will permit its passage from the gastrointestinal tract int
the critical organ. If this is not true, or if the uptake of the radioisotope by th
body is very small, such as the case of the trace elements, then the non-systemi
cally deposited activity will irradiate the gastrointestinal tract or the urinar
tract as it passes through the body. In the case of ^{60}Co, for example, if th
simplifying assumption is made that the amount of cobalt in the gastrointes
tinal tract remains constant throughout the day, then the activity in th
gastrointestinal tract that will deliver a radiation dose of 43 mrad/day ma
be calculated with the aid of equation (8.8). The gastrointestinal tract includ
ing the stomach may be approximated as a sphere of tissue whose weight i
3635 g (2000 g organ weight plus 1635 g contents). Assuming an averag
specific gravity of 1, the radius of this sphere is 9.54 cm. If this value is used t
compute g from equation (6.35), we find, after substituting the appropriat
values into equation (8.8), that q is 5 μCi. Since this total activity is ingeste
in 2200 ml water, the concentration in the water is 2.3×10^{-3} μCi/ml.

The values calculated above for the maximum allowable concentrations i
drinking are very close to those recommended by the ICRP. However, i
should be pointed out that they are not exactly the same. The difference i
due to the fact that, although the general principles on which the abov
calculations are based are the same as those of the ICRP, the assumption
here were simplified for purposes of clarity in the explanation of these basi
principles. For example, in computing the maximum allowable concentratio
in drinking water when the gastrointestinal tract is the critical organ, th
ICRP considers the separate portions of the tract separately. This is necessar
because of the different rates with which the contents move through th
various parts of the gastrointestinal tract. For more detailed formulae an
other pertinent biological and physical data for computation of maximun
allowable concentrations, the reader is referred to the report of ICR
Committee II on Permissible Dose for Internal Radiation, 1959†.

† *Health Physics*, vol. 3, June 1960.

Concentration in Drinking Water Based on Comparison with Radium

We have a great deal of experience with human exposure to radium. For this reason, and because radium is a "bone seeker", that is, it deposits in the bone, the maximum permissible body burdens of all bone seekers are established by comparison of the rem dose of the other isotope to that delivered to the bone by radium. On the basis of data on humans, 0.1 μg radium in equilibrium with its decay products has been recommended as the maximum permissible body burden of ^{226}Ra. The dose equivalent to the bone from 0.1 μg radium and its daughters is 0.56 rem/week. In computing this DE, a QF of 10 is assigned to the ^{226}Ra alpha particles. However, it has been found that the DE for some other bone seekers is less than that of radium for a given degree of damage. This is attributed to the greater degree of non-uniformity of deposition of the other bone seekers. For this reason, another factor, called the relative damage factor (DF), is introduced as a multiplier of the QF. This factor has a value of 5 for all corpuscular (alpha or beta) radiation, except for those cases where the corpuscular radiations are due to a decay chain whose first member is radium. When radium is the first member of the chain, then the relative damage factor is 1, since the distribution of the radionuclides will be determined by the ^{226}Ra. For example, the relative damage factor for

$$^{228}\mathrm{Th} \xrightarrow{\beta} {}^{224}\mathrm{Ra} \xrightarrow{\alpha}$$

is 5 for each alpha, while the same alphas are weighted with a relative damage factor of 1 in the chain

$$^{228}\mathrm{Ra} \xrightarrow{\beta} {}^{228}\mathrm{Ac} \xrightarrow{\beta} {}^{228}\mathrm{Th} \xrightarrow{\alpha} {}^{224}\mathrm{Ra} \xrightarrow{\alpha}$$

The energy dissipated in the bone by ^{226}Ra and the daughters that remain in the bone is 11 MeV per disintegration. Applying the QF value of 10 brings the effective energy to 110 MeV per disintegration. Since 99% of the ^{226}Ra body burden is in the skeleton, we have for the maximum permissible body burden of any other bone seeker

$$q = \frac{0.1\ \mu\mathrm{Ci} \times 0.99}{f_2} \times \frac{110\ \mathrm{MeV/d}}{E\ \mathrm{MeV/d}} = \frac{11}{f_2 E}, \tag{8.9}$$

where E is the effective corpuscular energy per disintegration of any other bone seeker, and f_2 is the fraction of the total body burden of bone seeker that is in the skeleton. In the case of ^{90}Sr–^{90}Y, for example,

Average energy = 0.194 (^{90}Sr) + 0.93 (^{90}Y) = 1.12 MeV/dis.
QF = 1,
Relative damage factor = 5,
Effective energy = 5(1.12) = 5.6 MeV/dis.

Substituting these values into equation (8.9), we find that $q = 2\ \mu$Ci. For ^{90}Sr, the effective half-life is 6.4×10^3 days, and $\lambda_e = 1.08 \times 10^{-4}$ day^{-1}. Since 9% of the ingested dose is deposited in the bone, the maximum allowable concentration to maintain a body burden of 2 μCi is

$$C \frac{\mu\text{Ci}}{\text{ml}} \times 2.2 \times 10^3 \frac{\text{ml}}{\text{day}} \times 9 \times 10^{-2} = 2\ \mu\text{Ci} \times 1.08 \times 10^{-4}\ \text{day}^{-1},$$

$$C = 1.09 \times 10^{-6} \frac{\mu\text{Ci}}{\text{ml}}.$$

Ingestion of water at the rate assumed in the calculation above will result in the maximum permissible body burden *when equilibrium is attained*. Because of the very long effective half-life of ^{90}Sr in the bone, the maximum allowable body burden is not attained during the 50-year occupational exposure time assumed for the purpose of computing values for the radiation protection guide. After 50 years of continuous ingestion at the above rate, the amount of ^{90}Sr in the skeleton will be

$$\begin{aligned} q &= q_m (1 - e^{-\lambda_e t}) \\ &= 2 (1 - e^{-1.08 \times 10^{-4} \times 50 \times 365}) \\ &= 1.72\ \mu\text{Ci}, \end{aligned}$$

or only 86% of the maximum allowable body burden. It is thus clear that the average body burden, and consequently the average dose rate to the skeleton during a 50-year period of maximum permissible ingestion, will be considerably less than the maximum permissible body burden. The mean body burden during a period of ingestion T starting at time zero when there is no radioisotope of the species in question in the body, and assuming the effective elimination rate for the radioisotope to be λ_e, is given by

$$\bar{q} = \frac{1}{T} \int_0^T q_m (1 - e^{-\lambda_e t})\, dt. \tag{8.10}$$

Integrating equation (8.9), we find that

$$\bar{q} = q_m \left[1 + \frac{1}{\lambda_e T} (e^{-\lambda_e t} - 1)\right]. \tag{8.11}$$

For ^{90}Sr, whose $\lambda_e = 0.0395$ year^{-1}, we have for a 50-year exposure period

$$\bar{q} = 1.130\ \mu\text{Ci}.$$

TABLE 8.2. RADIOISOTOPES THAT DO NOT REACH EQUILIBRIUM WITHIN 50 YEARS

Z	Radioisotope	T_e, years	Per cent equilibrium reached in 50 years
38	^{90}Sr	18	86
88	^{226}Ra	44	56
89	^{227}Ac	20	83
90	^{230}Th	200	16
90	^{232}Th	200	16
91	^{231}Pa	200	16
93	^{237}Np	200	16
94	^{238}Pu	62	43
94	^{239}Pu	200	16
94	^{240}Pu	190	16
94	^{241}Pu	12	94
94	^{242}Pu	200	16
95	^{241}Am	140	22
95	^{243}Am	200	16
96	^{243}Cm	30	69
96	^{244}Cm	17	87
96	^{245}Cm	200	16
96	^{246}Cm	190	16
98	^{249}Cf	140	22
98	^{250}Cf	10	97

Several other radioisotopes do not attain their equilibrium values in the ody during 50 years of continuous ingestion at the maximum recommended oncentrations. These radioisotopes are listed in Table 8.2.

irborne Radioactivity

In considering airborne radioactivity, the lung is considered from two oints of view: first, as a portal of entry for inhaled substances that are systemi-ally deposited, and second, as a critical organ that may suffer radiation amage. For purposes of computing permissible atmospheric concentrations f radioactivity, airborne contaminants may be broadly classified as gaseous nd particulate. For gaseous radioactive contaminants, the possible hazards be considered are immersion and inhalation; in the case of particulate atter, the main hazard is due to deposition of the radioactive particulates the lung.

The calculation for the permissible concentration of a gas may be illustrated r ^{41}A, a biologically inert gas. Argon-41 decays to ^{41}K by the emission of a .2-MeV beta particle and a 1.3-MeV gamma-ray. The half-life for ^{41}A is 00 min, or 0.076 days. For the case of immersion, it is assumed that a man exposed in an infinite hemisphere of the gas. In an infinite medium contain-g a uniformly distributed isotope, the density of emitted energy is equal to the density of absorbed energy. Using a value of 1.1 for the stopping power o tissue relative to air, and an effective energy of $1.3 + \frac{1}{3}(1.2) = 1.7$ MeV, the concentration of radioactivity that results in an absorbed dose rate of 0.0143 rad/day is computed from

$$0.0143 \frac{\text{rad}}{\text{day}} = \frac{q\mu\text{Ci/cm}^3 \times 3.7 \times 10^4 \text{ dps}/\mu\text{Ci} \times 1.7 \text{ MeV/d} \times 1.6 \times 10^{-6} \text{ erg/MeV} \times 8.64 \times 10^4 \text{ sec/day} \times 1.1}{1.293 \times 10^{-3} \text{ g/cm}^3 \times 100 \text{ ergs/g/rad}}$$

$$q = 1.91 \times 10^{-7} \ \mu\text{Ci/cm}^3.$$

The above calculation was made for a complete sphere of radioactive gas Since man is covered by a hemisphere of radioactive gas, the permissible concentration is twice that calculated above, or 3.8×10^{-7} μCi/cm³.

When a gas is inhaled, it may dissolve in the body fluids and fat afte diffusing across the capillary bed in the lung. In the case of an inert gas absorption into the body stops after the body fluids and fat are saturated with the dissolved gas. The saturation quantity of dissolved ^{41}A in the body fluids due to inhalation of air contaminated at the maximum level for immersion must now be calculated to ascertain that it does not lead to overexposure. As a first step in this calculation, the molar concentration of ^{41}A that corresponds to 3.8×10^{-7} μCi/cm³ is determined. The specific activity of ^{41}A is

$$\frac{1.6 \times 10^3 \text{ yr} \times 226 \times 365 \text{ day/year}}{0.076 \text{ day}^{-1} \times 41} = 4.24 \times 10^7 \text{ Ci/g},$$

$$\therefore \frac{3.82 \times 10^{-13} \text{ Ci/cm}^3}{4.24 \times 10^7 \text{ Ci/g}} \times \frac{1 \text{ mole}}{41 \text{ g}} = 2.2 \times 10^{-22} \frac{\text{mole}}{\text{cm}^3}.$$

The molar concentration of air is

$$\frac{1 \text{ mole}}{2.24 \times 10^4 \text{ cm}^3} = 4.47 \times 10^{-5} \frac{\text{mole}}{\text{cm}^3}.$$

Since argon constitutes 0.94% of the air, the molar concentration of naturally occurring argon in air is

$$9.4 \times 10^{-3} \times 4.47 \times 10^{-5} \frac{\text{mole}}{\text{cm}^3} = 4.2 \times 10^{-7} \frac{\text{mole A}}{\text{cm}^3 \text{ air}}.$$

The ^{41}A corresponding to the maximum permissible concentration is thus seen to be insignificant relative to the argon already in the air, and the molar concentration of argon in the air may be assumed to be unchanged as a result f adding 3.8×10^{-2} μCi ^{41}A per cm^3 air. With this ^{41}A in the air, the specific ctivity of the argon in the air is

$$\frac{3.8 \times 10^{-13}\ \text{Ci/cm}^3}{4.2 \times 10^{-7}\ \text{moles/cm}^3} = 4.55 \times 10^{-7}\ \frac{\text{Ci}}{\text{mole}}.$$

Now calculate the concentration of argon in the body fluids when the fluids re in equilibrium with the argon in the air. According to Henry's law, the mount of a gas dissolved in a liquid is proportional to the partial pressure of he gas above the liquid

$$p_{\text{gas}} = KN = K\frac{n_g}{n_g + n_s}, \tag{8.12}$$

here p_{gas} is the partial pressure of the gas in mm Hg, K is Henry's constant, nd N is the mole fraction of the dissolved gas; n_g is the molar concentration f the dissolved gas and n_s is molar concentration of the solvent, in moles olvent per liter solution. The solubilities of several gases in water at 38°C, xpressed in terms of Henry's constant, are given in Table 8.3. At body emperature, K for argon is 3.41×10^7, and the partial pressure of argon in he atmosphere is

$$p_{\text{gas}} = 0.0094 \times 760 = 7.15\ \text{mm Hg}.$$

he total body water in a 70 kg man is 43 liters. Assuming the body fluid to e all water, the molar concentration of solvent is

$$n_s = \frac{1000\ \text{g/liter}}{18\ \text{g/mole}} = 55.6\ \frac{\text{moles}}{\text{liter}}.$$

quation (8.12) may now be solved for the concentration of dissolved argon

$$7.15 = 3.41 \times 10^7 \left(\frac{n_g}{n_g + 55.6}\right),$$

$$n_g = 1.17 \times 10^{-5}\ \frac{\text{moles}}{\text{liter}}.$$

Since the specific activity of the argon is 4.55×10^{-7} Ci/mole, the activity oncentration in the body fluid is 5.3×10^{-6} μCi/liter and the total activity n the body is 2.28×10^{-4} μCi. Assuming the body fluids to be uniformly dis- ibuted throughout the body, the dose rate due to the dissolved ^{41}A is found, om equation (8.8), to be 1.04×10^{-4} mrad/day. Argon is more soluble in at than in water. At equilibrium, the partition coefficient, which is the con- entration ratio of A in fat to A in water, is 5.4 : 1. The amount of ^{41}A in the 0 kg of fat in a standard man therefore is $5.4 \times 5.3 \times 10^{-6}$ μCi/kg $\times$ 10 kg = 2.86×10^{-4} μCi, and the resulting dose rate to the body fat is about .8 $\times 10^{-4}$ mrad/day.

According to these calculations, the dose rate due to dissolved argon withi the body is negligible in comparison to the immersion dose.

Inhaled particulate matter may be either soluble or insoluble. If solubl it may be absorbed into the body fluids, or it may form an insoluble precipi tate in the lung, or it may interact with the tissue protein. "Insoluble" particu lates that are deposited in the deep respiratory tract may remain there fo

TABLE 8.3. SOLUBILITY OF CERTAIN GASES AT 38°C

$$K = \frac{\text{partial pressure of gas in millimeters Hg}}{\text{mole fraction of gas in solution}}$$

Gas	K
H_2	5.72×10^7
He	11.0
N_2	7.51
O_2	4.04
A	3.41
Ne	9.76
Kr	2.13
Xe	1.12
Em	0.651
CO_2	0.168
C_2H_2	0.131
C_2H_4	1.21
N_2O	0.242

long periods of time, until they are cleared either by phagocytosis, by respira tory excursions that propel them to ciliated surfaces in the bronchioles, or by slow solution. Unless specific biological data are available for specific particu lates, recommended protection guides for airborne particulates are based o the following assumptions:

Soluble particles

1. 25% exhaled
2. 50% deposited in the upper respiratory tract and swallowed within 24 hr
3. 25% dissolved and absorbed into the body fluids

Insoluble particles

1. 25% exhaled
2. 50% deposited in the upper respiratory tract and swallowed within 24 hr
3. 12½% deposited in the deep respiratory tract, but cleared into the throat and swallowed within 24 hr
4. 12½% deposited in the deep respiratory tract, and retained with a biological retention half-time of 120 days

The application of these assumptions may be illustrated by calculating th maximum permissible atmospheric concentration of insoluble particles of a dust containing ^{35}S. Sulfur-35 is a pure beta emitter whose mean energy is 0.0492 MeV, and whose half-life is 87.2 days. According to the assumptions above, 25% of the inhaled dust is immediately exhaled, 62.5% is deposited in the lung, and cleared out within 24 hr of deposition, and 12.5% is retained in the deep respiratory tract with an effective half retention time of

$$T_e = \frac{T_B \times T_p}{T_B + T_p}$$

$$= \frac{120 \times 87.2}{120 + 87.2} = 50.5 \text{ days},$$

corresponding to a clearance rate, λ_e, of 0.0138 per day. The quantity of activity, Q, in the lung (weight 1000 g) that will deliver an absorbed dose of 0.043 rad/day is calculated from the equation

$$0.043 \frac{\text{rad}}{\text{day}} =$$

$$\frac{Q\,\mu\text{Ci} \times 3.7 \times 10^4 \text{ dps}/\mu\text{Ci} \times 8.64 \times 10^4 \text{ sec/day} \times 4.9 \times 10^{-2} \text{ MeV/d} \times 1.6 \times 10^{-6} \text{ erg/MeV}}{10^3 \text{ g} \times 10^2 \text{ erg/g/rad}},$$

$$Q = 17.1\ \mu\text{Ci}.$$

When the lung burden is in equilibrium with the environment of concentration C μCi/cm^3,

amount inhaled × fraction deposited = amount eliminated,

$$2 \times 10^7\, C \frac{\mu\text{Ci}}{\text{day}} \times 0.75 = \lambda_e \text{ day}^{-1} \times q\,\mu\text{Ci} + 0.625 \times 2 \times 10^7\, C \frac{\mu\text{Ci}}{\text{day}}, \quad (8.13)$$

where q is the burden of activity in the deep respiratory tract. However,

$$q + 0.625 \times 2 \times 10^7\, C = Q. \quad (8.14)$$

Substituting the appropriate values for λ_e ($\lambda_e = 0.0138$ day^{-1}) and for Q ($Q = 17.1$ μCi) into equations (8.13) and (8.14), we find that $C = 8.8 \times 10^{-8}$ μCi/cm^3 is the concentration to be used as a radiation protection guide for inhaled insoluble ^{35}S particulates.

The radiation-absorbed dose to the gastrointestinal tract (weight, including contents, 3635 g) due to the fraction of the inhaled activity assumed to be brought up from the lung and swallowed is found, by equation (6.12), to be 0.76 mrad/day.

Suggested References

International Commission on Radiological Protection (ICRP)

1 X-ray and Radium Protection. Recommendations of the 2nd International Congress of Radiology, 1928. Circular No. 374 of the Bureau of Standards, U.S. Government Printing Office (January 23, 1929). *Br. J. Radiology*, **1**, 359 (1928).

2 Recommendations of the International X-ray and Radium Protection Commission. Alterations to the 1928 Recommendations of the 2nd International Congress of Radiology. 3rd International Congress of Radiology, 1931. *Br. J. Radiology*, **4**, 485 (1931).

3 International Recommendations for X-ray and Radium Protection. Revised by the International X-ray and Radium Protection Commission and adopted by the 3rd International Congress of Radiology, Paris, July 1931. *Br. J. Radiology*, **5**, 82 (1932).

4 International Recommendations for X-ray and Radium Protection. Revised by the International X-ray and Radium Protection Commission and adopted by the 4th International Congress of Radiology, Zurich, July 1934. *Radiology*, **23**, 682–5 (1934). *Br. J. Radiology*, **7**, 695 (1934).

5 *International Recommendations for X-ray and Radium Protection.* Revised by the International X-ray and Radium Protection Commission and adopted by the 5th International Congress of Radiology, Chicago, September 1937. British Institute of Radiology (1938).

6 International Recommendations on Radiological Protection. Revised by the International Commission on Radiological Protection at the 6th International Congress of Radiology, London, 1950. *Radiology*, **56**, 431–9 (March 1951). *Br. J. Radiology*, **24**, 46–53 (1951).

7 Recommendations of the International Commission on Radiological Protection (Revised December 1, 1954). *Br. J. Radiology*, Supplement 6 (1955).

8 Report on Amendments during 1956 to the Recommendations of the International Commission on Radiological Protection (ICRP). *Radiation Research*, **8**, 539–42 (June 1958). *Acta Radiol.* **48**, 493–5 (December 1957). *Radiology*, **70**, 261–2 (February 1958). *Fortschritte a.d. Gebiete d Rontgenstrahlen u.d. Nuklearmedizin*, **88**, 500–2 (1958).

9 *Recommendations of the International Commission on Radiological Protection* (adopted September 9, 1958). ICRP Publication 1, Pergamon Press (1959).

10 Report on Decisions at the 1959 Meeting of the International Commission on Radiological Protection (ICRP). *Radiology*, **74**, 116–19 (1960). *Am. J. Roentg.* **83**, 372–5 (1960). *Strahlentherapie*, Band 112, Heft (3 1960). *Acta Radiol.* **53**, fasc. 2 (February 1960). *Br. J. Radiology*, **33**, 189–92 (1960).

11 *Recommendations of the International Commission on Radiological Protection* (as amended 1959 and revised 1962). ICRP Publication 6, Pergamon Press (1964).

12 *Recommendations of the International Commission on Radiological Protection, Report of Committee II on Permissible Dose for Internal Radiation* (1959). ICRP Publication 2, Pergamon Press (1960). *Health Physics*, **3** (June 1960).

13 *Recommendations of the International Commission on Radiological Protection, Report of Committee III on Protection against X-rays up to Energies of* 3 *MeV and Beta- and Gamma-rays from Sealed Sources* (1960). ICRP Publication 3, Pergamon Press (1960).

14 *Recommendations of the International Commission on Radiological Protection, Report of Committee IV* (1953–1959) *on Protection against Electromagnetic Radiation above* 3 *MeV and Electrons, Neutrons and Protons* (adopted 1962, with revisions adopted 1963). ICRP Publication 4, Pergamon Press (1964).

15 *Recommendations of the International Commission on Radiological Protection, Report of Committee V on the Handling and Disposal of Radioactive Materials in Hospitals and Medical Research Establishments.* ICRP Publication 5, Pergamon Press (1965).
16 Report of the RBE Committee to the International Commissions on Radiological Protection and on Radiological Units and Measurements. *Health Physics*, **9**, no. 4, 357–84 (1963).
17 Exposure of Man to Ionizing Radiation Arising from Medical Procedures with Special Reference to Radiation-induced Diseases: An enquiry into methods of evaluation. A report of the International Commissions on Radiological Protection and on Radiological Units and Measurements. *Physics in Medicine and Biology*, **6**, no. 2 (1961).
18 Exposure of Man to Ionizing Radiation arising from Medical Procedures: An enquiry into methods of evaluation. A report of the International Commissions on Radiological Protection and on Radiological Units and Measurements. *Physics in Medicine and Biology*, **2**, no. 2 (1957).
19 Radiobiological Aspects of the Supersonic Transport: A report prepared by a Task Group of Committee 1. *Health Physics*, **12**, 209–26 (1966).
20 *Principles of Environmental Monitoring Related to the Handling of Radioactive Materials: A report prepared by a Task Group of Committee* 4. ICRP Publication 7, Pergamon Press (1966).
21 *The Evaluation of Risks from Radiation: A report prepared by a Task Group of Committee I.* ICRP Publication 8, Pergamon Press (1966). *Health Physics*, **12**, 239–302 (1966).
22 *Recommendations of the International Commission on Radiological Protection.* ICRP Publication 9. Pergamon Press (1966).
23 Deposition and Retention Models for Internal Dosimetry of the Human Respiratory Tract: A report prepared by a Task Group of Committee 2. *Health Physics*, **12**, 173–207 (1966).
24 A Review of the Physiology of the Gastro-Intestinal Tract in Relation to Radiation Doses from Radioactive Materials: A report prepared by a consultant to Committee 2. *Health Physics*, **12**, 131–61 (1966).
25 Calculation of Radiation Dose from Protons and Neutrons to 400 MeV: A report prepared by a Task Group of Committee 3. *Health Physics*, **12**, 227–37 (1966).
26 The Standard Man, as applied to Internal Dose Calculations: A report prepared by a Task Group of Committee 2. In preparation.
27 Evaluation of Radiation Doses to Body Tissues from Internal Contamination Due to Occupational Exposure: A report prepared by a Task Group of Committee 4. In preparation.

International Atomic Energy Agency (IAEA)

Safety Series No.	Title
1	*Safe Handling of Radioisotopes* (1962).
2	*Safe Handling of Radioisotopes—Health Physics Addendum* (1960).
3	*Safe Handling of Radioisotopes—Medical Addendum* (1960).
4	*Safe Operation of Critical Assemblies and Research Reactors* (1961).
5	*Radioactive Waste Disposal into the Sea* (1961).
6	*Regulations for the Safe Transport of Radioactive Materials* (1967).
7	*Regulations for the Safe Transport of Radioactive Materials: Notes on Certain Aspects of the Regulations* (1961).
8	*The Use of Film Badges for Personnel Monitoring* (1962).
9	*Basic Safety Standards for Radiation Protection* (1967).
10	*Disposal of Radioactive Wastes into Fresh Water* (1963).
11	*Methods of Surveying and Monitoring Marine Radioactivity.*
12	*The Management of Radioactive Wastes Produced by Radioisotope Users* (1965).
13	*The Provision of Radiological Protection Services* (1965).
14	*The Basic Requirements for Personnel Monitoring* (1965).
15	*Radioactive Waste Disposal into the Ground* (1965).

16 *Manual on Environmental Monitoring in Normal Operation* (1966).
17 *Techniques for Controlling Air Pollution from the Operation of Nuclear Facilities* (1966).
18 *Environmental Monitoring in Emergency Situations* (1966).
19 *The Management of Radioactive Wastes Produced by Radioisotope Users: Technical Addendum* (1966).
20 *Guide to the Safe Handling of Radioisotopes in Hydrology* (1966).
21 *Risk Evaluation for Protection of the Public* (1967).
22 *Respirators and Protective Clothing* (1967).
23 *Radiation Protection Standards for Radioluminous Timepieces* (1967).

International Labour Organization (ILO)

1 *Manual of Industrial Radiation Protection.* Part I: Convention and recommendation concerning the protection of workers against ionizing radiations. 1963.
2 *Manual of Industrial Radiation Protection.* Part II: Model code of safety regulations (ionising radiations). 1959.
3 *Manual of Industrial Radiation Protection.* Part III: General guide on protection against ionising radiations. 1963.
4 *Manual of Industrial Radiation Protection.* Part IV: Guide on radiation protection in industrial radiography and fluoroscopy.
5 *Manual of Industrial Radiation Protection.* Part V: Guide on radiation protection in the use of luminous compounds.

Recommendations of International Commission on Radiological Units and Measurements (ICRU)

ICRU
Report
No.

1 Discussion on International Units and Standards for X-ray Work. *Brit. J. Radiol.* **23,** 64 (1927).
2 International X-ray Unit of Intensity. *Brit. J. Radiol.* (new series) **1,** 363 (1928).
3 Report of Committee on Standardization of X-ray Measurements. *Radiology,* **22,** 289 (1934).
4 Recommendations of the International Committee for Radiological Units. *Radiology,* **23,** 580 (1934).
5 Recommendations of the International Committee for Radiological Units. *Radiology,* **29,** 634 (1937)
6 *Report of International Commission on Radiological Protection and International Commission on Radiological Units.* National Bureau of Standards Handbook 47, Washington, D.C. (1951).
7 Recommendations of the International Commission for Radiological Units. *Radiology,* **62,** 106 (1954).
8 *Report of International Commission on Radiological Units and Measurements (ICRU)* 1956. National Bureau of Standards Handbook 62, Washington, D.C. (1957).
9 *Report of International Commission on Radiological Units and Measurements (ICRU)* 1959. National Bureau of Standards Handbook 78, Washington, D.C. (1961).
10a *Radiation Quantities and Units.* National Bureau of Standards Handbook 84, Washington, D.C. (1962).
10b *Physical Aspects of Irradiation.* National Bureau of Standards Handbook 85, Washington, D.C. (1964).
10c *Radioactivity.* National Bureau of Standards Handbook 86, Washington, D.C. (1963).
10d *Clinical Dosimetry.* National Bureau of Standards Handbook 87, Washington, D.C. (1963).
10e *Radiobiological Dosimetry.* National Bureau of Standards Handbook 88, Washington, D.C. (1963).
10f *Methods of Evaluating Radiobiological Equipment and Materials.* National Bureau of Standards Handbook 89, Washington, D.C. (1963).

CHAPTER 9

HEALTH PHYSICS INSTRUMENTATION

Radiation Detectors

Man possesses no biological sensors of ionizing radiation. As a consequence he must depend entirely on instrumentation for the detection and measurement of radiation. Instruments used in the practice of health physics serve a wide variety of purposes. It is logical, therefore, to find a wide variety of instrument types. We have, for example, instruments such as the Geiger counter that measure particles; film badges and pocket dosimeters that measure accumulated radiation doses; and ionization chamber type instruments that measure dose rate. In each of these categories, one finds instruments that are designed principally for measurement of a certain type of radiation—such as low-energy X-rays, gamma-rays, fast neutrons, etc.

Although there are many different instrument types, the operating principles for most radiation-measuring instruments are relatively few. The basic requirement of any radiation-measuring instrument is that the instrument's detector interact with the radiation in such a manner that the magnitude of the instrument's response is proportional to the radiation effect or radiation property

TABLE 9.1. RADIATION EFFECTS USED IN THE DETECTION AND MEASUREMENT OF RADIATION

Effect	Type of Instrument	Detector
Electrical	1. Ionization chamber 2. Proportional counter 3. Geiger counter 4. Solid state	1. Gas 2. Gas 3. Gas 4. Semiconductor
Chemical	1. Film 2. Chemical dosimeter	1. Photographic emulsion 2. Solid or liquid
Light	1. Scintillation counter 2. Cerenkov counter	1. Crystal or liquid 2. Crystal or liquid
Thermo-luminescence	Thermoluminescent dosimeter	Crystal
Heat	Calorimeter	Solid or liquid

under measurement. Some of the physical and chemical radiation effects that are applied to radiation detection and measurement for health physics purposes are listed in Table 9.1.

Particle-counting Instruments

Particle-counting instruments are frequently used by health physicists to determine the radioactivity of a sample taken from the environment, such as an air sample, or to measure the activity of a biological fluid from someone suspected of being internally contaminated. Another important application of particle-counting instruments is in portable radiation-survey instruments. Particle-counting instruments may be very sensitive—they literally respond to a single ionizing particle. They are, accordingly, widely used for searching for unknown radiation sources, for leaks in shielding, and for areas of contamination. The detector in particle-counting instruments may be either a gas or a solid. In either case, passage of an ionizing particle through the detector results in energy dissipation through a burst of ionization. This burst of ionization is converted into an electrical pulse that actuates a readout device, such as a scaler or a ratemeter, to register a count.

Gas-filled Particle Counters

Consider a gas detector system such as is shown in Fig. 9.1. This system consists of a variable voltage source V, a high valued resistor R, and a gas-filled counting chamber, D, which has two coaxial electrodes that are very

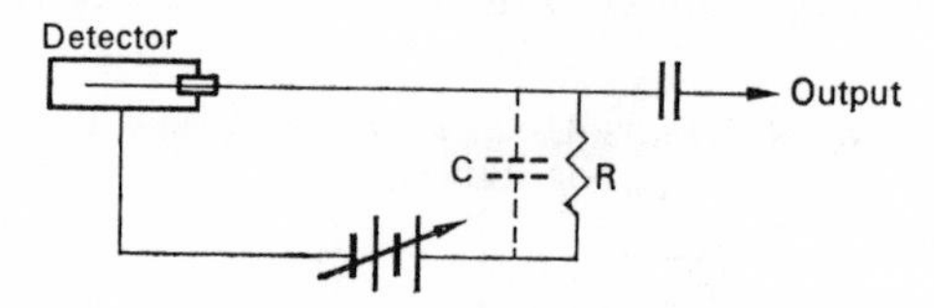

FIG. 9.1. Basic circuit for a gas-filled detector.

well insulated from each other. All the capacitance associated with the circuit is indicated by the capacitor C. Because of the production of ions within the detector when it is exposed to radiation, the gas within the detector becomes electrically conducting.

If the time constant RC of the detector circuit is much greater than the time required for the collection of all the ions resulting from the passage of a single particle through the detector, then a voltage pulse of magnitude

$$V = \frac{Q}{C}, \tag{9.1}$$

where Q is the total charge collected and C is the capacitance of the circuit, and of the shape shown by the top curve in Fig. 9.2, appears across the output of the detector circuit.

A broad output pulse would make it difficult to separate successive pulses. However, if the time constant of the detector circuit is made much smaller than the time required to collect all the ions, then the height of the developed

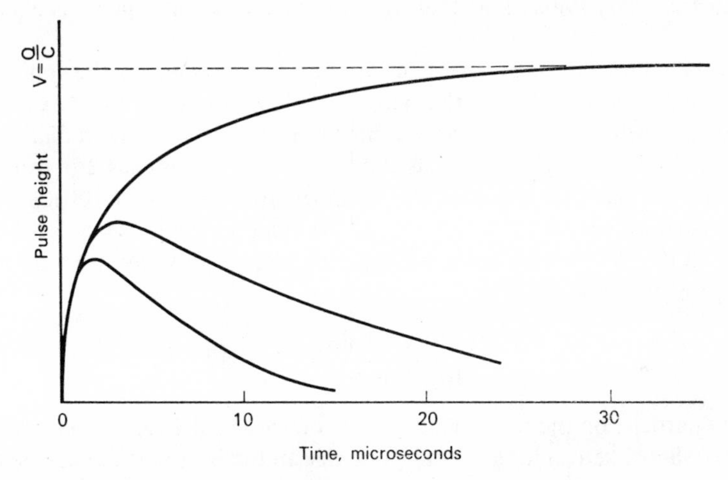

FIG. 9.2. Dependence of pulse shape on the time constant of the detector circuit. The top curve is for the case where $RC = \infty$, the center curve is for the case where RC is less than the ion collection time, while the lowest curve is for the case where RC is much less than the ion collection time.

voltage pulse is smaller, but the pulse is very much narrower, as shown by the curves in Fig. 9.2. This pulse "clipping", as it is called, allows individual pulses to be separated and counted.

Ionization chamber counter

If a constant flux of radiation is permitted to pass through the detector, and if the voltage V is varied, several well-defined regions of importance in radiation measurement may be identified. As the voltage is increased from zero through relatively low voltages, the first region, known as the *ionization chamber region*, is encountered. If the instrument has the electrical polarity shown in Fig. 9.1, then all the positive ions will be collected by the outer cathode, while the negative ions, or electrons, will be collected by the central anode. By "low voltages", in this case, is meant the range of voltage great enough to collect the ions before a significant fraction of them can recombine, yet not great enough to accelerate the ions sufficiently to produce secondary ionization by collision. The exact value of this voltage is a function of the type of gas, the gas pressure, and the size and geometric arrangement of the electrodes. In this region, the number of electrons collected by the anode will be equal to the number produced by the primary ionizing particle. The pulse size, accordingly, will be independent of the voltage, and will depend only on the number of ions produced by the primary ionizing particle during its passage through the detector. The ionization chamber region may be defined as the range of operating voltages in which there is no multiplication of ions due to secondary ionization; that is, the gas amplification factor is equal to one.

The fact that the pulse size from a counter operating in the ionization chamber region depends on the number of ions produced in the chamber makes it possible to use this instrument to distinguish between radiations of different specific ionization, such as alphas and betas or gammas. For example, an alpha particle that traverses the chamber produces about 10^5 ion pairs, which corresponds to 1.6×10^{-14} C. If the chamber capacitance is 10 $\mu\mu$F, and if all the charge were collected, then the voltage pulse resulting from the passage of this alpha will be

$$V = \frac{Q}{C} = \frac{1.6 \times 10^{-14}}{10 \times 10^{-12}} = 1.6 \times 10^{-3}\ \text{V}.$$

A beta particle, on the other hand, may produce about 1000 ion pairs within the chamber. The resulting output pulse due to the beta particle will be only 1.6×10^{-5} V. Amplification of these two pulses by a factor of 100 leads to pulses of 0.16 V for the alpha and 0.0016 V for the beta. With the use of a discriminator in the scaler (or other readout device), voltage pulses less than a certain predetermined size can be rejected; only those pulses that exceed this size will be counted. In the case of the example given above, a discriminator setting of 0.1 V would allow the pulses due to the alphas to be counted, but would not pass any of the pulses due to the beta-rays. This discriminator setting is often referred to as the input sensitivity of the scaler. Increasing the input sensitivity, in the example above, would allow both alphas and betas

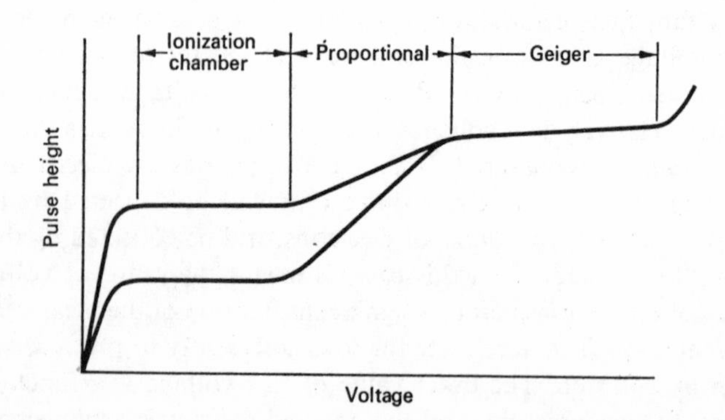

FIG. 9.3. Curve of pulse height versus voltage across a gas-filled pulse counter, illustrating the ionization chamber, proportional, and Geiger regions.

to be counted. This ability to distinguish between the two radiations is illustrated in Fig. 9.3, which shows the output pulse height as a function of voltage across the counting chamber.

Proportional counter

One of the main disadvantages of operating a counter in the ionization chamber region is the relatively feeble output pulse, which requires either much amplification or a high degree of input sensitivity in the scaler. To overcome this difficulty, and yet to take advantage of pulse size dependence on ionization for the purpose of distinguishing between radiations, the counter may be operated as a proportional counter. As the voltage across the counter is increased beyond the ionization chamber region, a point is reached where secondary electrons are produced by collision. This is the beginning of the proportional region. The voltage drop across resistor R will now be greater than it was in the ionization chamber region because of these additional electrons. The gas amplification factor is greater than 1. This multiplication of ions in the gas, which is called an avalanche, is restricted to the vicinity of the primary ionization. Increasing the voltage causes the avalanche to increase in size by spreading out along the anode. Since the size of the output pulse is determined by the number of electrons collected by the anode, the size of the output voltage pulse from a given detector is proportional to the high voltage across the detector. Besides the high voltage across the tube, the gas amplification depends on the diameter of the collecting electrode (the electric field intensity near the surface of the anode, which is given by equation (2.37), increases as the diameter of the collecting anode decreases) and on the gas pressure. Decreasing gas pressure leads to increasing gas multiplication, as shown in Fig. 9.4. Because of the dependence of gas multiplication, and consequently the size of the output pulse, on the high voltage it is important to use a very stable high-voltage power supply with a proportional counter.

An example of the use of a proportional counter to distinguish between alpha and beta radiation is shown below in Fig. 9.5. At point A, the "threshold" voltage, the pulses produced by the alpha particles that traverse the counter are just great enough to get by the discriminator. A small increase in voltage causes a sharp increase in counting rate because all the output pulses due to alphas now exceed the input sensitivity of the scaler. Further increase in high voltage has little effect on the counting rate, and results in a "plateau", a span of high voltage over which the counting rate is approximately independent of voltage. With the system operating on this alpha plateau, the pulses due to beta-rays are still too small to get by the discriminator. However, a point, B, is reached, as the high voltage is increased, where the gas amplification is great enough to produce output pulses from beta particles that exceed the input sensitivity of the scaler. This leads to another plateau where both

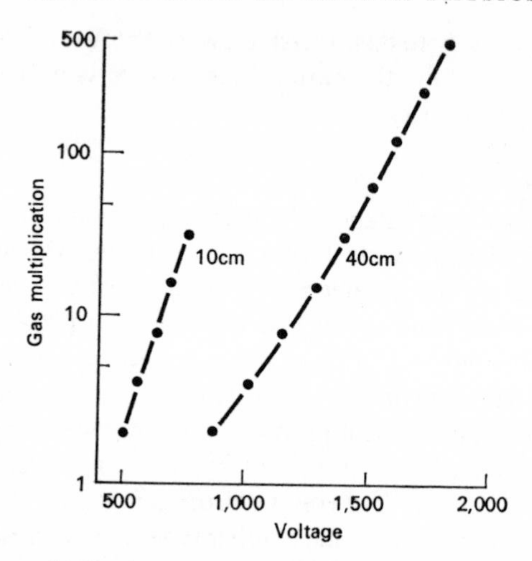

FIG. 9.4. Gas multiplication versus voltage for pressures of 10 and 40 cm Hg tank argon; anode diameter = 0.01 in., cathode diameter = 0.87 in. (From B. B. Rossi and H. H. Staub: *Ionization Chambers and Counters*, McGraw Hill, New York, 1949.)

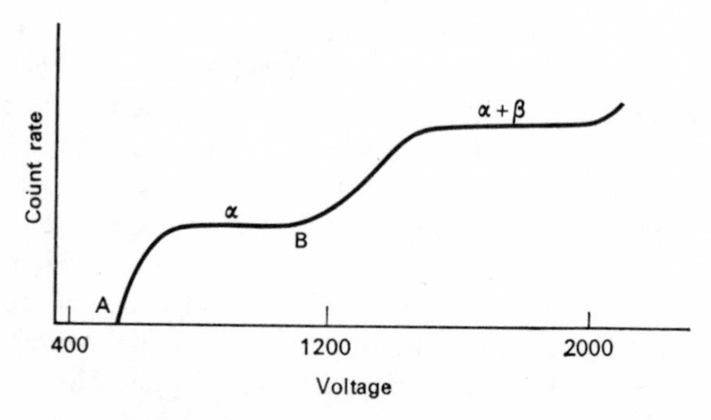

FIG. 9.5. Alpha and alpha–beta counting rates as a function of voltage in a proportional counter.

alpha and beta particles are counted. By subtracting the alpha count rate from the alpha–beta count rate, the beta-ray activity may be obtained.

Geiger counter

Continuing to increase the high voltage beyond the proportional region will eventually cause the avalanche to extend along the entire length of the anode. When this happens, the end of the proportional region is reached and the Geiger region begins. At this point, the size of all pulses—regardless of the nature of the primary ionizing particle—is the same. When operated in the Geiger region, therefore, a counter cannot distinguish among the several types of radiations. However, the very large output pulses (greater than $\frac{1}{4}$V) that result from the high gas amplification in a Geiger counter means either the complete elimination of a pulse amplifier or use of an amplifier that does not have to meet the exacting requirements of high pulse amplification.

Figure 9.5 shows the alpha and alpha–beta plateaus of a proportional counter. A Geiger counter too has a wide range of operating voltages over which the counting rate is approximately independent of the operating voltage. This plateau extends approximately from the voltage which results in pulses great enough to be passed by the discriminator to that which causes a rapid increase in counting rate that preceeds an electrical breakdown of the counting gas. In the Geiger region, the avalanche is already extended as far as possible axially along the anode. Increasing the voltage, therefore, causes the avalanche to spread radially, resulting in an increasing counting rate. We

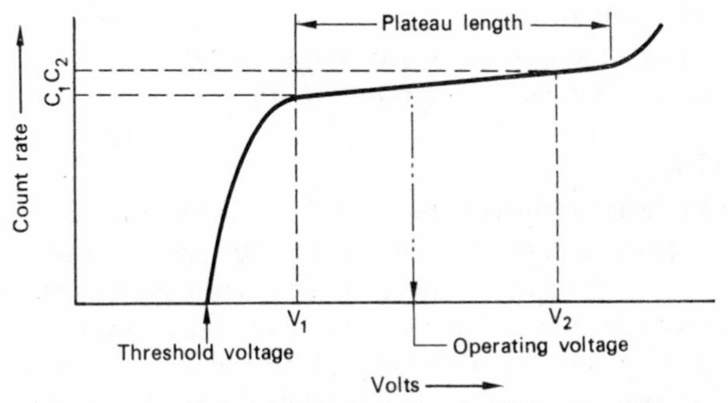

FIG. 9.6. Operating characteristics of a Geiger counter.

therefore have a slight positive slope in the plateau, as shown in Fig. 9.6. Figures of merit for judging the quality of a counter are the length of the plateau, the slope of the plateau, and the resolving time (discussed below). The slope is usually given as per cent increase in counting rate per 100 V:

$$\text{slope} = \frac{(C_2 - C_1)/C_1 \times 100}{0.01\,(V_2 - V_1)}. \tag{9.2}$$

A Geiger counter has a slope of about 3% per hundred volts. The operating voltage for a Geiger counter is about one-third to one-half the distance from the knee of the curve of count rate versus voltage.

Quenching a Geiger counter

When the positive ions are collected after a pulse, they give up their kinetic energy by striking the wall of the tube. Most of this kinetic energy is dissipated as heat. Some of it, however, excites the atoms in the wall. In falling back to the ground state, these atoms may lose their excitation energy by emitting U.V. photons. Since at this time the electric field around the anode is re-established to its full intensity, the interaction of a U.V. photon with the gas in the counter may initiate an avalanche, and thereby produce a spurious count. Prevention of such spurious counts is called quenching. Quenching may be accomplished either electronically, by lowering the anode voltage after a pulse until all the positive ions have been collected; or chemically, by using a self-quenching gas. A self-quenching gas is one that can absorb the U.V. photons without becoming ionized. One method of doing this is to introduce a small amount of an organic vapor, such as alcohol or ether, into the tube. The energy from the U.V. photon is then dissipated by dissociating the organic molecule. Such a tube is useful only as long as it has a sufficient number of organic molecules for the quenching action. In practice, an organic vapor Geiger counter has a useful life of about 10^8 counts. Self-quenching also results when the counting gas contains a trace of a halogen. In this case, the halogen molecule does not dissociate after absorbing the energy from the U.V. photon. The useful life of a halogen-quenched counter, therefore, is not limited by the number of pulses that had been produced in it.

Resolving Time

If two particles enter the counter in rapid succession, the avalanche of ions from the first particle paralyzes the counter, and renders it incapable of responding to the second particle. Because the electric-field intensity is greatest near the surface of the anode, the avalanche of ionization starts very close to the anode, and spreads longitudinally along the anode. The negative ions thus formed migrate towards the anode, while the positive ions move towards the cathode. The negative ions, being electrons, move very rapidly, and are soon collected, while the massive positive ions are relatively slow moving, and therefore travel for a relatively long period of time before being collected. The collection time for positive ions formed near the surface of the anode is given by the equation

$$t = \frac{(b^2 - a^2)\, p \ln b/a}{2V\mu} \text{ sec.} \tag{9.3}$$

where
b = radius of cathode, cm,
a = radius of anode, cm,
p = gas pressure in counter, mm Hg,
V = potential difference across counter, volts,
μ = mobility of positive ions (cm/sec)/(V/cm); for air, μ has a value of 1070 and for argon its value is 1040.

Example 9.1

How long will it take to collect all the positive ions in a Geiger counter filled with argon at a pressure of 100 mm Hg, if the operating voltage is 1000 V, and if the cathode and anode have radii of 1 cm and 0.01 cm respectively?

By substituting the appropriate numbers into equation (9.3), we have

$$t = \frac{(1 - 0.01^2) \times 100 \ln 1/0.01}{2 \times 1{,}000 \text{ to } 1{,}040}$$

$$= 221 \times 10^{-6} \text{ sec.}$$

These slow-moving positive ions form a sheath around the positively charged anode, thereby greatly decreasing the electric field intensity around the anode and making it impossible to initiate an avalanche by another ionizing particle. As the positive ion sheath moves towards the cathode, the electric field intensity increases, until a point is reached when another avalanche could be started. The time required to attain this electric field intensity is called the *dead time*. After the end of the dead time, however, when another avalanche can be

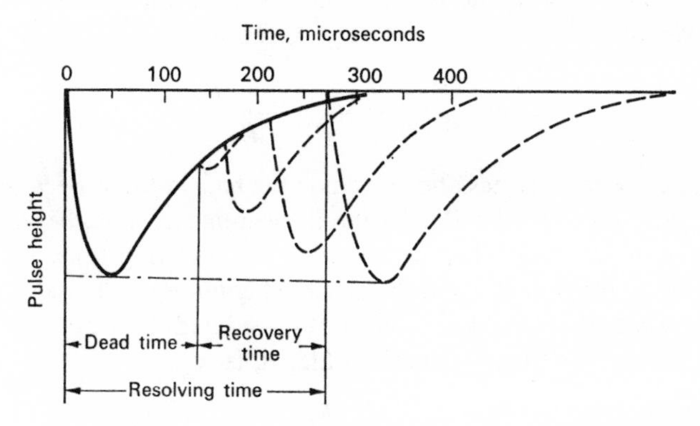

FIG. 9.7. Relationship among dead time, recovery time, and resolving time.

started, the output pulse from this avalanche is still relatively small, since the electric field intensity is still not great enough to produce a Geiger pulse. As the positive ions continue their outward movement, an output pulse resulting from another ionizing particle would increase in size. When the output pulse is large enough to be passed by the discriminator and be counted, the counter is said to have recovered from the previous ionization, and the time interval between the dead time and the time of full recovery is called the *recovery time*. The sum of the dead time and the recovery time is called the *resolving time*. Alternatively, the resolving time may be defined as the minimum time that must elapse after the detection of an ionizing particle before a second particle can be detected. The relationship between the dead time, recovery time, and resolving time is illustrated in Fig. 9.7. The resolving time of a Geiger counter is on the order of 100 μsec or more. A proportional counter is much faster than a Geiger counter. Since the avalanche in a proportional counter is limited to a short length of the anode, a second avalanche can be started elsewhere along the anode while the region of the first avalanche is completely paralyzed. The resolving time of a proportional counter, therefore, is on the order of several microseconds.

Measurement of resolving time

The resolving time of a counter may be conveniently measured by the two-source method. Two radioactive sources are counted—singly and together. If there were no resolving time loss, the counting rate of the two sources together would be equal to the sum of the two single source counting rates. However, because of the counting losses due to the resolving time of the counting system, the sum of the two single counting rates exceeds that of the two sources together. If R_1 is the counting rate of source 1, R_2 of source 2, R_{12} of the two sources together, and R_b the background counting rate, then the resolving time is given by

$$\tau = \frac{R_1 + R_2 - R_{12} - R_b}{R_{12}^2 - R_1^2 - R_2^2}. \tag{9.4}$$

All the source counting rates above include the background. Because $R_1 + R_2$ is only slightly greater than R_{12}, all the measurements must be made with a great degree of accuracy when determining the resolving time. For the case where the resolving time is τ and the observed counting rate of a sample is R_0, the counting rate that would have been observed had there been no resolving time loss, that is, the "true" counting rate, R, is

$$R = \frac{R_0}{1 - R_0 \tau}. \tag{9.5}$$

Scintillation Counters

A scintillation detector is a transducer that changes the kinetic energy of an ionizing particle into a flash of light. Historically, one of the earliest means of measuring radiation was by scintillation counting. Rutherford, in his classical experiments on scattering of alpha particles, used a zinc sulfide crystal as the primary detector of radiation; he used his eye to see the flickers of light that appeared when alpha particles struck the zinc sulfide. Today, the light is viewed electronically by photomultiplier tubes whose output pulses may be amplified, sorted by size, and counted. The various radiations may be detected with scintillation counters by using the appropriate scintillating material. Table 9.2 lists some of the substances used for this purpose.

Scintillation counters are widely used to count gamma-rays and low-energy beta-rays. The counting efficiency of Geiger or proportional counters for low-energy betas may be very low due to the dissipation of the beta energy within the sample. (This phenomenon is called self-absorption.) This disadvantage

TABLE 9.2. SCINTILLATING MATERIALS

Phosphor	Density (g/cm³)	Wavelength of maximum emission, Å	Relative pulse height	Decay time (μsec)
NaI (Tl)	3.67	4100	210	0.25
CsI (Tl)	4.51	Blue	55	1.1
KI (Tl)	3.13	4100	50	1.0
Anthracene	1.25	4400	100	0.032
Trans-Stilbene	1.16	4100	60	0.0064
plastic		3550–4500	28–48	0.003–0.005
liquid		3550–4500	27–49	0.002–0.008
p-Terphenyl	1.23	4000	40	0.005

From R. Swank, Characteristics of scintillators, *Annual Review of Nuclear Science*, vol. V, 1954.

can be overcome by dissolving the radioactive sample in a scintillating liquid, such as toluene. Such liquid scintillation counters result in detection efficiencies that approach 100%. They are widely used in research applications, especially in the field of biochemistry, where they are used to measure ^{14}C and ^{3}H. However, liquid scintillation counters find relatively little direct application in operational health physics.

Whereas the inherent detection efficiency of gas-filled counters is close to 100% for those alphas or betas that enter the counter, their detection efficiency for gamma-rays is very low—usually less than 1%. Solid scintillating crystals, on the other hand, have high detection efficiencies for gamma-rays. Furthermore, since the intensity of the flicker of light in the detector is proportional to the energy of the gamma-ray that produces the light, a scintillation detector can, with the aid of the appropriate electronics, be used as a gamma-ray spectrometer. (With a suitable detector, a scintillation counter may also be used as a beta-ray or an alpha-ray spectrometer.)

For gamma-ray measurement, the detector used most frequently is a sodium iodide crystal activated with thallium [NaI(Tl)], that is optically coupled to a photomultiplier tube. The thallium activator, which is present as an "impurity" in the crystal structure to the extent of about 0.2%, converts the energy absorbed in the crystal to light. The high density of the crystal, together with its high effective atomic number, results in a high detection efficiency, Fig. 9.8. Gamma-ray photons, passing through the crystal, interact with the atoms of the crystal by the usual mechanisms of photoelectric absorption, Compton scattering, and pair production. The primary ionizing particles resulting from the gamma-ray interactions—the photoelectrons, Compton electrons, and positron-electron pairs—dissipate their kinetic energy by exciting and

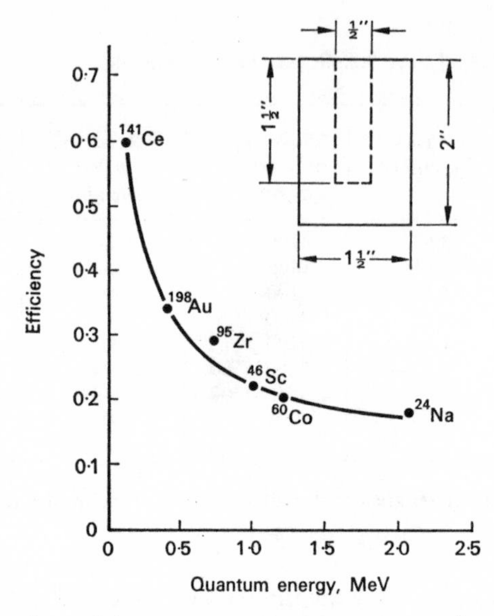

FIG. 9.8. Detection efficiency versus gamma-ray energy for a NaI(Tl) well crystal. (From C. J. Borkowski: O.R.N.L. Progress Report 1160, 1951.)

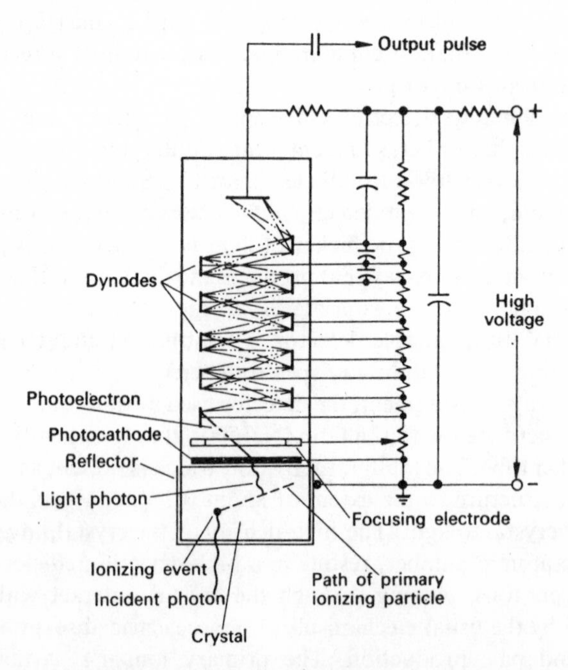

FIG. 9.9. Schematic representation of the sequence of events in the detection of a gamma-ray photon by a scintillation counter. An average of about four electrons are knocked out of a dynode by an incident electron.

ionizing the atoms in the crystal. The excited atoms return to the ground state by the emission of quanta of light. These light pulses, upon striking the photosensitive cathode of the photomultiplier tube, cause electrons to be ejected from the cathode. These electrons are accelerated to a second electrode, called a dynode, whose potential is about 100 V positive with respect to the photocathode. Each electron that strikes the dynode causes several other electrons to be ejected from the dynode, thereby "multiplying" the original photocurrent. This process is repeated about 10 times before all the electrons thus produced are collected by the plate of the photomultiplier tube. This current pulse, whose magnitude is proportional to the energy of the primary ionizing particle, can then be amplified and counted. Figure 9.9 illustrates schematically the sequence of events in the detection of a photon by a scintillation chamber.

A photoelectric interaction within the crystal produces essentially mono-energetic photoelectrons, which in turn produce light pulses of about the same intensity. These light pulses, being of equal intensity, lead to current output pulses of approximately the same magnitude. In Compton scattering, on the other hand, a continuous spectrum of energy results from the Compton electron—the most energetic electron being that which results from a 180° backscatter of the incident photon. This most energetic Compton electron is called the "Compton edge" in scintillation spectrometry. The scattered photon may pass out of the crystal, or it may interact again, either by photo-electric absorption (which would be the most likely interaction) or by another Compton scattering.

In pair production, a flicker of light representing the original quantum energy minus 1.02 MeV is produced as the positron and negatron simultaneously dissipate their energies in the crystal. After losing its energy, the positron combines with an electron, thus annihilating the two particles and producing two photons of 0.51 MeV. Depending on the time sequence, on the crystal size, and on the geometric location of the initial interaction, we may have two pulses representing 0.51 MeV each, one light pulse representing 1.02 MeV, or one light pulse representing the total energy of the original photon.

The scintillation spectrometer

The scintillation spectrometer is an instrument that separates the output pulses from a scintillation detector according to size. Since the size distribution is proportional to the energy of the gamma-ray photon, the scintillation spectrometer enables us to see the sharp line spectra from gamma-emitting isotopes, thus providing a useful tool for identifying unknown isotopes. The single-channel spectrometer consists essentially of the detector, a linear amplifier, a pulse height selector, and a readout device such as a scaler or a ratemeter, Fig. 9.10.

The pulse height selector is an electronic "slit", which may be adjusted

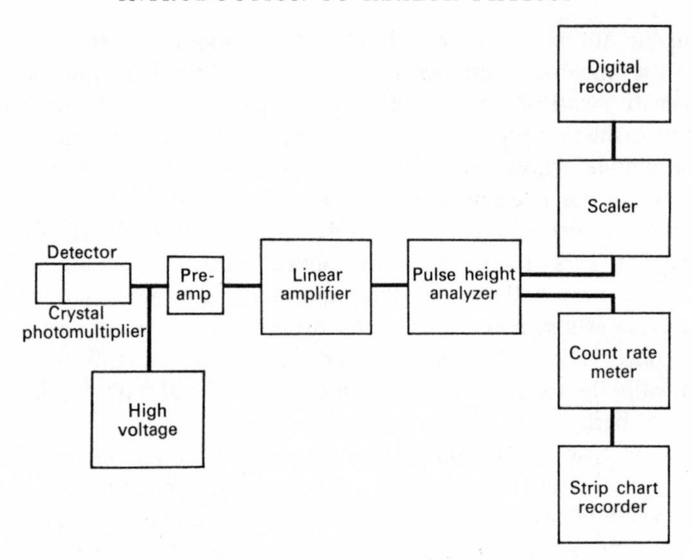

FIG. 9.10. Block diagram of a single-channel gamma-ray spectrometer.

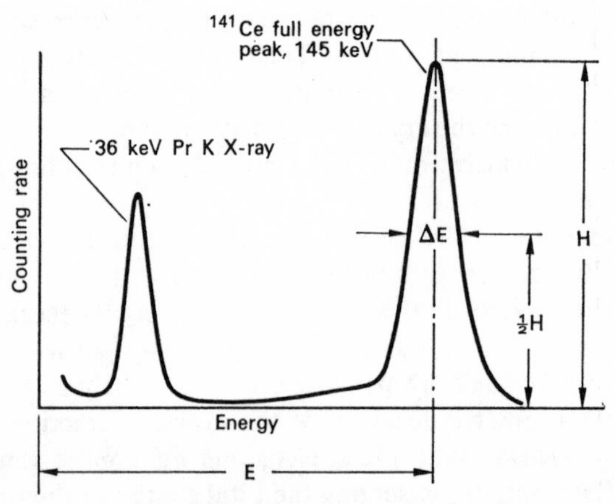

FIG. 9.11. Gamma-ray spectrum of ^{141}Ce. Per cent resolution of the detector is defined as 100 ($\Delta E/E$).

to pass pulses whose amplitude lies between any two desired limits of maximum and minimum. A gamma-ray spectrometer may be of two types: either a single-channel instrument, which has only one movable slit, or a multi-channel instrument, which has many fixed slits. The single slit may be moved either manually, or it may be driven synchronously with a strip chart to scan and record a gamma-ray spectrum. The multi-channel spectrometer can print out the counting rates in each channel as well as visually display the gamma-ray spectrum on a cathode-ray tube. A typical simple gamma-ray spectrum is shown in Fig. 9.11.

The spectrometer may be used for qualitative analysis by locating gamma-ray spectral lines resulting from the photoelectric absorption of the gamma-ray by the crystal. For this purpose, the resolution of the detector is important. The resolution is defined as the half width of the photoelectric absorption peak (commonly called the "photopeak") divided by the midpoint of the photopeak, Fig. 9.11.

$$\text{Per cent resolution} = \frac{\Delta E}{\bar{E}} \times 100. \tag{9.6}$$

The resolution of a detector is a function of energy, and improves with increasing quantum energy. For example, for a 100 KeV gamma-ray the resolution may be about 14%, while for a one MeV gamma it may be about 7%.

Cerenkov Detector

Cerenkov radiation is visible light that results from the passage of a charged particle through a medium at a velocity greater than the phase velocity of light in that medium. The emission of Cerenkov radiation is favoured by a high index of refraction of the medium. Accordingly, Cerenkov detectors are made of high-density glass whose index of refraction for the sodium D lines is on the order of 1.6–1.7. Liquids of high refractive index also are used. Cerenkov radiation, after being produced in the detector, is viewed by a photomultiplier tube for measurement. The Cerenkov detector is not usually used in operational health physics. Because of the dependence of Cerenkov radiation on the velocity of charged particles in the detector, the Cerenkov phenomenon is used mainly in high-energy physics research to measure the velocity of very energetic particles.

Semiconductor Detector

A semiconductor detector acts as a solid state ionization chamber. The ionizing particle—beta-ray, alpha particle, etc.—interacts with atoms in the sensitive volume of the detector to produce electrons by ionization. The collection of these ions leads to an output pulse. In contrast to the relatively high mean ionization energy of 30–35 eV for most counting gases, a mean energy expenditure of only 3.5 eV is required to produce an ionizing event in a semiconductor detector (silicon).

A semiconductor is a substance that has electrical conducting properties midway between a "good conductor" and an "insulator". Although many substances can be classified as semiconductors, the most commonly used semiconductor materials are silicon and germanium. These elements, each of which has four valence electrons, form crystals that consist of a lattice of atoms that are joined together by covalent bonds. Absorption of energy by the crystal leads to disruption of these bonds—only 1.12 eV are required to knock out one of the valence electrons in silicon—which results in a free electron and a "hole" in the position formerly occupied by the valence electron. This free electron can move about the crystal with ease. The hole too can move about in the crystal; an electron adjacent to the hole can jump into the hole, and thus leave another hole behind. Connecting the semiconductor in a closed electric ciruit results in a current through the semiconductor as the electrons flow towards the positive terminal and the holes flow towards the negative terminal.

The operation of a semiconductor radiation detector depends on its having either an excess of electrons or an excess of holes. A semiconductor with an excess of electrons is called an *n*-type semiconductor, while one with an excess of holes is called a *p*-type semiconductor. Normally a pure silicon crystal will have an equal number of electrons and holes. (These electrons and holes

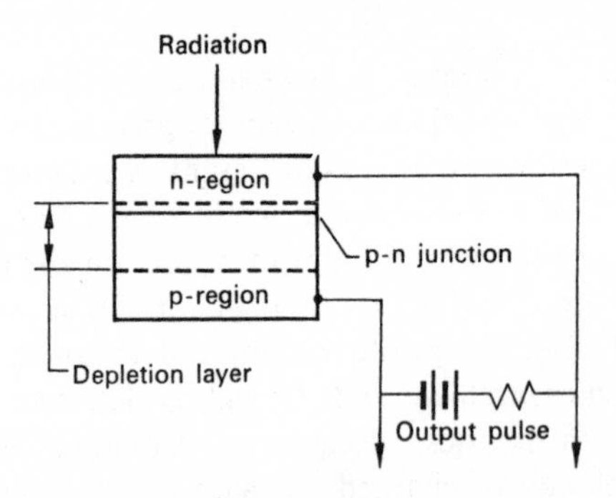

FIG. 9.12. Semiconductor junction detector. This detector is most useful for measuring electrons or other charged particles. For measuring gamma-rays, an electron radiator is interposed between the radiation and the detector, and the photo and Compton electrons thus produced are measured by the detector.

result from the rupture of the covalent bonds by absorption of heat or light energy.) By adding certain impurities to the crystal, either an excess number of electrons (an *n* region) or an excess number of holes (a *p* region) can be produced. Germanium and silicon both are in group IV of the periodic table. If atoms from one of the elements in group V, such as phosphorous, arsenic, antimony, or bismuth, each of which has five valence electrons, are added to the pure silicon or germanium, four of the five electrons in each of the added atoms are shared by the silicon or germanium atoms to form a covalent bond. The fifth electron from the impurity is thus an excess electron, and is free to move about in the crystal and to participate in the flow of electric current. Under these conditions, the crystal is of the *n* type. A *p*-type semiconductor, having an excess number of holes, can be made by adding an impurity from group III of the periodic table to the semiconductor crystal. Elements from group III, such as boron, aluminum, gadolinium, or indium, have three alence electrons. Incorporation of one of these elements as an impurity in he crystal, therefore, ties up only three of the four valence bonds in the crystal attice. This deficiency of one electron is a hole, and we have *p*-type silicon or germanium.

A *p* region in silicon or germanium that is adjacent to an *n* region is called an *n–p* junction. If a forward bias is applied to the junction, that is, if a voltage s applied across the junction such that the *p* region is connected to the positive erminal and the *n* region to the negative terminal, the impedance across the unction will be very low, and current will flow across the junction. If the polarity of the applied voltage is reversed, that is, if the *n* region is connected o the positive terminal and the *p* region to the negative, we have the condition nown as reverse bias. Under this condition, no current (except for a very mall current due to thermally generated holes and electrons) flows across the unction. The region around the junction is swept free, by the potential lifference, of the holes and electrons in the *p* and *n* regions. This region is called the depletion layer, and is the sensitive volume of the solid state detector. When an ionizing particle passes through the depletion layer, electron–hole pairs are produced as a result of ionizing collisions between the ionizing particles and the crystal. The electric field then sweeps the holes and electrons apart, giving rise to a pulse in the load resistor as the electrons flow through he external circuit.

A semiconductor detector is especially useful for measurement of charged particles, such as electrons, protons, and alphas. By coating the semiconductor with an appropriate material, such as ^{10}B, ^{235}U, or a hydrogenous radiator, either slow or fast neutrons can be detected. Among the advantages of a semiconductor detector are the following:

(a) High-speed counting due to the very low resolving time—on the order of nanoseconds.

(b) Response is linearly proportional to the energy deposited in the depletion layer by the ionizing particle. This results in a very sharp energy esolution for charged particles, thereby making the semiconductor an excellent detector for particle spectrometry.

(c) Operates on a very low voltage—about 25–300 V.

Dose-measuring Instruments

Radiation flux is only one of the several factors that determine radiation dose. That a flux measuring instrument does not necessarily measure dose is shown by the following example:

Example 9.2

Consider two radiation fields of equal energy density. In one case we have a 0.1 MeV photon flux of 2000 photons per cm²/sec. In the second case, the photon energy is 2 MeV and the flux is 100 photons per cm²/sec. The energy absorption coefficient for air for 0.1-MeV gamma radiation is 0.0233 cm²/g; for 2 MeV gammas, the energy absorption coefficient is 0.0238 cm²/g. The dose rates for the two radiation fields are given by:

$$D = \frac{\phi \text{ phot/cm}^2\text{/sec} \times \text{E MeV/phot} \times 1.6 \times 10^{-6} \text{ erg/MeV} \times \mu \text{ cm}^2\text{/g}}{87.8 \text{ ergs/g/}R} \quad (9.7)$$

$= 8.5 \times 10^{-8}$ R/sec for the 0.1 MeV radiation, and

8.7×10^{-8} R/sec for the 2 MeV photons.

The dose rates for the two radiation fields are about the same. A flux-measuring instrument, however, such as a Geiger counter, would register about 20 times more for the 0.1 MeV radiation than for the higher-energy radiation.

Pocket Dosimeters

To measure radiation dose, the response of the instrument must be proportional to absorbed energy. A basic instrument for doing this, the free air ionization chamber, was described in Chapter 6. In that chapter, too, it was shown that an "air wall" ionization chamber could be made on the basis of the operational definition of the roentgen, and that such an instrument could be used to measure exposure dose. Ionization chambers of this type, which are often called "pocket dosimeters", are widely used for personnel monitoring. Two types of pocket dosimeters are in common use. One of these is the condenser type, as illustrated in Fig. 6.2. This type pocket dosimeter is of the indirect reading type; an auxiliary device is necessary in order to read the measured dose. This device, which is in reality an electrostatic voltmeter that is calibrated in roentgens, is called a "charger-reader" (because it is also used to charge the chamber). The term minometer is often used synonymously with charger-reader. Figure 9.14 shows a photograph of a pocket dosimeter and its charger-reader. Commercially available condenser-type pocket dosimeters measure integrated X- or gamma-ray exposure doses up to 200 mR with an accuracy of about ±15% for quantum energies between about 0.05 and 2 MeV. For quantum energies outside this range, correction factors, which are supplied by the manufacturer, must be used. These dosimeters also respond to beta-rays whose energy exceeds 1 MeV. By coating the inside of the chamber with boron, the pocket dosimeter can also be made sensitive to thermal neutrons. The standard type of pocket dosimeter, however, is designed for measuring X and gamma radiation only. It is calibrated either with radium or ^{60}Co gamma-rays. Pocket dosimeters discharge slowly even when they are not in a radiation field because of cosmic radiation and because charge leaks across the insulator that separates the central electrode from the outer electrode. A dosimeter that leaks more than 5% of the full-scale reading per

FIG. 9.13. Photograph of a lithium-drifted solid state detector whose front window is $\frac{1}{2}\mu$ gold. This detector can be used to measure charged particle and electromagnetic radiation. Gamma-rays are detected with high resolution but with low efficiency, while particles are detected with both high efficiency and high resolution. (Courtesy Technical Measurement Corp.)

FIG. 9.14. Condenser-type pocket dosimeter and its charger-reader. The dosimeter measures gamma- and X-rays within ±15% from 30 keV to 1.2 MeV in the range 0–200 mR. (Courtesy Victoreen Instrument Co.)

day should not be used. Usually, two pocket dosimeters are worn. Since a malfunction will always cause the instrument to read high, the lower of the two readings is considered as more accurate. Because of leakage and possibility of malfunction due to being dropped, pocket dosimeters are usually worn for one day. Reading the instrument erases its information content. It is therefore necessary to recharge the indirect reading pocket dosimeter after each reading.

The second type pocket dosimeter is direct reading, and operates on the principle of the gold-leaf electroscope (Fig. 9.15). A quartz fiber is displaced electrostatically by charging it to a potential of about 200 V. An image of the fiber is focused on a scale and is viewed through a lens at one end of the instrument. Exposure of the dosimeter to radiation discharges the fiber, thereby

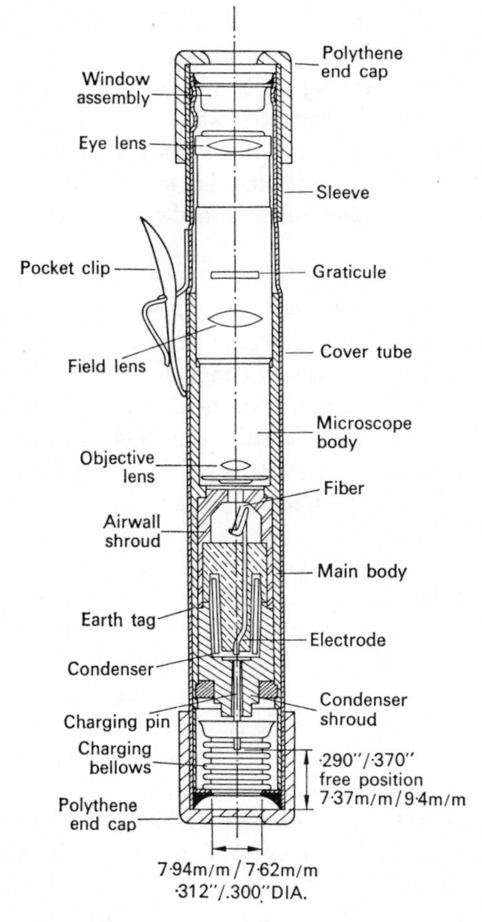

FIG. 9.15. Simplified cross-section of a direct reading quartz fiber electroscope-type pocket dosimeter. The energy dependence characteristics of this dosimeter is shown in Fig. 6.4. (Courtesy R. A. Stephen & Co., Ltd.).

allowing it to return to its original position. The amount discharged, and consequently the change in position of the fiber, is proportional to the radiation exposure. An advantage of the direct reading dosimeter is that it does not have to be recharged after being read. Commonly used direct reading dosimeters that are commercially available have a range of 0–200 mR, and read within about $\pm 15\%$ of the true exposure for energies from about 50 keV to 2 MeV. An auxiliary charger must be used with this dosimeter.

Pocket dosimeters, as the name implies, are mainly used as personnel monitoring devices, and are worn by persons who may be exposed to X- or gamma-radiation in order to measure the actual exposure of the wearer. However, these same instruments may also be used as area monitoring devices by locating them at the points where the exposure dose is to be measured. For this purpose, one or more dosimeters may be left in place for periods up to 1 week. For area monitoring applications, there are available chambers of larger volume (Fig. 9.16) and consequently more sensitive than the pocket dosimeter. Such chambers, which are often called stray radiation chambers, are designed to be used with a charger-reader in the same manner as a condenser-type pocket dosimeter. These chambers are especially useful in monitoring scattered radiation from medical and dental X-ray apparatus.

Film Badges

Another very commonly used personnel monitoring device is the film badge (Fig. 9.17), which consists of a packet of two (for X or gamma) or three (for X, gamma, and neutrons) pieces of dental-sized film wrapped in light-tight paper and worn in a suitable plastic or metal container. The two films for X- and gamma-radiation include a sensitive emulsion and a relatively insensitive emulsion. Such a film pack is useful over an exposure range of about 10 mR to about 1800 R of radium gamma-rays. The film is also sensitive to beta-radiation, and may be used to measure beta-ray dose, from betas whose maximum energy exceeds about 400 keV, from about 50 mrad to about 1000 rad. Using appropriate film and technics, thermal neutron doses of 10 mrem to 1000 rem, and fast neutron doses from about 40 mrem to 100 rem may be measured.

Film badge dosimetry is based on the fact that ionizing radiation exposes the silver halide in the photographic emulsion, which results in a darkening of the film. The degree of darkening, which is called the optical density of the film, can be precisely measured with a photoelectric densitometer whose reading is expressed as the logarithm of the intensity of the light transmitted through the film. The optical density of the exposed film is quantitatively related to the magnitude of the exposure (Fig. 9.18). By comparing the optical density of the film worn by an exposed individual to that of films exposed to known amounts of radiation, the exposure to the individual's film may be determined. Small variations in emulsions greatly affect their quantitative

FIG. 9.16. Stray radiation chambers. The smaller chamber has a range of 0–10 mR, while the larger one, which is more sensitive, has a range of 0–1 mR. These chambers are used with the charger-reader shown in Fig. 9.14. (Courtesy Victoreen Instrument Co.)

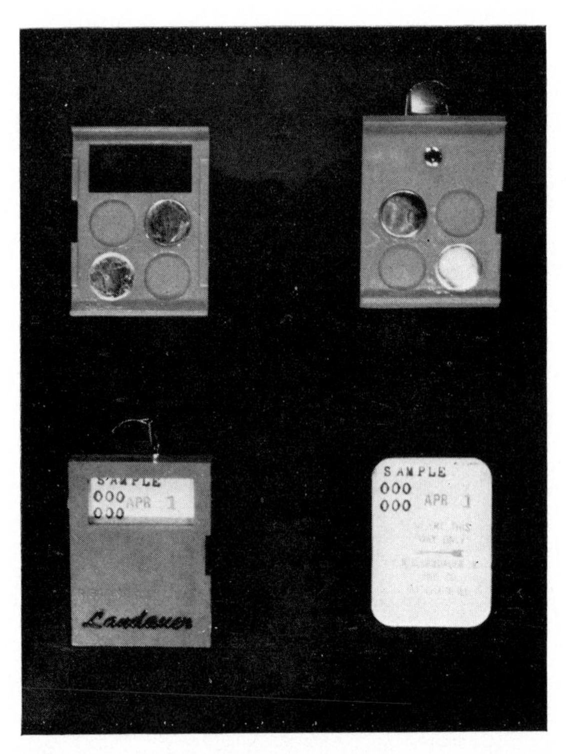

FIG. 9.17. Film badge dosimeter for X-rays and beta-rays. For neutron dosimetry, a second film pack sensitive to neutrons, as well as an additional filter, made of cadmium, is added to the badge. (Courtesy R. S. Landauer Co.)

esponse to radiation. Since the films used in film badges are produced in atches, and since slight variations from batch to batch may be expected, it s necessary to separately calibrate the film from each batch.

Films used in film badge dosimeters are highly energy dependent in the low nergy range, from about 0.2 MeV gamma radiation downward (Fig. 9.19). This energy dependence arises from the fact that the photoelectric cross section or the silver in the emulsion increases much more rapidly than that of air or issue as the photon energy decreases below about 200 keV. A maximum ensitivity is observed at about 30–40 keV. Below this energy, the sensitivity of the film decreases because of the attenuation of the radiation by the paper

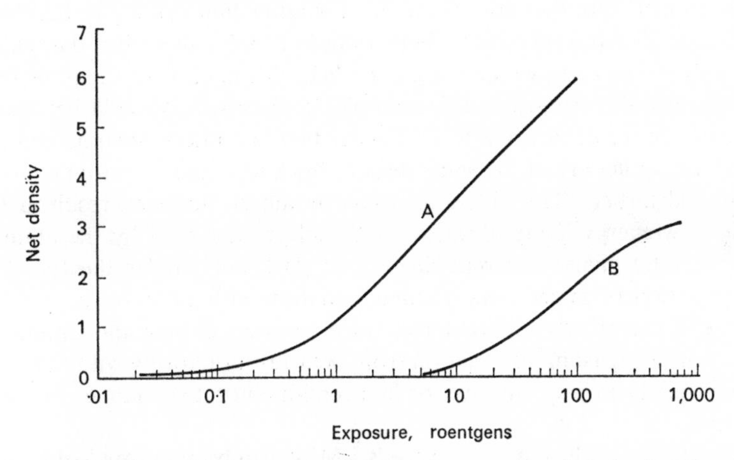

9.18. Relationship between radiation exposure and optical density. Curve he response of duPont type 555 and curve *B* is the response of duPont type 834 dosimeter film to ^{60}Co gamma-rays.

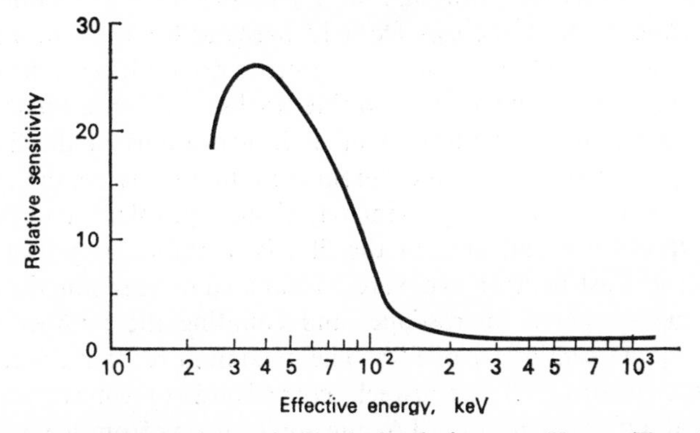

FIG. 9.19. Energy dependence of a film badge dosimeter to X-rays. (From N.B.S. Handbook 50, *Photographic Dosimetry of X- and Gamma-Rays*, 1954.)

wrapper. As a result of this very strong energy dependence, film dosimetry i useless for X-rays less than 200 keV unless the film was calibrated with radia tion of the same energy distribution as the radiation being monitored, o unless the energy dependence of the film is accounted for. This allowance fo energy dependence is made by selective filtration. The film badge holder i designed so that radiation may reach the film directly through an open window or the radiation may be filtered by the film badge holder, or by one of severa different filters, such as aluminum, copper, cadmium, tin, silver, and lead The exact design and choice of filter is governed by the type of radiation t be monitored. The evaluation of the exposure is then made by considerin the ratio of the film densities under each of the various filters. Beta-ray dos is determined from the ratio of the open window film reading to that behin the filters. If exposure was to beta radiation only, then film darkening i seen only in the open window area. To help distinguish between low-energy gamma-rays and beta-rays, for example, comparison is made between th darkening in the open window, and under two thin filters, such as aluminum and silver, which are of the same density thickness, and therefore equivalen beta-ray absorbers. The different atomic numbers, however, result in much greater low-energy X-ray filtration by the silver filter than by the aluminum filter, thereby giving different degrees of darkening under the two filters Interpretation of mixed beta–gamma radiation with a film badge is difficul because of the greatly different penetrating powers of beta and gamma radi ation. For this reason, information from beta-ray monitoring with film badge is used mainly in a qualitative, or in a semi-quantitative manner to evaluat exposure.

Fast neutrons, whose energy exceeds $\frac{1}{2}$ MeV, can be monitored with nuclea track film, such as Eastman Kodak NTA, which is added to the film badge Irradiation of the film by fast neutrons results in proton recoil tracks due to elastic collisions between hydrogen nuclei in the paper wrapper, in the emul- sion, and in the film base. Although the *n*, *p* scattering cross section decrease with increasing neutron energy—from 13 barns at 0.1 MeV to 4.5 barns a 1 MeV to 1 barn at 10 MeV—the recoil protons do not have sufficient energy below about $\frac{1}{2}$ MeV to make recognizable tracks, and hence the threshold a $\frac{1}{2}$ MeV. Because the concentration of hydrogen atoms in the film and its paper wrapper is not very much different from that of tissue, the response o the film to fast neutrons is approximately tissue equivalent, and the numbe of proton tracks per unit area of the film is therefore proportional to the absorbed dose. Fast-neutron exposure is estimated by scanning the developed film with high-powered microscope, and counting the number of proton tracks per square centimeter of film. The maximum recommended exposure rate to fast neutrons, 100 mrem/week, corresponds to a mean proton track density of about 2600/cm^2 of NTA film for neutrons from a Pu–Be source. Since the area seen by the oil immersion lens is about 2×10^{-4} cm^2, a fast eutron dose of 100 mrem corresponds to a mean track density of about 1 roton recoil track per two microscopic fields.

Thermal neutrons also produce proton recoil tracks in the neutron film s a result of their capture by nitrogen in the film according to the $^{14}N(n, p)$ ^{14}C reaction. Although the cross section for 2200 m/sec neutrons for this eaction is 1.75 barns, the concentration of nitrogen in the film is much less han that of hydrogen, thus making this reaction less sensitive, on a per eutron basis, than the n, p scattering reaction for fast neutrons. Nevertheless, ecause in practice fast neutrons are usually part of a mixed radiation field hat includes thermal neutrons (and gamma radiation), and because the ermissible flux for thermal neutrons is much higher than for fast neutrons, llowance must be made for the proton tracks due to thermal neutrons. A ilm badge designed for use in a mixed radiation field that includes neutrons lways has at least two metal filters of equal density thickness—one of cad-nium and the other usually of tin. Cadmium has a very high cross section, 500 barns for the $^{113}Cd(n, \gamma)$ ^{114}Cd for 0.025 eV neutrons, and 7400 barns for .179 eV neutrons. The capture cross section of tin for thermal neutrons s insignificantly small. As a result, a thermal neutron field will show a high rack density under the tin filter, but no tracks under the cadmium. Fast eutrons, on the other hand, will produce the same track density under both ilters. Furthermore, because of the n, γ reaction in the cadmium, a thermal eutron field will produce a darker area on the gamma-ray film under the admium than under the tin. In the absence of any neutrons, gamma radiation vould expose the film under each of these filters to the same degree. By count-ng the tracks and measuring the gamma-ray film density, we determine the hermal neutron flux as well as allowing for the thermal neutron background rack density in the determination of the fast neutron flux. It should be re-mphasized that the ordinary neutron film badges are not sensitive to neutrons n the energy range between epithermal and $\frac{1}{2}$ MeV. However, if the spectral listribution of the neutron field is known, then allowance for neutrons in the ilm badge insensitive range can be made.

Thermoluminescent Dosimeter

Many crystals, including CaF_2, containing Mn as an impurity and LiF emit ight if they are heated after having been exposed to radiation; they are called hermoluminescent crystals. Absorption of energy from the radiation excites he atoms in the crystal, which results in the production of free electrons and oles in the thermoluminescent crystal. These are trapped by impurities or mperfections in the crystalline lattice thus locking the excitation energy into he crystal. Heating the crystal releases the excitation energy as light. Figure .20 shows a characteristic glow curve for LiF, which is obtained by heating he irradiated crystal at a uniform rate, and measuring the light output as the emperature increases. The temperature at which the maximum light output occurs is a measure of the binding energy of the electron on the hole in th trap. More than one peak on a glow curve indicates different trapping site each with its own binding energy. The total amount of light is proportiona to the number of trapped, excited electrons, which in turn is proportional t the amount of energy absorbed from the radiation. The intensity of the ligh emitted from the thermoluminescent crystals is thus directly proportiona to the radiation dose. In use, a very small quantity, on the order of abou 50 mg, is placed into a small capsule (Fig. 9.21), and exposed to radiatior

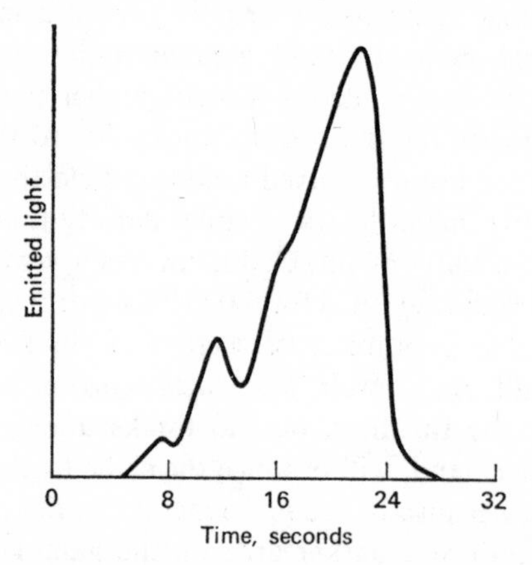

FIG. 9.20. Glow curve for LiF that had been exposed to 100 R. The area under the curve is proportional to the total exposure. (From J. R. Cameron, *et al.*, Thermoluminescent Dosimetry Utilizing LiF, *Health Physics*, **10**, 25–29, 1967.)

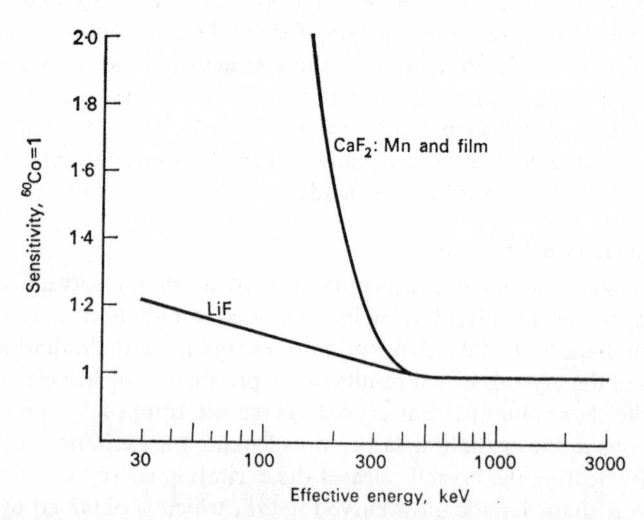

FIG. 9.22. Energy dependence of LiF compared with that of other unshielded dosimeters. (From J. R. Cameron, *et al.*, Thermoluminescent Dosimetry Utilizing LiF, *Health Physics*, **10**, 25–29, 1964).

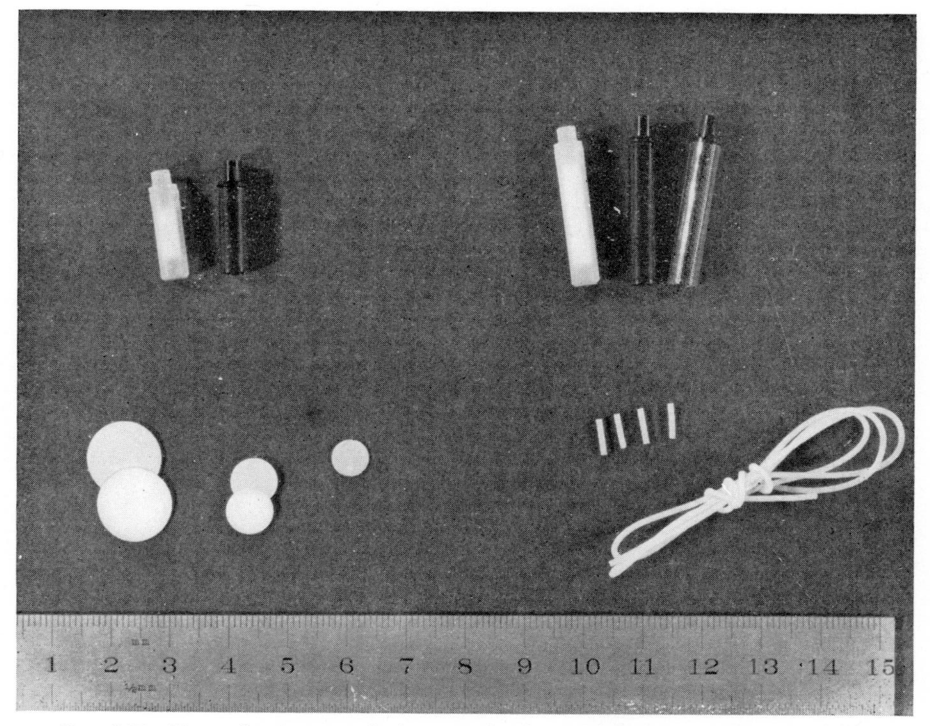

FIG. 9.21. Thermoluminescent dosimeters. At the top left are capsules containing 43 mg LiF powder; the capsules on the top right contain 140 mg LiF, the capsule on the extreme right is surrounded with a low-energy filter. From left to right, on the bottom, are LiF–Teflon discs of two sizes, a CaF_2: Mn–Teflon disc, microrods with LiF, and a long strand of LiF–Teflon from which the microrods are cut. (Courtesy Controls for Radiation Inc.)

For readout, the phosphor is heated electrically and the intensity of the esulting luminescence is measured by a photomultiplier tube whose output ignal, after amplification, is applied to a suitable readout instrument, such as digital voltmeter. The instrument is calibrated by measuring the intensity f light from phosphors that had been exposed to known doses of radiation. ince the intensity of luminescence is proportional to the quantity of the hosphor as well as to the radiation absorbed dose, the amount of phosphor sed in making a measurement must be kept as close as possible to the amount sed in calibrating the instrument.

Thermoluminescent dosimeters respond quantitatively to X-rays, gamma-ays, beta-rays, electrons, and protons over a range that extends from about 0 mrad to about 100,000 rad. LiF thermoluminescent dosimeters are pproximately tissue equivalent, since the effective atomic number of the LiF hosphor is 8.1, while the effective atomic number of soft tissue is about 7.4. The response of a LiF thermoluminescent dosimeter is almost energy inde-endent from about 100 keV to 1.3 MeV gamma-rays. Below 100 keV, the ensitivity increases somewhat, as shown in Fig. 9.22.

on Current Chamber

Ion current chambers have a response whose magnitude is proportional to bsorbed energy, and hence are widely used by health physicists in making ose measurements. A current ion chamber consists basically of a chamber vith two electrodes across which is placed a potential low enough to prevent as multiplication (Fig. 9.23). The ions that are generated in the chamber by adiation are collected, and flow through an external circuit. The ion chamber hus acts as a current source of infinite internal resistance. Although in rinciple an ammeter can be placed in the external circuit to read the ion urrent, in practice this is usually not done because the current is very small. nstead, a high-valued load resistor, R, on the order of 10^{10} ohms is placed in he circuit, and the voltage drop across the resistor is measured with a sensitive lectrometer. Because of the capacitance of the counter and the associated

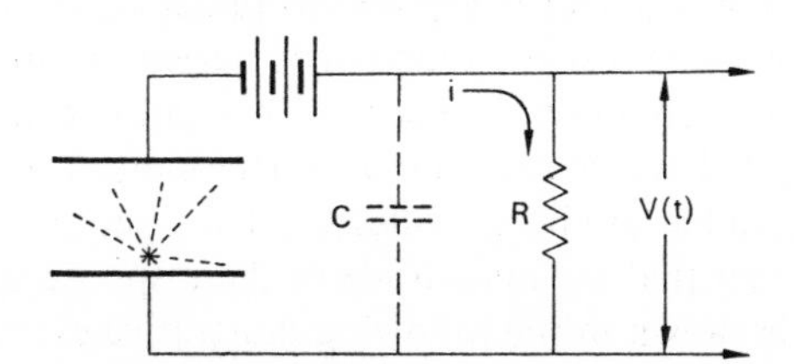

FIG. 9.23. Operating principle of a current ionization chamber. The radiation-produced ions are collected from the chamber, thus causing a current i to flow through the external circuit, resulting in a voltage drop $V(t)$ across the high-valued resistor R. C represents all the capacitance associated with the chamber.

circuit, C, the voltage across the load resistor varies with time, t, after closin
the circuit according to the equation

$$V(t) = IR(1 - e^{t/RC}). \quad (9.8$$

The product RC is called the time constant of the detector circuit, and deter
mines the speed with which the detector responds. When t is equal to RC
the exponent in equation (9.8) becomes 1, and the voltage attains 63% of it
final value. As t increases beyond several time constants the instrument read
the final steady-state voltage. It should be noted that the sensitivity of
detector increases with increasing resistance of the load resistor. Since th
capacitance of the detector is fixed, this means that, in an instrument wit
several ranges—which is accomplished by varying the value of R—the mor
sensitive ranges have longer time constants, and hence are slower to respon
than the less sensitive ranges. The time constants for health physics surveyin

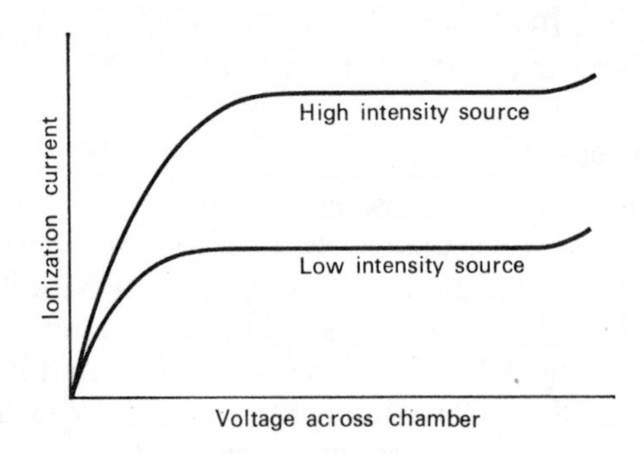

FIG. 9.24. Variation of ionization current with voltage across the ionization chamber for different levels of radiation. The plateau represents the saturation current.

instruments vary up to about 10 sec. Laboratory instruments, where fas
response is not important, may have time constants on the order of 100 sec

When a current ion chamber is exposed to radiation levels of different in
tensity, and the voltage across the chamber is varied, a family of curves, a
shown in Fig. 9.24, is obtained. The current plateau is called the *saturatio
current*. When operated at a voltage that lies on the plateau, all the ions that ar
produced in the chamber are being collected. The operation of a current io
chamber, and the fact that the magnitude of the response is proportional t
absorbed energy, is shown in the following illustrative example.

Example 9.3

A large air-filled ionization chamber has a window whose thickness is 1 mg
cm^2. (a) What ionization current will result if 1200 alpha particles from ^{210}P

nter the chamber per minute? (b) What would be the ionization current if he window thickness were increased to 3 mg/cm²?

The ionization current within the chamber may be calculated from the quation

$$I = \frac{N \text{ particles/sec} \times \bar{E} \text{ eV/particle} \times 1.6 \times 10^{-19} \text{ C/ion}}{W \text{ eV/ion}}. \tag{9.9}$$

(a) The energy of the alpha particle after it penetrates the window into the hamber, is equal to the difference between its initial kinetic energy, 5.3 MeV, nd the energy lost in penetrating the window. Assuming the plastic window o be equivalent to tissue in regard to its stopping power, we calculate, from quations (5.14) and (5.16), that the range of a 5.3 MeV alpha particle in the lastic of which the window is made is 5.1 mg/cm². After passing through the mg/cm² window, therefore, the remaining kinetic energy of the alpha particle s

$$\frac{5.1 - 1}{5.1} \times 5.3 = 4.25 \text{ MeV},$$

nd the resulting ion current is, from equation (9.9),

$$I = \frac{1.2 \times 10^3\ \alpha/\text{min} \times 4.25 \times 10^6 \text{ eV}/\alpha \times 1.6 \times 10^{-19} \text{ C/ion}}{60 \text{ sec/min} \times 35 \text{ eV/ion}}$$

$$= 3.9 \times 10^{-13} \text{ amps.}$$

(b) If the window thickness were increased to 3 mg/cm², the energy of the lpha particle entering into the ionization chamber would be

$$\frac{5.1 - 3}{5.1} \times 5.3 = 2.18 \text{ MeV},$$

nd the ion current would be only 2×10^{-13} A. It should be pointed out that n both instances, alpha particles were entering into the ion chamber at the ame rate. If individual pulses had been counted, therefore, the counting rate vould have been the same in both cases.

As a result of electronic design considerations, the basic circuit of the ioniza-ion chamber survey instrument usually is a Wheatstone bridge (Fig. 9.25), vith an electrometer tube being the resistive element in one of the arms of he bridge. The grid of the tube is connected to one of the electrodes of the on chamber. The bridge is balanced after by-passing the ion chamber current round the bridge. Switching the ion chamber to produce a voltage change on he grid of the electrometer tube as the ionization current flows through the oad resistor R_1 unbalances the bridge and results in a deflection of the micro-mmeter A. The amount of deflection of the meter is proportional to the signal oltage applied to the grid of the tube; this in turn is proportional to the onization current, which is proportional to the radiation dose rate. Because of the characteristics of an electrometer tube, the input signal to the electrometer tube must be on the order of about 1 volt. The current from the ionization chamber is usually very low. The ion current, which depends on the volume of the chamber, V, and on the radiation dose rate, D, is given by the equation

$$I = \frac{V \text{ cm}^3 \times D \text{ mR/hr} \times 10^{-3} \text{ R/mR} \times 1 \text{ sC/cm}^3\text{/R} \times 1 \text{ A/C/sec}}{3.6 \times 10^3 \text{ sec/hr} \times 3 \times 10^9 \text{ sC/C}}. \quad (9.10)$$

For a chamber whose volume is 400 cm³, and a dose rate of 25 mR/hr, the ionization current i is 9.25×10^{-13} A. To produce a voltage drop of about 1 volt across a resistor with this current requires a resistor of about 10^{11} ohms. If a resistor of this value is used as R_1 on the most sensitive scale of the instrument, then we need resistors of 10^{10} and 10^9 ohms respectively for full-scale

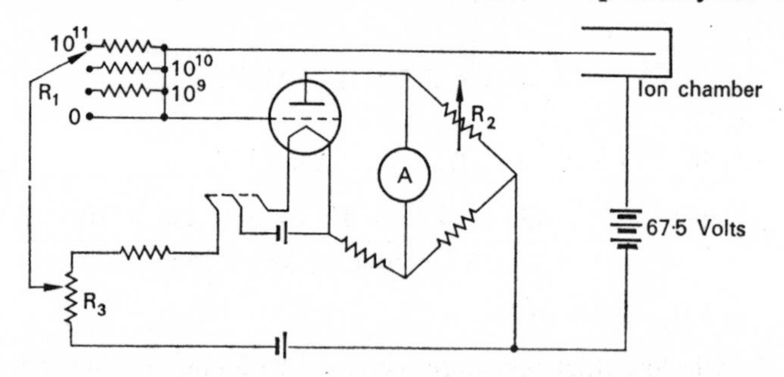

FIG. 9.25. Cutie-pie-type ionization chamber survey instrument circuit diagram. The bridge is balanced, with $R_1 = 0$, by adjusting R_2 and R_3. R_2 is usually inside the case and R_3 is the "zero" knob.

dose rate measurements 10 and 100 times greater than that of the most sensitive scale. These very high-valued resistors must be sealed in glass if their resistance is to remain unchanged. Touching with the hands or otherwise dirtying the connections, or in any way lowering the grid to plate or grid to cathode resistance of the electrometer tube, will very greatly affect the sensitivity of the instrument. In making repairs on the instrument, therefore, great care must be taken to assure the maintenance of the proper resistance in the electrometer circuit.

Certain of the ion chamber instruments, such as the Juno type, are designed to respond selectively to alpha-, beta-, and gamma-radiation, while others, such as the cutie pie, can respond to betas and gammas. The span of dose rates on the commonly used ion chamber survey meters is up to several thousand mR per hour—usually in three ranges. Full-scale readings of 25, 250, and 2500 mR per hour gamma-radiation may be considered typical. However, less sensitive as well as more sensitive ionization-chamber-type survey instruments are commercially available.

Neutron Measurements

Detection Reactions

Neutrons, like gamma-rays, are not directly ionizing; they must react with another medium to produce a primary ionizing particle. Because of the strong dependence of neutron reaction rate on the cross section for that particular reaction, we either use different detection media, depending on the energy of the neutrons that we are trying to measure, or we modify the neutron energy distribution in order that it be compatible with the detector. Some of the basic neutron detection reactions used in health physics instrumentation include:

1. $^{10}B(n, \alpha)\,^{7}Li$. Boron, which may be enriched in the ^{10}B isotope, is introduced into the counter either as BF_3 gas or as a thin film on the inside surfaces of the detector tube. The ionization due to the alpha particle and the ^{7}Li recoil nucleus is counted.

2. Elastic scattering of high-energy neutrons by hydrogen atoms. The scattered protons are the primary ionizing particles, and the ionization they produce is detected and measured.

3. Nuclear fission: fissile material (n, f) fission fragments. The fissile material is deposited as a thin film on the inside surface of a counter tube. Capture of a neutron and splitting of the fissile nucleus results in highly ionizing fission fragments, which can easily be detected. Fission reactions are energy dependent, and several fissile isotopes have a threshold energy below which fission cannot occur.

4. Neutron activation: *threshold detectors*. Many neutron reactions produce radioactive isotopes. The degree of neutron-induced radioactivity of any given substance depends on the total neutron irradiation. By measuring the induced activity, and allowing for decay time between exposure and measurement, the integrated neutron exposure can be calculated. Furthermore, since many of these activation reactions have energy thresholds, they can be useful in determining the neutron energy distribution. This technic is especially useful in measuring neutron dose at very high dose rates, such as those that would be encountered in a criticality accident. (A criticality accident is an accidental uncontrolled chain reaction in which a very large amount of energy is liberated during a very brief time interval.) For measuring high neutron fluxes for health physics purposes, a series of threshold detectors of various threshold energies is packaged in a single unit. Exposure to neutrons activates the detectors. Since the induced activity in the threshold detectors depends on the neutron flux whose energy exceeds the threshold energy, the relative counting rates of the threshold detectors after exposure is used as a measure of the spectral distribution of the neutrons, while the absolute activity of the detector is a measure of exposure. A number of different substances may be used as threshold detectors. One type of pocket criticality dosimeter (Fig. 9.26)

uses a combination of indium, cadmium-covered indium, gold, cadmium-covered gold, sulfur, and cadmium-covered copper. By combining the results of the activity in each of these foils, the neutron spectrum may be broken down to the following energy intervals:

Thermal	to	0.4 eV
0.4 eV	to	2 eV
2 eV	to	10 eV
10 eV	to	1 MeV
1 MeV	to	2.9 MeV
above		2.9 MeV

The total neutron flux is thus divided into six energy groups, thereby permitting a reasonably accurate means for computing absorbed dose. The threshold detectors are calibrated by exposing the pack to a known beam of neutrons, and then measuring the induced activity. (A thermoluminescent dosimeter may be added to measure the gamma-ray exposure from a criticality accident.)

Neutron Counting with a Proportional Counter

Counting in the proportional region makes it simple to measure neutrons in the presence of gamma radiation. A neutron counter used for this purpose uses BF_3 gas to take advantage of the n, α reaction on ^{10}B:

$$^{10}B(n,\alpha)\ ^{7}Li.$$

Boron-10 has a high cross section for this reaction, 4010 barns for thermal neutrons, and consequently makes a very sensitive detecting medium. The alpha particles resulting from this reaction are produced inside the detector. Because of the great difference between the output pulses resulting from the alpha particles and Li ions and those due to gamma-rays, it is a simple matter to discriminate electronically against all pulses except those due to the alphas and Li ions when a neutron is captured by ^{10}B.

The sensitivity of a BF_3 neutron detector may be increased by using ^{10}B-enriched BF_3 gas. Naturally occurring boron contains about 19.8% ^{10}B. However, it is possible to concentrate the ^{10}B isotope to the extent that BF_3 gas containing 96% ^{10}B is routinely available from commercial suppliers. The BF_3-filled counter is a sensitive and simple thermal neutron detector, and is widely used by health physicists for measuring thermal neutron flux. Furthermore, since 667 thermal neutrons per cm^2/sec for 40 hr corresponds to 100 mrem/week, it is a simple matter to measure dose rate by converting counts per minute of a calibrated counter to neutron flux.

The capture cross section for the ^{10}B (n, α) ^{7}Li reaction, and hence the counting rate of a BF_3 detector, depends on the neutron energy. Consider

FIG. 9.26. Pocket criticality dosimeter for measuring neutron fluxes. Inside the dosimeter are six different foils whose induced radioactivity, following exposure to neutrons, depends on the neutron energy and fluence. (Courtesy Reactor Experiments, Inc.)

the case of a thermal neutron flux of ϕ neutrons per cm²/sec having a Maxwell–Boltzman energy distribution and a BF_3 counter containing a total of N atoms of ^{10}B. If there is negligible neutron absorption by the counter wall, and if the intrinsic counting efficiency is 100%, the counting rate is given by

$$CR = N \int \phi(v)\, \sigma(v)\, dv, \tag{9.11}$$

where $\phi(v)$ is the flux of neutrons of velocity v, and $\sigma(v)$ is the capture cross section for neutrons of velocity v. Substituting $\phi(v) = n(v)v$, where $n(v)$ is the density, in neutrons per cm³, of neutrons whose velocity is v cm/sec; and, from equation (5.53), substituting $\sigma(v) = \sigma_0(v_0/v)$, into equation (9.11), we have

$$CR = N\sigma_0\, v_0 \int n(v)\, dv, \tag{9.12}$$

$$CR = N\sigma_0\, v_0\, n. \tag{9.13}$$

From equation (9.13), we see that the counting rate of a BF_3 counter is proportional to the total neutron density within the range of energies where the $1/v$ law is valid (up to about 1,000 eV). If v is the mean neutron velocity (not the velocity corresponding to the mean neutron energy), then

$$\phi = n\bar{v} \tag{9.14}$$

or, $n = \phi/\bar{v}$. Substituting this value for n into equation (9.13), and solving for ϕ, we have

$$\phi = \frac{\bar{v}}{v_0} \times \frac{CR}{N\sigma_0}. \tag{9.15}$$

In the Maxwell–Boltzman distribution, $\bar{v}/v_0 = 2/\sqrt{\pi} = 1.128$. For thermal neutrons, therefore, the flux is related to the counting rate by

$$\phi = \frac{1.128}{N\sigma_0} \times CR. \tag{9.16}$$

Example 9.4

A thermal neutron counting rate of 600 counts per min is measured with a BF_3 counter whose inside dimensions are 2 cm diameter × 20 cm long, and is filled with 96% enriched $^{10}B\,F_3$ at a pressure of 20 cm Hg. What is the thermal flux? The volume of the counter is $\pi r^2 l$, or 62.83 cm³. The molar quantity of BF_3 in the counter is

$$m = \frac{PV}{RT} = \frac{20/76 \text{ atm} \times 0.06283 \text{ liter}}{0.082 \text{ liter atm/mole}^\circ \times 300^\circ} = 6.72 \times 10^{-4} \text{ moles}$$

and the number ^{10}B atoms in the counter is

$$N = 0.96 \frac{^{10}\text{B atoms}}{\text{molecule}} \times 6.03 \times 10^{23} \frac{\text{molecules}}{\text{mole}} \times 6.72 \times 10^{-4} \text{ moles}$$

$$= 3.89 \times 10^{20}\ ^{10}\text{B atoms.}$$

From equation (9.16), we have

$$\phi = \frac{1.128}{3.89 \times 10^{20} \text{ atoms} \times 4.010 \times 10^{-21} \text{ cm}^2/\text{atom}} \times 10 \frac{\text{counts}}{\text{sec}}$$

$$= 7.2 \frac{\text{neutrons}}{\text{cm}^2 \text{ sec}}.$$

The sensitivity of a radiation detector may be defined as

$$\text{sensitivity} = \frac{\text{counting rate}}{\text{flux}}. \quad (9.17)$$

The sensitivity, S, of the BF_3 counter in Example 9.4 is

$$S = \frac{CR}{\phi} = \frac{600 \text{ counts/min}}{7.2 \text{ neutrons/cm}^2 \text{ sec}} = 83.3 \frac{\text{counts/min}}{\text{neutron/cm}^2 \text{ sec}}.$$

Long Counter

The BF_3 counter can also be used for fast neutrons if it is surrounded by paraffin or another moderator. Fast neutrons are sufficiently slowed down by the moderator to allow them to be captured by the ^{10}B. The counting rate of a moderated BF_3 counter in a field of fast neutrons increases with increasing moderator thickness until the paraffin is sufficiently thick to absorb a significant fraction of the thermalized neutrons. Beyond this optimum thickness, the counting rate decreases with further increase in thickness. The exact thickness at which this occurs depends on the energy of the neutrons. A paraffin moderator $2\frac{3}{8}$ in. thick results in an approximately flat response over a neutron energy span from about 10 keV to better than 1 MeV. The outside of the paraffin may be covered with a thin sheet of cadmium to absorb thermal neutrons while allowing fast neutrons to pass through into the paraffin. The capture gammas due to absorption of thermal neutrons by the cadmium produce very small pulses in the counter, pulses small enough to be rejected by the discriminator. Such a paraffin surrounded BF_3 counter is called a *long counter* because of its long energy-independent response. In health physics work it is most useful in the energy range of about 10 keV to 500 keV. Measurement of higher energy neutrons with a BF_3 counter requires relatively large amounts of paraffin for slowing down the neutrons, thus making it impractical for many health physics surveying applications. Such a counter, which is reported as having a fairly uniform response from 10 keV to 5 MeV

(A. D. Hanson and J. L. McKibben, A Neutron Detector Having Uniform Sensitivity from 10 keV to 5 MeV, *Phys. Rev.* **72**, 673 (1947), and R. A. Nobles, *et al.*, *Rev. Sci. Instr.* **25**, 334 (1954)), is shown below in Fig. 9.27. The counter response is highly directional, and is designed to measure neutrons that are

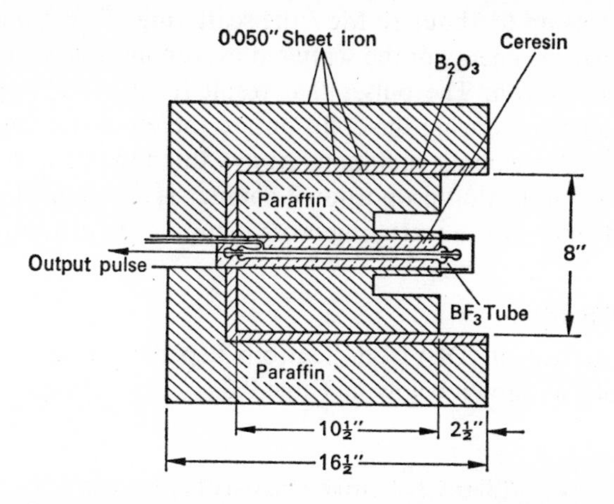

FIG. 9.27. Diagram of a *long counter*, a neutron counter whose response is approximately uniform from about 10 keV to 5 MeV. (Hanson and McKibben, *Physical Review* **72**, 673, 1947.)

incident only on the front face. The layer of paraffin outside the B_2O_3 is a shield designed to remove neutrons that are incident on the sides. The series of 8 holes in the front face permit the lower-energy neutrons to reach the detector tube after being thermalized. The sensitivity of this counter is about 1 count per sec per neutron per square centimeter.

Proton recoil counter

A proportional counter that responds to recoil protons resulting from the collision of fast neutrons with hydrogen atoms may be used for the detection of neutrons whose energy exceeds 500 keV. (Below this energy, the output pulses from the counter are very weak, and high gain amplification would be required to record them.) Such a counter may be made simply by using a hydrogenous gas, such as methane, in the counter. The counter is enclosed in a thin sheet of cadmium to absorb thermal neutrons in order to prevent pulses due to deuteron recoils following absorption of thermal neutrons by the hydrogen. The hydrogenous material may also be a solid, such as paraffin or polyethylene, incorporated into the wall of the counter. The fast neutrons "knock out" protons from these solids, and the protons dissipate their energy in the counter gas. When used in this way as a source of protons, the hydrogenous substance is called a "proton radiator". The sensitivity for fast neutrons of a proton recoil counter is very much less than the sensitivity of a

BF_3 counter for thermal neutrons. This is due to two reasons: the cross section of hydrogen for scattering of fast neutrons is very much less than the slow neutron capture cross section of ^{10}B, and also the energy distribution of the scattered protons includes a large fraction of very low-energy protons. For neutron energies up to about 10 MeV the scattering of neutrons is isotropic. This means that the energy of the scattered proton may vary from zero to the energy of the neutron. The pulses that result from protons to which little energy was imparted during the collision are therefore not counted because of the bias against gamma radiation. Above this threshold, the energy response of the recoil proton proportional counter is determined mainly by the energy dependence of the scattering cross section of hydrogen.

Neutron Dosimetry

The dose equivalent (DE) from neutrons depends strongly on the energy of the neutrons as shown below in Table 9.3.

TABLE 9.3. NEUTRON QUALITY FACTOR AND FLUX CORRESPONDING TO 100 MILLIREMS PER WEEK

Neutron energy MeV	QF	$n/cm^2/sec$ 100 mrem/wk
Thermal	3	670
0.0001	2	500
0.005	2.5	570
0.02	5	280
0.1	8	80
0.5	10	30
1.0	10.5	18
2.5	8	20
5.0	7	18
7.5	7	17
10.0	6.5	17
10–30		10

We therefore cannot simply convert neutron flux into millirems unless we know the energy spectral distribution of the neutrons. For pure thermal neutrons, of course, the energy distribution is known, and a BF_3 counter could therefore be calibrated to read directly in millirems per hour. For higher energy neutrons, neither the moderated BF_3 long counter with its "flat" energy independent response to fast neutrons, nor the simple proton recoil proportional counter, whose response to fast neutrons depends strongly on the scattering cross section, is suitable for dosimetry.

Fast neutrons: Hurst counter

The proton recoil proportional counter can be modified to measure fast neutron dose rate. By using a combination of several different proton radiators and filling gases, as shown schematically in Fig. 9.28, the energy distribution of the recoil protons that enter into the gas-filled cavity of a proportional counter is such that the resulting count rate is proportional to the variation of tissue dose with the neutron energy. Hence, the counting rate meter read out from the proportional counter can be calibrated directly in millirads per hour of fast neutrons. Figure 9.29 compares the energy depen-

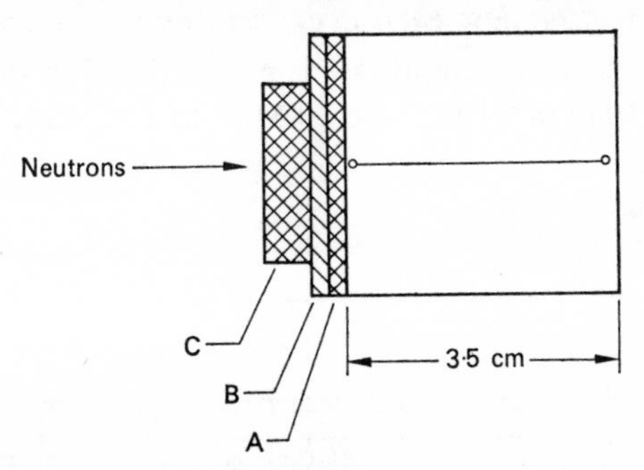

FIG. 9.28. Schematic representation of a count-rate fast-neutron dosimeter. *A*—paraffin (13 mg/cm²), *B*—aluminum (29 mg/cm²), *C*—paraffin (100 mg/cm²). Ratio of the paraffin areas, $a_A/a_C = 2.9$. The gas is methane at 30 cm Hg pressure. The counter has a highly directional response, and must be oriented towards the neutron beam as shown in the diagram. (G. S. Hurst, R. H. Ritchie, and H. N. Wilson, A count rate method of measuring fast neutron dose, *R.S.I.* **22**, 981–6, 1951.)

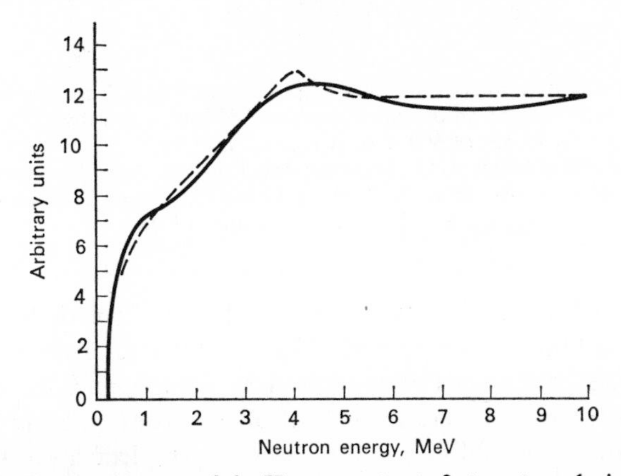

FIG. 9.29. Energy response of the Hurst count-rate fast-neutron dosimeter. The solid curve shows the relationship between counting rate per unit flux and neutron energy; the broken curve relates dose rate per unit flux and neutron energy. (G. S. Hurst, R. H. Ritchie, and H. N. Wilson, A count rate method of measuring fast neutron dose, *R.S.I.* **22**, 981–6, 1951.)

dence of the dose with the response of the count rate dosimeter. Gamma-insensitive fast neutron dosimeters based on this design principle are commercially available. Because of the design, this instrument is limited to measurement of neutrons in the energy range from 0.2 to 14 MeV. The discriminator settings necessary to obtain the correct dose response rejects all counts due to neutrons whose energy is less than 0.2 MeV. Above 14 MeV, the response of this fast neutron dosimeter is not proportional to the absorbed dose rate.

Thermal and fast neutron dose equivalent: neutron rem counter

A neutron counter whose count rate is proportional to the dose equivalent across the energy span of 0.025 eV to 10 MeV can be made by surrounding a

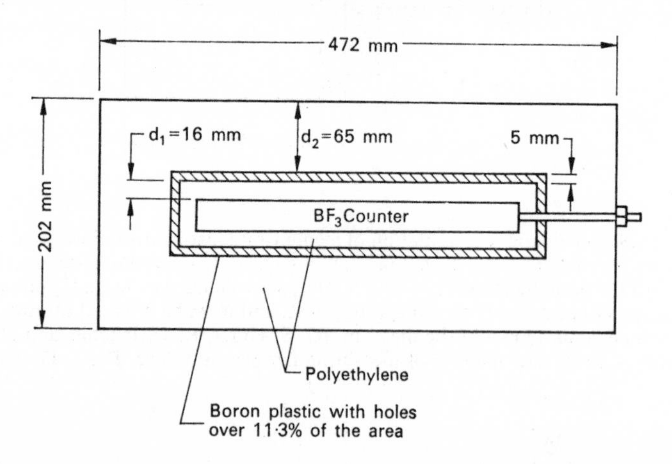

FIG. 9.30. Construction of the neutron rem counter. The BF_3 counter embedded in the polyethylene has an outside diameter of 30 mm, a sensitive length of 200 mm, and a total length of 300 mm; it is filled with 94% ^{10}B-enriched BF_3 at a pressure of 600 mm Hg. (I. O. Anderson and J. Braun, A neutron rem counter with uniform sensitivity from 0.025 eV to 10 MeV, *Neutron Dosimetry*, vol. II, pp. 87–95, I.A.E.A., Vienna, 1963.)

BF_3 proportional counter with two cylindrical layers of polythylene moderator separated by a boron-loaded plastic (Fig. 9.30). Circular discs of the same materials make up the ends of the cylinders. In designing the counter, the simplifying assumptions were made that all neutrons entering the counter and making their first collision in the outer polyethylene layer are absorbed in the boron plastic shield and thus lost for the purpose of detection, while all neutrons making their first collision in the polyethylene inside the boron plastic are considered as having a detection probability k. On the basis of these assumptions, the likelihood of obtaining a pulse from a neutron of energy E that is incident normally to the long axis of the counter is given by

$$P = ke^{-\Sigma(E)d_2}(1 - e^{-2\Sigma(E)d_1}), \tag{9.18}$$

where $\Sigma(E)$ is the macroscopic total scattering cross section for neutron energy E, and d_1 and d_2 are the thicknesses of the inner and outer polyethylene moderators. Varying d_1 and d_2 changes the energy response. With $d_1 =$ 16 mm and $d_2 = 32$ mm, the energy response of the counter very closely approximates, within $\pm 15\%$, the curve of rem per neutron per cm^2 versus neutron energy (Fig. 9.31). A commercial survey meter based on this design has a sensitivity of 120 cpm per mrem/hr, and a range of 0.1–100 m/rem/hr.

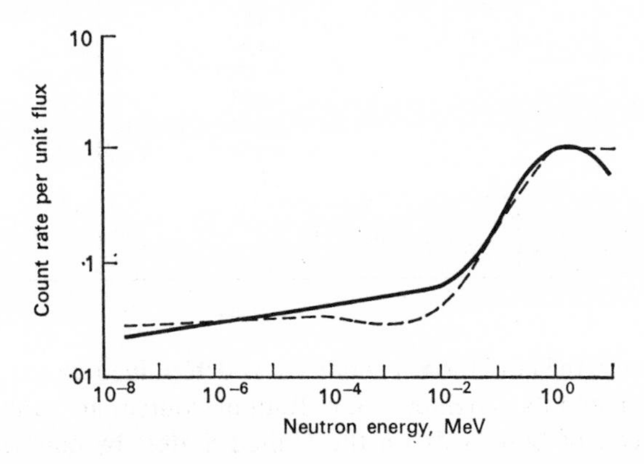

FIG. 9.31. Energy response curve (solid line) of the neutron rem counter shown in Fig. 9.30. The broken line is the dose equivalent per unit flux. Ideally, the response should be 7.6 counts per second per mrem per hour. (I. O. Anderson and J. Braun, A neutron rem counter with uniform sensitivity from 0.025 eV to 10 MeV, *Neutron Dosimetry*, vol. II, pp. 87–95, I.A.E.A., Vienna, 1963.)

Calibration

Health physics survey instruments are calibrated by exposing the instrument in a known radiation field, and then comparing the meter reading to the known radiation field.

Gamma-rays

For calibrating gamma ray measuring instruments, the most widely accepted source is radium (usually a radium salt such as RaBr), in equilibrium with its daughters, sealed into a platinum–iridium capsule. The gamma radiation from such a source is very heterochromatic, as shown below in Table 9.4; it originates mainly from the Ra B (^{214}Pb) to Ra C (^{214}Bi) and Ra C′ (^{214}Po) transitions. Since the half-lives of these two transitions are 26.8 and 19.7

TABLE 9.4. APPROXIMATE GAMMA-RAY SPECTRUM OF RADIUM AND ITS EQUILIBRIUM DECAY PRODUCTS

Transition	Quanta per radium disintegration	MeV/photon
Ra → Rn	0.012	0.184
Ra B → Ra C	0.115	0.241
	0.258	0.294
	0.450	0.350
Ra C → Ra C′	0.658	0.607
	0.065	0.766
	0.067	0.933
	0.206	1.120
	0.063	1.238
	0.064	1.379
	0.258	1.761
	0.074	2.198
	Total = 2.3	$\bar{E}$ = 0.7

min, respectively, the equilibrium condition is determined by the equilibrium between radium and its daughter radon. Radium sources are calibrated at the National Bureau of Standards (in the United States) by comparing, under standard conditions of measurement, the gamma-radiation output of the unknown with that of a standard source.

Knowing the quantity of radium in the capsule and the physical dimensions of the capsule permits the calculation of the gamma ray dose rate at various distances from the source. The dose rate at a distance d cm from the source is given by

$$R = \frac{9.3\, me^{-ux}}{d^2} \tag{9.19}$$

where R is the dose rate in roentgens per hour, m is the corrected (according to the note at the bottom of the calibration certificate) weight of the radium, u is the linear absorption coefficient (for Pt, u is 0.19 per mm), and x is the capsule wall thickness. For a capsule whose wall thickness is 0.5 mm Pt, the radiation output is 840 mR/hr/g Ra at a distance of 1 m. The distance from the source for any other radiation dose rate may be calculated with the aid of the inverse square law. Thus, if we had a source containing 10 mg Ra in a capsule of 0.5 mm Pt wall thickness, the dose rate at 1 m would be 84 mR/hr. To calibrate the 12.5 mR/hr midpoint on the scale of an instrument that measures up to 25 mR/hr, the detector would be placed at a distance from the source calculated as follows:

$$\frac{I_1}{I_2} = \frac{d_2^2}{d_1^2} \tag{9.20}$$

$$\frac{84}{12.5} = \frac{d_2^2}{1}$$

$$d_2 = 2.6 \text{ meters.}$$

The inverse square law, which assumes the radium to be a point source, is applicable to distances that exceed about 20 times the largest linear dimension of the source. Scatter from the floor, walls, or ceiling of the room in which the calibration is performed will result in a slower decrease in radiation dose rate than that predicted by the inverse square law. This point may be checked by measuring the gamma-ray dose rate at several distances with a calibrated ion chamber or counter. A plot of the logarithm of the dose rate versus the logarithm of distance (or dose rate versus distance on log paper) will result in a straight line of slope −2 if the inverse square law is applicable. A smaller slope would suggest the presence of significant amounts of scattered radiation.

Other sources besides radium may be employed to calibrate gamma-ray measuring instruments. Among the most frequently used sources in this category are ^{60}Co and ^{137}Cs. Such sources must be of high specific activity in order to have a source sufficiently small (in physical dimensions) in order that it could be considered a "point" source. The nominal source strengths for ^{60}Co and ^{137}Cs are 1.32 and 0.356 mR/hr/mCi at 1 m. However, since these sources too are required to be encapsulated, the nature of the capsule must be known in order to account for radiation attenuation by the capsule. In using such a source, the radiation output should be measured with a previously calibrated instrument. It should be pointed out that since both these radioisotopes have relatively short half-lives (5.3 years for ^{60}Co and 33 years for ^{137}Cs), appropriate correction factors must be applied to the original source strength measurements or calculations to account for radioactive decay.

Beta-rays

Dose-response characteristics of most portable survey instruments for beta-rays are strongly energy dependent. As a consequence, survey instruments are usually used only to detect beta radiation, but not to measure beta-ray dose rate. Such instruments, therefore, are normally not calibrated for beta-ray dose measurements. If beta-ray dose rates are required, an infinitely thick source of known specific activity may be made. The surface dose rate, which is equal to one-half the dose rate in an infinitely large medium of that specific activity, may be easily calculated. The beta-ray source used most frequently to calibrate film badges is an infinitely thick (about 5 mm) slab of metallic uranium; the beta-ray dose rate at the surface is 240 mrad/hr.

Alpha-rays

Portable survey instruments used to measure alpha contamination are usually designed to read disintegrations per minute. Calibration sources for these instruments are most often metal discs on which are electroplated known amounts of radioactivity. For this purpose, either ^{210}Po (half-life = 138 days) or ^{239}Pu (half-life = 24,000 years) is commonly used. Both these isotopes are essentially uncontaminated with gamma radiation.

Neutrons

For fast neutrons, commonly used calibration sources include Po–Be, Ra–Be, and Pu–Be. The approximate neutron yield for each of these sources is given in Table 5.3. The exact output of fast neutrons, in neutrons per sec, is usually given by the supplier of the source. For intercomparison of sources, the long counter of Hanson and McKibben may be used with an accuracy of ±5%. The neutron flux at any distance, *d*, from the source, assuming the source to be a "point", is given by

$$\phi \frac{\text{neutrons}}{\text{cm}^2\ \text{sec}} = \frac{N\ \text{neutrons/sec}}{4\pi\, d^2\ \text{cm}^2}. \tag{9.21}$$

All neutron sources supply fast neutrons. For the three α, n sources listed above, the neutron energies span a range from about ½ MeV to about 10 MeV, with a mean energy of about 4–5 MeV (Figs. 5.20, 9.32, and 9.33). In using a fast neutron source for calibration of a dose equivalent neutron survey meter, attention must be paid to the neutron energy spectrum. For example,

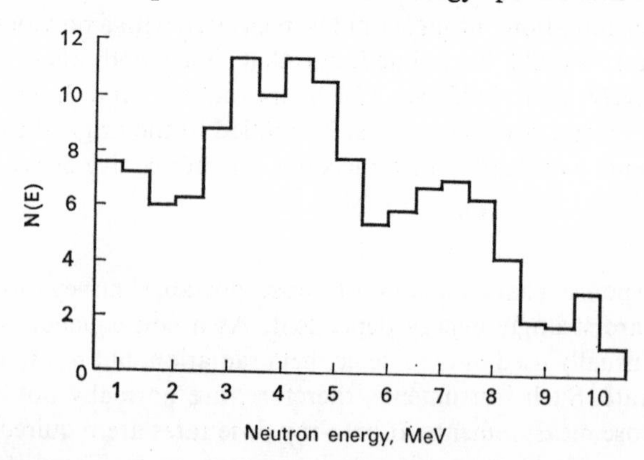

FIG. 9.32. Neutron energy spectrum for Ra–Be (α, n). The neutron spectrum was determined by measuring track lengths in nuclear emulsion. (N.B.S. Handbook 85, *Physical Aspects of Irradiation*, 1964.)

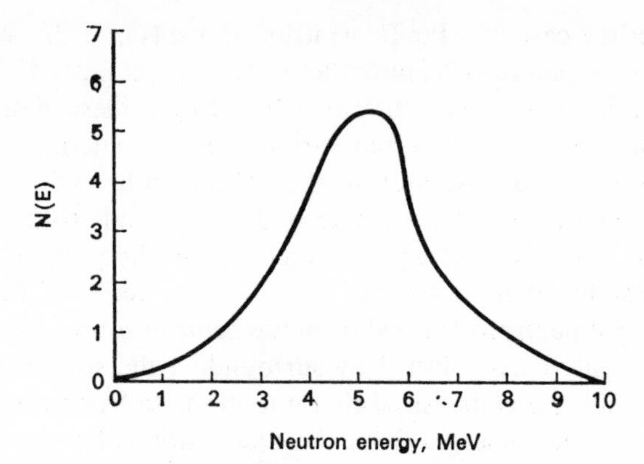

FIG. 9.33. Neutron energy spectrum for Pu–Be (α, n). Nuclear emulsions weie used by Steward (*Phys. Rev.* **98**, 740–3, 1955) for the energy measurements. (From *Neutron Sources and their Characteristics*, Tech. Bulletin NS-1, AECL, Ottawa.)

TABLE 9.5. DATA FOR COMPUTATION OF REM DOSE RATE FROM Pu–Be NEUTRONS

Energy interval E_i	Mean energy $\bar{E}_i$	f_i, fraction of neutrons in E_i	$E_i \times f_i$	Flux ϕ_i, for 2.5 mrems/hr	$\phi_i \times f_i$
0–0.5	0.25	0.038	0.0095	60	2.29
0.5–1	0.75	0.049	0.0368	22	1.05
1–1.5	1.25	0.045	0.0563	18	0.82
1.5–2	1.75	0.042	0.0735	19	0.81
2–2.5	2.25	0.046	0.1035	20	0.92
2.5–3	2.75	0.062	0.1705	20	1.24
3–3.5	3.25	0.077	0.2503	20	1.52
3.5–4	3.75	0.083	0.3113	19	1.61
4–4.5	4.25	0.082	0.3485	19	1.56
4.5–5	4.75	0.076	0.3610	19	1.43
5–5.5	5.25	0.057	0.2993	18	1.05
5.5–6	5.75	0.042	0.2415	18	0.76
6–6.5	6.25	0.042	0.2625	18	0.74
6.5–7	6.75	0.052	0.3510	17	0.89
7–7.5	7.25	0.054	0.3915	17	0.93
7.5–8	7.75	0.051	0.3953	17	0.84
8–8.5	8.25	0.038	0.3135	17	0.64
8.5–9	8.75	0.017	0.1488	17	0.29
9–9.5	9.25	0.018	0.1665	17	0.30
9.5–10	9.75	0.022	0.2145	17	0.37
10–10.5	10.25	0.007	0.0718	17	0.11
		1.000	4.5774		$\Sigma = 20.15$

$$\text{Flux corresponding to } 2.5\ \frac{\text{mrem}}{\text{hr}} = \frac{\Sigma\, \phi_i f_i}{\Sigma f_i} = 20\ \frac{\text{neutrons}}{\text{cm}^2/\text{sec}},$$

$$\text{mean neutron energy} = \frac{\Sigma\, E_i f_i}{\Sigma f_i} = 4.6\ \text{MeV}.$$

let us consider the case of a Pu–Be neutron source (Fig. 9.33) and determine the flux that corresponds to 2.5 mrem/hr, or to 100 mrem per 40-hr week. The spectral distribution is given in Table 9.5. From these data, and using the appropriate QF values for the various neutron energies, expressed as flux that will result in a dose equivalent of 2.5 mrem/hr, we calculate that 20 neutrons per square centimeter per second corresponds to 2.5 mrem/hr. If monoenergetic neutrons are required, then one of the γ, n sources listed in Table 5.3 must be used. If thermal neutrons are required for calibration purposes, the fast neutrons emitted from the neutron source must be thermalized. This is easily accomplished by surrounding the source with paraffin. However, it should be emphasized that not all the fast neutrons are thermalized. The beam that emerges from the moderator is heterochromatic, the moderator thus merely extends the lower end of the neutron energy spectrum to thermal energies. If the moderator is too thin, the neutron beam is too rich in fast neutrons; if the moderator is too thick, the ratio of thermal to fast neutrons increases, but the intensity of the neutron beam is decreased. The thickness of paraffin for maximizing the intensity of thermal neutrons from a calibration source is about 4 in. With this thickness, the thermal flux within a distance of 2 m from the source has been found to be about 40% of the fast neutron flux that would have been obtained without the paraffin moderator. If a relatively constant ratio of thermal to fast neutrons is required, the paraffin moderator thickness must be increased beyond four inches.

The exact ratio of thermal to fast neutrons may be measured by using the neutrons to activate a bare gold foil and a cadmium-covered gold foil. Because of the sharp cadmium absorption cut off at about 0.4 eV, the induced activity in the cadmium shielded foil is due only to neutrons whose energy exceeds 0.4 eV, while the activity of the bare foil is due to thermal neutrons as well as to the higher energy neutrons. The cadmium ratio, which is defined

$$\text{Cd ratio} = \frac{\text{activity of bare foil}}{\text{activity of Cd-covered foil}}, \tag{9.22}$$

is a measure of the purity of a thermal neutron field. The thermal flux may be computed from the gold foil activity measurements with the aid of equation (5.60).

Counting Statistics

Distributions

Radioactive decay and other nuclear reactions are randomly occurring events, and must therefore be described quantitatively in statistical terms. The sampling distribution of a population of randomly occurring events is called the binomial distribution and is given by the expansion of the binomial

$$(p+q)^n = p^n + np^{n-1}q + \frac{n(n-1)}{2!}p^{n-2}q^2 + \frac{n(n-1)(n-2)}{3!}p^{n-3}q^3 + \ldots, \tag{9.23}$$

where p is the mean probability of the occurrence of an event, q is the mean probability of non-occurrence of the event, $p + q = 1$, and n is the number of chances of occurrence. The probability of the occurrence of exactly n events is given by the first term of the binomial expansion, the probability of occurrence of $n - 1$ events is given by the second term, and so on. For example, the likelihood of throwing 3 ones in three consecutive throws of a die, in which the mean probability of throwing a one is 1/6, is given by the first term of the expansion, according to equation (9.23), of $(1/6 + 5/6)^3$:

$$(1/6 + 5/6)^3 = (1/6)^3 + 3\,(1/6)^2\,(5/6) + \frac{3\times 2}{2!}(1/6)\,(5/6)^2 + \frac{3\times 2\times 1}{3!}(5/6)^3$$

$$= 1/216 + 15/216 + 75/216 + 125/216 = 1.$$

The probabilities of 2 ones, 1 one, and no ones are given by the second, third, and fourth terms as 15/216, 75/216, and 125/216, respectively. A plot of these probabilities (Fig. 9.34, curve *A*), shows the distribution to be very asymmetrical. If we were to make similar calculations for the probability of throwing 6, 5, 4,

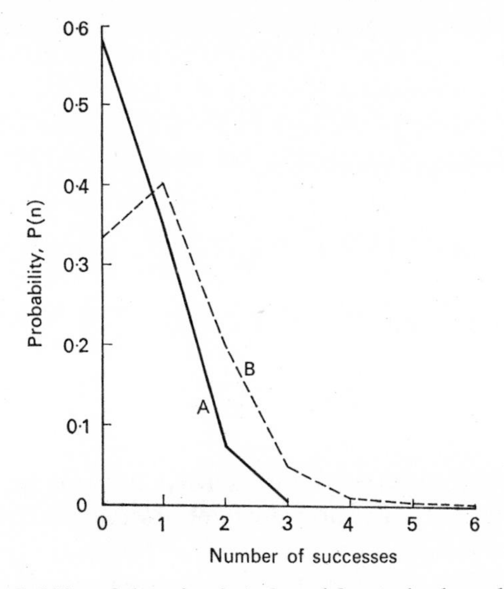

FIG. 9.34. Probability of throwing 0, 1, 2, and 3 ones in three throws of a die, curve *A*; and the probability of throwing 0, 1, 2, 3, 4, 5, and 6 ones in six throws of a die, curve *B*.

3, 2, 1 or 0 ones in six throws, that is $(1/6 + 5/6)^6$, we would find the distribution given in Fig. 9.34, curve *B*. By comparing curves *A* and *B* we see that the latter curve is more symmetrical. As n increases, the distribution curve becomes increasingly symmetrical around the center line. For the case where n is infinite, we have the familiar bell-shaped *normal curve*. For cases where $n \geq 30$ the binomial sampling distribution curve is for practical purposes indistinguishable from a normal curve. The normal distribution, which is given by

$$p(n) = \frac{1}{\sigma\sqrt{(2\pi)}}\, e^{-(n-\bar{n})^2/2\sigma^2}, \tag{9.24}$$

where $p(n)$ = probability of finding exactly n,
$\bar{n}$ = mean value,
σ = standard deviation (also called standard error),

must be fitted by two parameters: the mean and the standard deviation.

For the case where $p <<< 1$, that is where the occurrence of an event is highly improbable, the binomial distribution approaches the *Poisson distribution*. Since radioactive decay of a particular atom is a highly unlikely event (in ^{32}P, for example, whose half-life is 14.3 days, the probability of decay of any particular atom is given by the decay constant as 5.6×10^{-7}/sec), radioactive processes are described by Poisson statistics. According to the Poisson distribution, the probability of the occurrence of exactly n events per unit time, if the true average rate is $\bar{n}$, is given by the $(n + 1)$ term of the expansion of

$$e^{-\bar{n}} \times e^{\bar{n}} = 1, \tag{9.25}$$

which gives, upon series expansion of $e^{\bar{n}}$:

$$e^{-\bar{n}}\left(1 + \frac{\bar{n}}{1!} + \frac{\bar{n}^2}{2!} + \frac{\bar{n}^3}{3!} + \ldots\right), \tag{9.26}$$

or by the general term for the Poisson distribution

$$p(n) = \frac{(\bar{n})^n \times e^{-\bar{n}}}{n!}. \tag{9.27}$$

For example, if we have 0.001 μCi of activity, the mean number of disintegrations per second is 37. The probability of observing exactly 37 disintegrations in 1 sec is calculated from equation (9.27) as

$$p(37) = \frac{37^{37} \times e^{-37}}{37!}.$$

Using Sterling's approximation

$$n! = (2\pi n)^{\frac{1}{2}} \times \left(\frac{n}{e}\right)^n \tag{9.28}$$

to evaluate 37!, we find $p(37)$ to be equal to 0.066. It thus follows that a wide range of observations around the true mean of 37 would be made in a series of measurements of this activity. This distribution of observations is given by the standard deviation of the distribution of measurements. As in the case of the normal distribution, in a large number of measurements, 68 % of all the observations would lie between plus and minus one standard deviation of the mean, 96 % between plus and minus two standard deviations, and so on. The normal distribution, equation (9.24), contains, as one of the parameters, σ, the standard deviation. The Poisson distribution, equation (9.27), contains only one parameter, the mean. The standard deviation of the Poisson distribution is equal to the square root of the mean number of observations made during a given measurement interval:

$$\sigma \text{ (Poisson)} = \sqrt{n}. \tag{9.29}$$

The "size" of the sample when we are counting events such as radioactive decay or other nuclear events is the arbitrary length of time over which we are making the measurement. In equation (9.29), n is equal to the total number of events that occurred during that time of observation, and is thus the average rate for that time interval. Thus, if we observe 10,000 counts during a 10-min counting interval, the standard deviation of the observation is $\sqrt{10{,}000} = 100$ counts per 10 min. The measurement represents a mean value of 10,000 counts for the 10-min measurement interval. One of the main virtues of the Poisson distribution is that, for practical purposes, when $n \geq 20$, it is indistinguishable from a normal distribution of the same mean and of standard deviation equal to the square root of the mean. Under these conditions, all statistical tests that are valid for normal distribution, such as the t-test, the chi-square criterion, and the variance ratio test (which is called the F-test) may also be used for Poisson distributions. Although all these tests are applicable to radioactivity measurements, a full discussion of them is beyond the scope of this book. Details and applications of these tests may be found in the suggested references at the end of this chapter.

In the example cited above, where 10,000 counts were recorded during 10 min of counting, the mean counting rate, in counts per minutes and the standard deviation of the mean rate is given by

$$r \pm \sigma_r = \frac{n}{t} \pm \frac{\sqrt{n}}{t}. \tag{9.30}$$

Since $$\sigma_r = \frac{\sqrt{n}}{t} = \sqrt{\left(\frac{n}{t} \cdot \frac{1}{t}\right)} = \sqrt{\left(r \cdot \frac{1}{t}\right)},$$

we have

$$r \pm \sigma_r = r \pm \sqrt{\frac{r}{t}}, \tag{9.31}$$

which gives 1000 ± 10 cpm. If the activity had been measured over a 1-min interval, and had given 1000 counts, we would have $1000 \pm \sqrt{1000}$ or 1000 ± 32 cpm. The relative error, which is defined as

$$\text{relative error} = \frac{\sqrt{n}}{n}, \tag{9.32}$$

is 3.2% in the case of the 1-min count, but only 1% in the case of the 10-min count. The degree of precision of a given counting measurement is thus seen to depend on the total number of counts.

When making radioactivity measurements, we usually must account for background, and are thus interested in the net counting rate; that is, the difference between the gross counting rate of the sample, which includes background, and the background counting rate. Each of these counting rates has its own standard deviation. The standard deviation of the net counting rate is given as

$$\sigma_n = \sqrt{(\sigma_g^2 + \sigma_{bg}^2)} = \sqrt{\left(\frac{r_g}{t_g} + \frac{r_{bg}}{t_{bg}}\right)}, \tag{9.33}$$

where σ_g = standard deviation of gross counting rate,
σ_{bg} = standard deviation of background counting rate,
r_g = gross counting rate,
r_{bg} = background counting rate,
t_g = time during which gross count was made,
t_{bg} = time during which background count was made.

Example 9.5

A 5-min sample count gave 510 counts, while a 1-hr background measurement yielded 2400 counts. What is the net sample counting rate and the standard deviation of the net counting rate?

$$r_n = \frac{510 \text{ counts}}{5 \text{ min}} - \frac{2400 \text{ counts}}{60 \text{ min}} = 102 \text{ cpm} - 40 \text{ cpm}$$

$$= 62 \text{ cpm}.$$

$$\sigma_n = \sqrt{\left(\frac{102}{5} + \frac{40}{60}\right)} = 4.6$$

62 ± 4.6 cpm.

The standard deviation is a measure of the dispersion of randomly occurring events around the mean. If a large number of replicate measurements were made, it was pointed out above that 68% of the observations would fall within ±1 standard deviation of the mean. For this reason, we say that one standard deviation is the 68% confidence limit. Similarly two standard deviations represent the 96% confidence limit. If we report data within the limits of two standard deviations, this means that we are 96% certain that the true value lies within the limits given. Several levels of confidence, together with their corresponding number of standard deviations, are given in the table below:

Confidence level	Number of standard deviations
50%	0.6745
68%	1.0
90%	1.645
95%	1.960
96%	2.0
99%	2.575

Example 9.6

A preliminary measurement made during a short counting time suggested a gross counting rate of 55 cpm. The background counting rate, determined by a 1 hr measurement, is 25 cpm. How long should the sample be counted in order to be 96% certain that the measured net counting rate will be within 10% of the true counting rate?

The estimated net counting rate is 30 cpm; 10% of this is 3 cpm. Since we want to be at the 96% confidence level, this allowable error of ±3 cpm represents two standard deviations; one standard deviation therefore is 1.5 cpm. Using this value in equation (9.33), and substituting the other given values leads to

$$1.5 = \sqrt{\left(\frac{55}{t_g} + \frac{25}{60}\right)},$$

$$tg = 30 \text{ min.}$$

Difference between Means

In making counting measurements, we are almost always interested in the difference between two counting rates—as, for example, the difference between the sample counting rate and the background counting rate, or the difference between the net counting rates of two samples. If this difference is very great, then we know intuitively that there is a real difference between the two samples. However, if the difference is small, then, because of the fact that radioactive decay is a random process in which we know that 96% of the measurements will lie between the true mean and plus or minus two standard deviations, we cannot decide intuitively whether or not the two counting rates are really different, or whether the observed difference is merely due to errors of random sampling, and that the two measurements are in fact two samples of the same population. To help in our decision, we make use of an objective statistical test based on the *Null Hypothesis*. The null hypothesis assumes that there is no difference between the two measurements. On this assumption, we calculate the probability of finding a difference between the two mean counting rates as great as or greater than that actually observed. Furthermore, we arbitrarily set a limit to this probability. If the calculated likelihood of randomly finding the measured difference is greater than this limit, then we say that there is no difference between the two counting rates, and that the two samples come from the same population. If, on the other hand, the probability of finding the measured difference among samples of the same population is less than the calculated value, then we reject the null hypothesis and we say that the difference between the two means is statistically significant; the samples are in fact different. The two arbitrary levels of significance most frequently used in statistical calculations is 1% and 5%. That means, if we choose the 1% level, that if the probability of randomly finding a difference between two samples of the same population as great as that observed in the two sample measurements is greater than 1 in 100, we accept the null hypothesis, and we say that there is no difference between the two samples. If the calculated probability is 1 in 100 or less, then we say that the difference is significant. It should be emphasized that the significance level is purely arbitrary, and is set by the experimenter. The experimenter is not bound to the 1% or 5% level; he may, if he wishes, be more liberal, and use 10%, or any other significance level. However, if he uses a 10% level, he is more likely to accept an apparent difference, which in fact is not real, than he would if he used 5% or 1% criteria. In reading and interpreting experimental data, therefore, it is important to know the probability level that the experimenter used as a criterion for significance.

Determination of the significance of the difference between means is based on the fact that, in a normally distributed population, not only are means of population samples normally distributed about the true means of the population, but differences between means are also normally distributed. If we should draw a very large number of duplicate samples from a population, compute the mean for each sample, then subtract the mean of the second sample, M_2, from the mean of the first sample, M_1, and plot the differences, we will obtain a normal curve about a mean of zero, whose standard deviation is called the standard error of the difference between means. To estimate the standard error of the difference between means from two samples, we use equation (9.33), which can be written in a more general form as

$$\sigma_{\text{diff}} = \sqrt{(\sigma_{M_1}^2 + \sigma_{M_2}^2)}, \tag{9.34}$$

where σ_M^2 is the square of the standard error of the mean. In counting measurements, the standard error of the mean is given in equation (9.31) which gives the standard error of the mean counting rate as

$$\sigma_M = \sqrt{\frac{r}{t}}. \tag{9.35}$$

The "t" test is used to tell us by how many units of the standard error of the difference between means the difference between two measured means differs from zero:

$$t = \frac{|M_1 - M_2|}{\sigma_{diff}} = \frac{|M_1 - M_2|}{\sqrt{r_1/t_1 + r_2/t_2}}. \tag{9.36}$$

Example 9.7

A sample of drinking water was counted for 10 min, and gave 530 counts. A 30-min background count gave a background rate of 50 cpm. At the 95% confidence level, was there any activity in the water?

$$M_1 = \frac{530 \text{ counts}}{10 \text{ min}} = 53 \text{ cpm} \qquad t_1 = 10 \text{ min},$$

$$M_2 = 50 \text{ cpm} \qquad t_2 = 30 \text{ min},$$

$$t = \frac{53 - 50}{\sqrt{(53/10 + 50/30)}} = 1.13.$$

This result shows that the measured difference differs from zero by 1.13 standard deviations. From a table of areas under a normal curve, we find that 1.13 standard deviations includes 37% of the area on each side of the mean, or 74% of the area between ± 1.13 standard deviations. We interpret this result as saying that we would expect a difference this great or greater 26 times out of 100 if the two samples came from the same population. Therefore, the difference between the two samples is not significant. To be significant at the 95% confidence level, the difference between the two means would have to be great enough to be observed only 5 times or less out of 100.

Problems

1. If a certain counting standard has a mean activity of 1000 cpm,

(a) What is the probability of observing exactly 400 counts in 1 min?

(b) What is the probability of measuring 390–410 counts in 1 min?

2. A sample counted 560 counts in 10 min, while the background counted 390 counts in 15 mins.

(a) What is the standard deviation of the gross and background counting rates?

(b) What is the standard deviation of the net counting rate?

(c) What are the 90% and 99% confidence limits for the net counting rate?

3. A background counting rate of 30 cpm was determined by a 60-min count. A sample that was counted for 5 min gave a gross count of 170.

(a) At the 90% confidence level, is there activity in the sample?

(b) Is there activity in the sample at the 95% confidence level?

4. As a test of the operation of a certain counter, two measurements were made on the same long-lived sample. The first gave 10,210 counts in 10 min, and the second gave 4995 counts in 5 min. Is the counter operating satisfactorily?

5. A 1 min count shows a gross activity of 35 counts. If the background is 1560 counts in 60 min, how long must the sample be counted in order to be within $\pm 10\%$ of the true activity at the 95% confidence level?

6. An ionization chamber has a window thickness of 2 mg/cm^2. If a 0.01 μCi ^{210}Po source is located 1 cm in front of the window, so that the counting geometry is 25%, calculate the saturation ionization current.

7. An "air" wall, air-filled ionization chamber, whose volume is 100 cm^3, gives a saturation current of 10^{-12} A when placed in an X-ray field. If the temperature was 27°C, and the atmospheric pressure was 740 mm Hg, what was the radiation exposure rate?

8. (a) What value resistor, to be placed in series with the ion chamber, problem 7, is required to generate a voltage drop of 10 mV?

(b) If the capacity of the chamber is 250 $\mu\mu$F, what is the time constant of detector circuit?

(c) How much time is required before the meter will read 99% of the saturation current?

9. A pocket dosimeter has a capacitance of 5 $\mu\mu$F and a sensitive volume of 1.5 cm^3. What is the charging voltage if it is to be used in the range 0–200 mR and the voltage across the dosimeter should be one-half the charging voltage when the dosimeter reads 200 mR?

10. A Geiger tube has a capacitance of 25 $\mu\mu$F. The time required to collect all the positive ions is 221×10^{-6} sec. In order to produce sharp output pulses, it is desired to limit the time constant of the detector circuit to 50 μsec.

(a) What is the value of the series resistor?

(b) If 10^8 ion pairs are formed per Geiger pulse, what is the upper limit of the output voltage pulse?

11. A Geiger counter has a resolving time of 250 μsec. What fraction of the counts is lost due to the counter's dead time if the observed counting rate is 30,000 cpm?

12. The fact that the gas multiplication in a proportional counter is very much less than that in a Geiger counter means that a pulse amplifier for use with a proportional counter must have a lower input sensitivity than one used with a Geiger counter. Calculate the input sensitivity for an amplifier to be used with a 2-in. dia hemispherical windowless gas-flow proportional counter whose capacitance is 20 $\mu\mu$F and which is operated to give a gas amplification of 5×10^3. Assume that the output pulse is "clipped" to one-half its maximum height.

13. What is the sensitivity of a thermal neutron detector whose volume is 50 cm^3, and is filled with 96% enriched $^{10}BF_3$ to a total pressure of 70 cm Hg?

14. If the BF_3 tube of problem 13 is used as a current ionization chamber, what saturation current would result from a thermal flux of 10^9 neutrons per cm^2/sec?

15. How long would it take for the sensitivity of the BF_3 detector of problem 13 to decrease by 10%?

16. What is the sensitivity for 1 MeV and for 10 MeV neutrons (amps per neutron per cm^2/sec) of an ion chamber that is filled with CH_4 gas to a pressure of 760 mm Hg, if its volume is 500 cm^3?

17. A 25 mCi ^{60}Co source is lost. At what distance can the lost source be detected with a survey meter whose sensitivity is 0.05 mR/hr above background?

Suggested References

1. HANDLOSER, JOHN S.: *Health Physics Instrumentation*, Pergamon Press, Oxford, 1959.
2. SNELL, ARTHUR, H. (Ed.): *Nuclear Instruments and Their Uses*, Wiley, New York, 1962.
3. PRICE, WILLIAM J.: *Nuclear Radiation Detection*, McGraw-Hill, New York, 1965.
4. *Metrology of Radionuclides*, I.A.E.A., Vienna, 1960.
5. *Measurement of Absorbed Dose of Neutrons, and of Mixtures of Neutrons and Gamma Rays*, N.B.S. Handbook 75, Government Printing Office, Washington, 1961.

6. *Measurement of Neutron Flux and Spectra for Physical and Biological Applications*, N.B.S. Handbook 72, Government Printing Office, Washington, 1960.
7. *A Manual of Radioactivity Procedures*, N.B.S. Handbook 80, Government Printing Office, Washington, 1961.
8. *Photographic Dosimetry of X- and Gamma-Rays*, N.B.S. Handbook 57, Government Printing Office, Washington, 1954.
9. Unruk, C. M., Baumgartner, W. V., Kocher, L. F., Brackenbush, L. W. and Endres, G. W. R.: *Personnel Neutron Dosimeter Developments*, BNWL-SA-537 Batelle —Northwest, Richland, Washington, 1966.
10. Li, J. C. R.: *Introduction to Statistical Inference*, Edwards Bros., Ann Arbor, 1957.
11. Li, C. C.: *Numbers From Experiments*, Boxwood Press, Pittsburgh, 1959.
12. Natrella, M. G.: *Experimental Statistics*, N.B.S. Handbook 91, Government Printing Office, Washington, 1963.

CHAPTER 10

EXTERNAL RADIATION PROTECTION

Basic Principles

Radiation protection practice is a special aspect of the control of environmental health hazards by engineering means. In the industrial environment, the usual procedure is first to try to eliminate the hazard, as was done when benzene was replaced with carbon tetrachloride, and then the carbon tetrachloride was replaced with trichlorethylene as a solvent for degreasing machined parts. If elimination of the hazard is not feasible, an attempt is made to enclose the hazard, thereby isolating the hazard from man. If neither of these two solutions can be achieved, exposure to the hazard can usually be prevented by isolating the man. The exact manner of application of these general principles to radiation protection depends on the individual situation. It is convenient, in radiation-protection practice, to break down the problem to protection against external radiation and protection against personal contamination resulting from inhaled, ingested, or tactilly transmitted radioactivity.

Technics of External Radiation Protection

External radiation originates in X-ray machines and in other devices specifically designed to produce radiation; in devices in which production of X-rays is a side effect, as in the case of the electron microscope; and in radioisotopes. If it is not feasible to do away with the radiation source, then exposure of personnel to external radiation may be controlled by concurrent application of one or more of the following three technics:

1. minimizing exposure time,
2. maximizing distance from the radiation source,
3. shielding the radiation source.

Time

Although many biological effects of radiation are dependent on dose rate, it may be assumed, for purposes of environmental control, that the reciprocity relationship

$$\text{dose rate} \times \text{exposure time} = \text{total dose}$$

is valid. For total dose within one or two orders of magnitude of the RPG value, we have no data, either clinical or experimental, to contraindicate this assumption. Thus, if work must be performed in a relatively high radiation field, such as the repair of a cyclotron made radioactive by the absorption of neutrons, or manipulation of a radiographic source in a complex casting, restriction of exposure time so that the product of dose rate and exposure time does not exceed the maximum allowable total dose permits the work to be done in accordance with radiation safety criteria. For example, in the case of a radiographer who must make radiographs 5 days per week while working in a radiation field of 25 mR/hr, overexposure can be prevented by limiting his daily working time in the radiation field to 48 min. His total daily exposure would then be only 20 mR. If the volume of work requires a longer exposure, then either another radiographer must be used or the operation must be redesigned in order to decrease the intensity of the radiation field in which the radiographer must work.

Distance

Intuitively, it is clear that radiation exposure decreases with increasing distance from a radiation source. When translated to quantitative terms, this fact becomes a powerful tool in radiation safety. We will consider three cases, viz: a point source, a line source, and a surface source.

In the case of a point source, the variation of dose rate with distance is given simply by the inverse square law, equation (9.20). Cobalt 60, for example, which emits one photon of 1.17 MeV and one photon of 1.31 MeV per disintegration, has a source strength which is approximated by

$$\begin{aligned} I &= 6\,\Sigma f_i\,E_i \\ &= 6\,(1 \times 1.17 + 1 \times 1.31) \\ &= 14.9 \text{ Rhf per curie.} \end{aligned}$$

For a 100-mCi source, the exposure dose rate at a distance of 1 ft is about 1490 mR/hr. If a radiographer were to manipulate this source for 1 hr per day, his maximum dose rate should not exceed 20 mR/hr. This restriction could be attained through the use of a remote-handling device whose length, as calculated from the inverse square law, is at least 8.65 ft.

If the radiography is to be done at one end of the shop, which is exclusively set aside for this purpose, then either a barricade must be erected outside of which the dose rate does not exceed the maximum allowable weekly rate, or, if this is not possible because of limitation of space, a shield must be erected. If the barricade is used, its distance from the source must be such that the dose rate at the barricade will not exceed

$$\frac{100 \text{ mR/week}}{40 \text{ hr/week}} = 2.5 \text{ mR/hr.}$$

Using the inverse square law, this distance is found to be 23.8 ft.

In the case of a line source of radiation, such as a pipe carrying contaminated liquid waste, the variation of dose rate with distance is somewhat more complex mathematically than in the case of the point source. If the linear concentration of activity in the line is C_l curies per unit length of a gamma

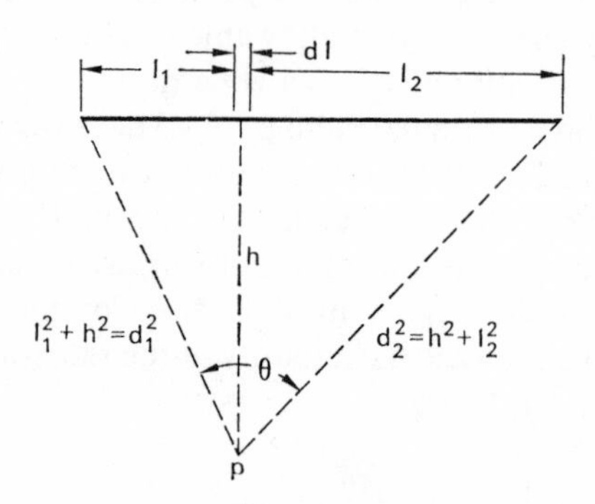

FIG. 10.1. Geometry for computing the gamma-ray dose rate at a finite distance, h, from a line source of uniform activity C_1 mCi per unit length.

emitter whose source strength is Γ, then the dose rate at point p (Fig. 10.1), at a distance h from the infinitesimal length dl, is given by:

$$dD_p = \frac{\Gamma \times C_l \times dl}{l^2 + h^2} \tag{10.1}$$

and for the dose rate due to the activity in the total length of pipe, we have

$$D_p = \Gamma C_l \int_0^{l_1} \frac{dl}{l^2 + h^2} + \Gamma C_l \int_0^{l_2} \frac{dl}{l^2 + h^2}$$

$$= \frac{\Gamma C_l}{h}\left[\tan^{-1}\frac{l^1}{h} + \tan^{-2}\frac{l^2}{h}\right]$$

$$= \frac{\Gamma \times C_l \times \theta}{h}. \tag{10.2}$$

Example 10.1

Induced ^{24}Na activity in a cooling water line passes through a small diameter pipe in an access room 20 ft wide. The door to the room is in the center of the 20-ft wall, at a distance of 10 ft from the pipe, as shown in Fig. 10.2. If the linear concentration of activity is 1 mCi per foot,

(a) What is the dose rate in the doorway (D_1)?

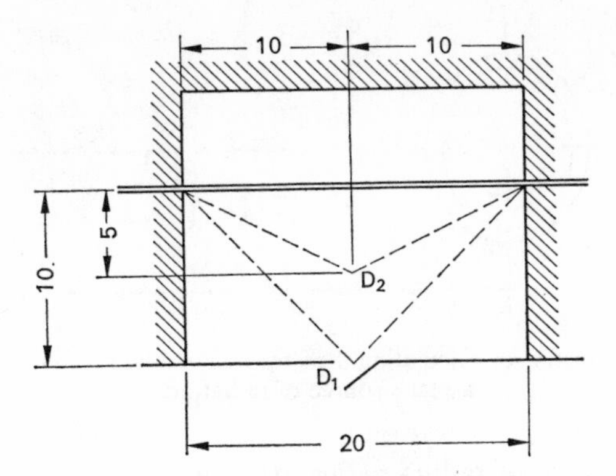

FIG. 10.2. Layout of the room described in Example 10.1.

For ^{24}Na, $\Gamma = 20.6$ R per hour per curie at 1 ft. The dose rate at a distance of 10 ft from the pipe is

$$D_1 = \frac{2\,\Gamma\,C_l}{h}\tan^{-1}\frac{l}{h}$$

$$= \frac{2 \times 20.6 \times 1 \times 10^{-3}}{10}\tan^{-1}\frac{10}{10}$$

$$= 0.412 \times \frac{\pi}{4} = 3.24 \text{ mR per hour.}$$

(b) What is the dose rate midway between the pipe and the door, point D_2, at a distance of 5 ft from the pipe?

The ratio of the two dose rates, D^1 at a distance of 10 ft and D_2 at a distance of 5 ft, is

$$\frac{D_1}{D_2} = \frac{2\,\Gamma\,C_l/h_1 \times \theta_1}{2\,\Gamma\,C_l/h_2 \times \theta_2}$$

$$= \frac{h_2}{h_1} \times \frac{\theta_1}{\theta_2}. \qquad (10.3)$$

In this example, $\theta_1 = 2\tan^{-1} 10/10 = \pi/2$ radians, while $\theta_2 = 2\tan^{-1} 10/5 = 0.7\pi$ radians. Substituting into equation (10.3), we have

$$\frac{3.24}{D_2} = \frac{0.25\;\pi}{0.35\;\pi} \times \frac{5}{10}$$

$$D_2 = 9.15 \text{ mR per hour.}$$

Frequently, the health physicist may find it useful to know the quantitative relationship between dose rate and distance from a plane radiation source. If

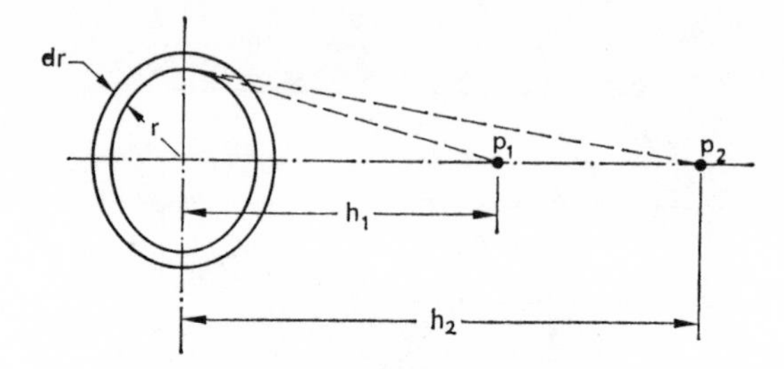

FIG. 10.3. Geometry for calculating the variation of dose rate with distance from a plane source of radiation.

we have a thin source of radius r (Fig. 10.3) and a surface concentration of C_a curies per unit area of a gamma emitter of source strength Γ, the dose rate at point p_1, a distance h_1 along the central axis, is given by:

$$D = \int_0^R \frac{\Gamma \times C_a \times 2\pi r\, dr}{r^2 + h^2} = \pi \times \Gamma \times C_a \times \ln \frac{R^2 + h^2}{h^2}. \tag{10.4}$$

The ratio of the dose rate at a distance h_1 to the dose rate at any other distance h_2 is given by:

$$\frac{D_1}{D_2} = \frac{\ln (R^2 + h_1^2/h_1^2)}{\ln (R^2 + h_2^2/h_2^2)}. \tag{10.5}$$

Example 10.2

A one mCi solution of ^{24}NaCl was spilled over a circular area 2 ft in diameter.

(a) What is the gamma-ray exposure rate at a height of 1 ft over the ground? $\Gamma = 20.6$ R per hour per curie at 1 ft, and $C_a = 10^{-3}/\pi$ Ci/ft^2.

Substituting these values into equation (10.4), we have

$$D = \pi \times 20.6 \times \frac{10^{-3}}{\pi} \ln \left(\frac{1^2 + 1^2}{1^2}\right)$$
$$= 14.25 \times 10^{-3} \text{ R per hour}$$
$$= 14.25 \text{ mR per hour.}$$

(b) What is the exposure rate at a distance of 3 ft above the ground? From equation (10.5), we have

$$D_2 = \frac{\ln (1^2 + 3^2/3^2)}{\ln (1^2 + 1^2/1^2)} \times 14.25$$
$$= 2.14 \text{ mR per hour.}$$

The radiation exposure rate from a thick source containing a uniformly distributed gamma emitting isotope may be estimated from the effective surface activity after allowing for self absorption within the slab. Consider a large slab of thickness t cm (Fig. 10.4), containing C_v μCi/cm^3 of uniformly distributed radioactivity. The linear absorption coefficient of the slab material

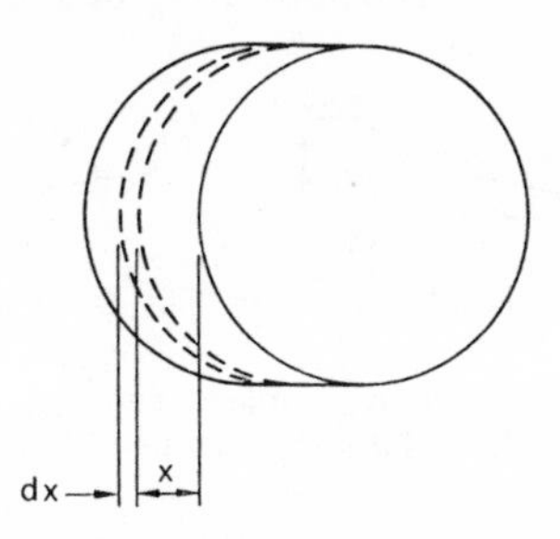

FIG. 10.4. Conditions for setting up equation (10.6).

is μ/cm. The activity on the surface due to the radioactivity in the layer dx, at a depth of x, is

$$d\,(C_a) = C_v \,.\, dx \,.\, e^{-\mu x}. \tag{10.6}$$

Integrating equation (10.6) over the total thickness t yields the effective surface activity:

$$C_a = \int_0^t c_v e^{-\mu x} dx = \frac{C_v}{\mu}(1 - e^{-\mu t}). \tag{10.7}$$

Substituting equation (10.7) into equation (10.4) yields

$$D = \pi\Gamma \frac{C_v}{\mu}(1 - e^{-\mu t}) \ln \frac{R^2 + h^2}{h^2}. \tag{10.8}$$

Shielding

In Chapter 5 we saw that, under conditions of good geometry, the attenuation of a beam of gamma radiation is given by:

$$I = I_0\, e^{-\mu t}. \tag{5.19}$$

However, under conditions of poor geometry, i.e. for a broad beam or for a very thick shield, equation (5.19) underestimates the required shield thickness because it assumes that every photon that interacts with the shield will be removed from the beam, and thus will not be available for counting by the detector. Under conditions of poor geometry (Fig. 10.5) this assumption is not valid; a significant number of photons may be scattered by the shield into the detector, or photons that had been scattered out of the beam may be scattered back in after a second collision. This effect may be illustrated by Fig. 10.6, which shows the broad beam and narrow beam attenuation of ^{60}Co gamma-rays by concrete. To transmit 10% of the incident ^{60}Co radiation,

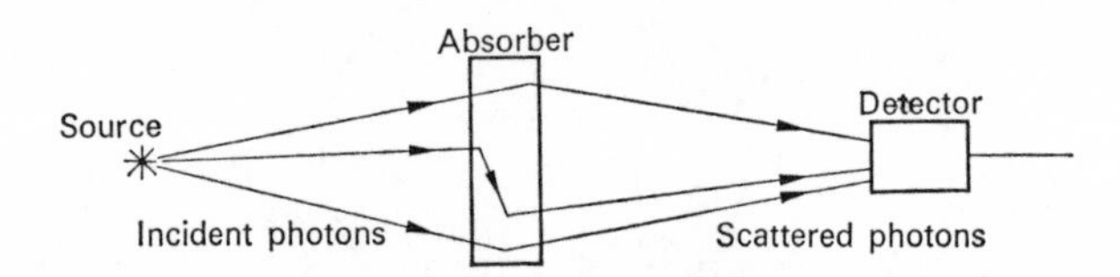

FIG. 10.5. Gamma-ray absorption under conditions of "poor geometry" showing the effect of photons scattered into the detector.

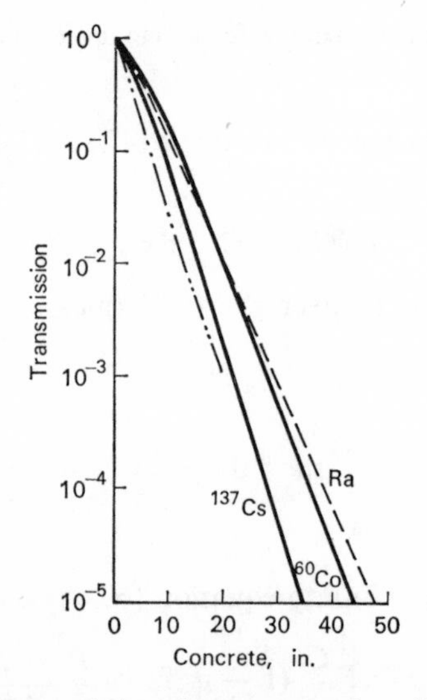

FIG. 10.6. Fractional transmission of gamma rays from ^{137}Cs, ^{60}Co, and Ra (in equilibrium with its decay products) through concrete. The short broken curve represents transmission of ^{60}Co gamma-rays under conditions of good geometry. The other curves represent transmission of broad beams.

Fig. 10.6 shows that, under conditions of good geometry, about 7 in. of concrete shielding is required. For a broad beam, on the other hand, this thickness of concrete will transmit about 25% of the radiation incident on it. To transmit only 10% of a broad beam requires about 11 in. of concrete. When designing a shield against a broad beam of radiation, experimentally determined shielding data for the radiation in question should be used whenever they are available. (Broad beam attenuation curves for radium, ^{60}Co, and ^{137}Cs for concrete, iron, and lead are given in Figs. 10.6, 10.7 and 10.8). When such data are not available, a shield thickness for conditions of poor geometry may be estimated by modification of equation (5.19) through the use of a *build-up factor*, B:

$$I = B \times I_0 \, e^{-\mu t}. \qquad (10.9)$$

The build-up factor, which is always greater than 1, may be defined as the ratio of the intensity of the radiation, including both the primary and scattered radiation, at any point in a beam, to the intensity of the primary radiation only at that point. Build-up factor may apply either to radiation flux or to radiation dose. Build-up factors have been calculated for various gamma-ray energies and for various absorbers.† Some of these values are given in Figs. 10.9 and 10.10.

In these curves the shield thickness is given in units of *relaxation lengths.* One relaxation length is that thickness of shield that will attenuate a narrow beam to $1/e$ of its original intensity. One relaxation length, therefore, is numerically equal to the reciprocal of the absorption coefficient. The use of the build-up factor in the calculation of a shield thickness may be illustrated with the following examples:

Example 10.3

A 1-Ci source of ^{137}Cs is to be stored in a spherical lead container when it is not in use. How thick must the lead be if the exposure at a distance of 1 m from the source is not to exceed 2.5 mR/hr? Assume the source to be sufficiently small to be considered a "point".

Ninety-two per cent of the ^{137}Cs disintegrations are accompanied by a 0.662 MeV gamma-ray. From equation (6.10), we find the source strength of ^{137}Cs to be closely approximated by:

$$\begin{aligned} \Gamma &= 0.48 \, \Sigma f_i E_i \\ &= 0.48 \, (0.92 \times 0.662) \\ &= 0.292 \text{ Rhm per curie.} \end{aligned}$$

In Figs. 10.9 and 10.10, we see that the build-up factor is a function of the shield thickness. Since the shield thickness is not yet known, equation (10.9) has two unknowns, the build-up factor B and the shield thickness t. To determine the proper shield thickness, we assume a thickness, then substitute this estimated value into equation (10.9) to determine whether it will yield the required reduction in radiation dose rate. A minimum shield thickness can be estimated by assuming narrow beam attenuation, and then increasing the thickness thus obtained by about one half-value layer (HVL). The lead

† H. Goldstein and J. E. Wilkins, *Calculation of the Penetration of Gamma Rays*, U.S.A. E.C. report NYO-3075, 1954.

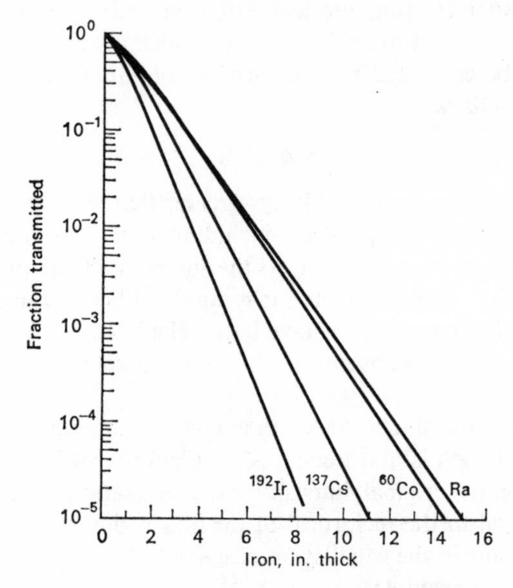

FIG. 10.7. Broad beam attenuation of gamma-rays from ^{192}Ir, ^{137}Cs, ^{60}Co, and radium by iron. (From N.B.S. Handbook 73, *Protection Against Radiations from Sealed Gamma Sources*, 1960.)

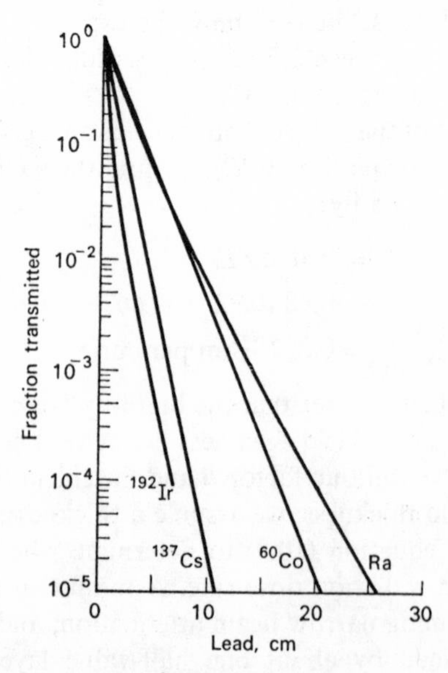

FIG. 10.8. Broad beam attenuation by lead of gamma-rays from ^{192}Ir, ^{137}Cs, ^{60}Co, and radium. (From N.B.S. Handbook 73, *Protection Against Radiations from Sealed Gamma Sources*, 1960.)

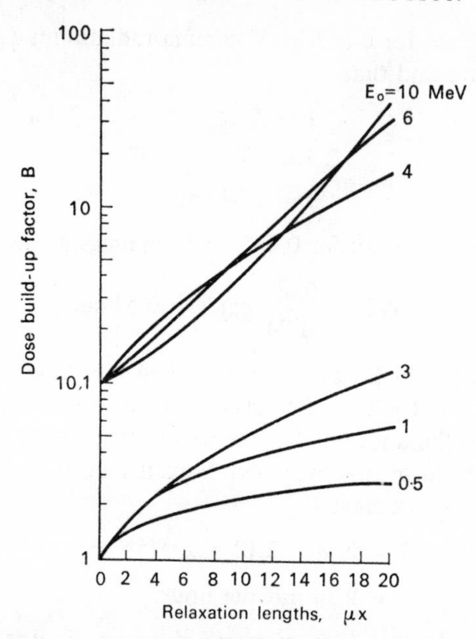

FIG. 10.9. Dose build-up factor in lead for a point isotropic gamma-ray source of energy E_0. (From *Radiological Health Handbook*, 1960.)

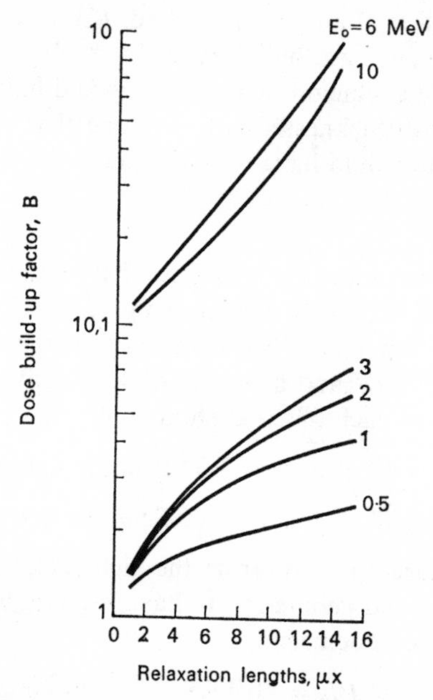

FIG. 10.10. Dose build-up factor in lead for a plane monodirectional gamma-ray source of quantum energy E_0. (From *Radiological Health Handbook*, 1960.)

absorption coefficient for 0.662-MeV gamma-radiation is 1.24 per cm. From equation (5.19), we find that

$$I = I_0 \times e^{-\mu t},$$
$$2.5 = 292 \times e^{-1.24t},$$
$$t = 3.84 \text{ cm}.$$

The half-value layer of lead for 0.662-MeV gamma-rays is

$$\text{HVL} = \frac{0.693}{1.24} \text{cm}^{-1} = 0.59 \text{ cm}.$$

The estimated shield thickness therefore is 3.84 plus 0.59, or 4.43 cm, which corresponds to $1.24 \times 4.43 = 5.5$ relaxation lengths. From Fig. 10.9 we find (by interpolation) the dose build-up factor for 0.662-MeV gamma-rays to be 2.12 for a lead shield of this thickness. Substituting these values for B and t into equation (10.9), we have

$$I = 292 \times 2.12 \times e^{-1.24 \times 4.43}$$
$$= 2.34 \text{ mR per hour}.$$

This calculated reduction in gamma-ray dose rate is just slightly less than the desired value of 2.5 mR/hr. The thickness of 4.43, as calculated above, is therefore just about correct. In this example, we found the correct thickness after one attempt in a trial-and-error method. If the calculated reduction in radiation dose rate would not have turned out to be so close to the design value with the estimated shield thickness, we would have continued, by trial and error, to estimate thicknesses until the one that results in the desired reduction of dose rate would have been obtained.

Example 10.4

Design a spherical lead storage container that will attenuate the radiation dose rate from 1 Ci of ^{24}Na to 10 mR/hr at a distance of 1 m from the source.

In each disintegration of a ^{24}Na atom, two gamma-rays are emitted in cascade: one of 2.75 MeV and a second of 1.37 MeV. The dose rate, at a distance of 1 m, due to each of these photons is

$$\Gamma_{2.75} = 0.48 \times 2.75 = 1.32 \text{ Rhm per curie, and}$$
$$\Gamma_{1.37} = 0.48 \times 1.37 = 0.65 \text{ Rhm per curie.}$$

To reduce the exposure dose rate from the high-energy photon to 10 mR/hr for the condition of good geometry, we have, using a value of 0.475 per cm for the total absorption coefficient,

$$I/I_0 = 10/1320 = e^{-0.475t}$$
$$t = 10.25 \text{ cm}.$$

The half-value layer of lead for this quantum energy is 1.46 cm. Let us add one half HVL, or 0.73 cm, to the thickness calculated above; and then calculate the attenuation of the 2.75-MeV gamma-ray.

$$I = 1320 \times e^{-0.475 \times 10.98}$$
$$= 7.14 \text{ mR per hour.}$$

With this thickness of lead, the 1.37 MeV gamma-ray will be attenuated to

$$I = 657 \times e^{-0.621 \times 10.98}$$
$$= 0.72 \text{ mR per hour.}$$

and the total dose rate at a distance of 1 m will be the sum of the dose rates due to the two quantum energies, or 7.86 mR/hr.

The calculation above we based on good geometry. Let us now account for build-up in the shield. Consider at this time only the high-energy photon, let us add another half of one HVL to the shield thickness, which gives us 11.71 cm. Since one relaxation length of lead for 2.75 MeV gammas is 2.1 cm, the shield thickness corresponds to 5.6 relaxation lengths. From Fig. 10.7 we find the build-up factor to be 3.13. The attenuation of the shield, therefore, according to equation (10.9), is

$$I = 1320 \times 3.13 \times e^{-0.475 \times 11.71}$$
$$= 15.3 \text{ mR per hour.}$$

This exposure dose rate is too high; the shield thickness must therefore be increased. If we add another HVL to the shield to give us 13.17 cm, or 6.27 relaxation lengths, and using the corresponding dose build-up factor of 3.42, we find the exposure dose rate to be 8.5 mR/hr for the high-energy photon. For the lower-energy photon, whose relaxation length in lead is 1.61 cm, this shield thickness corresponds to 8.18 relaxation lengths, and the dose build-up factor is 3.61. Using these values in equation (10.9), the exposure dose rate due to the 1.37-MeV photons is calculated as 0.7 mR/hr. The total exposure dose rate at a distance of 1 m from the shielded ^{24}Na source is thus 9.2 mR/hr. Since this rate may be considered, for most practical purposes, to be equivalent to the design value of 10 mR/hr, the required shield thickness is 13.17 cm. Since the shield may be interposed anywhere between the source and the point where the desired attenuated dose rate is located, and since the volume of lead, for a given wall thickness in a spherical shield, increases rapidly with increasing outer radius according to the expression

$$\text{Volume} = \frac{4}{3}\pi (r_0^3 - r_i^3),$$

the inner radius of the shield is kept as small as possible consistent with the space requirements set by the physical dimensions of the source. The outside radius then is equal to the sum of the inside radius and the shield thickness.

X-ray shielding

Shielding for protection against X-rays is considered under two categories: source shielding and structural shielding. Source shielding is usually supplied by the manufacturer of the X-ray equipment as a lead shield in which the X-ray tube is housed. The safety regulations recommended by the National Committee on Radiation Protection and published by the U.S. Department of Commerce in the pertinent National Bureau of Standards handbooks specify the following two types of protective tube housings for medical X-ray installations:

1. *Diagnostic type*. One that reduces the leakage radiation to at most 0.10 R/hr (100 mR/hr) at a distance of 1 m from the target when the tube is operating at its maximum continuous-rated current for the maximum-rated voltage.
2. *Therapeutic type*. One that reduces the leakage radiation to at most 1.0 R/hr (1000 mR/hr) at a distance of 1 m from the tube target and 30 R/hr (30,000 mR/hr) at any point 5 cm from the surface of the housing when the tube is operating at its maximum continuous-rated current for the maximum-rated voltage.

For non-medical X-rays, a protective tube housing is one which surrounds the X-ray tube itself, or the tube and other parts of the X-ray apparatus (for example, the transformer) and is so constructed that the leakage radiation at a distance of 1 m from the target cannot exceed 1 R in 1 hr when the tube is operated at any of its specified ratings. Leakage radiation, as used in these specifications for tube housings, means all radiation, except the useful beam, coming from the tube housing.

Structural shielding is designed to protect against the useful X-rays, leakage radiation, and scattered radiation. It encloses both the X-ray tube (with its protective tube housing) and the space in which is located the object being irradiated. Structural shielding may vary considerably in form. It may, for example, be either a lead-lined box in the case of an X-ray tube used by a radiobiologist to irradiate small organisms, or it may be the shielding around a room in which a patient is undergoing radiation therapy. In any case, structural shielding is designed to protect people in an occupied area outside an area of high radiation intensity. The structural shielding requirements for a given installation are determined by:

1. The maximum kilovoltage at which the X-ray tube is operated.
2. The maximum milliamperes of beam current.

3. The workload (W), which is a measure, in suitable units, of the amount of use of an X-ray machine. For X-ray shielding design, workload is usually expressed in units of milliampere-minutes per week.
4. The use factor (U), which is the fraction of the workload during which the useful beam is pointed in the direction under consideration.
5. The occupancy factor (T), which is the factor by which the workload should be multiplied to correct for the degree or type of occupancy of the area in question. When adequate occupancy data are not available, the values for T given in Table 10.1 may be used as a guide in planning shielding.

TABLE 10.1. OCCUPANCY FACTORS

Full occupancy $T = 1$	Control space, wards, workrooms, darkrooms, corridors large enough to hold desks, waiting rooms, restrooms used by occupationally exposed personnel, children's play areas, living quarters, occupied space in adjacent buildings
Partial occupancy $T = 1/4$	Corridors too narrow for desks, utility rooms, rest rooms not used routinely by occupationally exposed personnel, elevators using operators, and uncontrolled parking lots
Occasional occupancy $T = 1/16$	Stairways, automatic elevators, outside areas used only for pedestrians or vehicular traffic, closets too small for future workrooms, toilets not used routinely by occupationally exposed personnel

For design purposes, structural shielding is divided into the two categories of primary and secondary protective barriers. The primary barrier is designed to protect against the useful beam, while the secondary barrier is designed to protect against leakage and scattered radiation. Methods for the calculation of the thickness of each of these protective barriers are given below.

1. *Primary protective barrier*

The maximum dose rate at any occupied point at a distance d meters from the target in the X-ray tube is given by

$$D_m = \frac{P}{T} \text{ R/week,} \tag{10.10}$$

where P is the maximum permissible weekly dose rate (0.1 R/week for controlled areas and 0.01 R/week for uncontrolled areas) and T is the occupancy factor. By applying the inverse square law, we find this radiation field to have a dose rate at one meter from the target that is given by

$$D_1 = d^2 \times D_m = \frac{d^2 P}{T} \text{ R/week at 1 meter.} \tag{10.11}$$

This dose rate is due to the workload WU mA minutes per week. Let us now define the ratio

$$K = \frac{D_1}{WU} = \frac{d^2 P}{WUT} \frac{R/\text{mA} - \text{min}}{\text{week}} \text{ at 1 meter,} \tag{10.12}$$

and measure this value for broad beams of X-rays of various energies that have been transmitted through lead or concrete shields of varying thicknesses. The results of these measurements are shown in Figs. 10.11 through 10.14. The transmission of X-rays through thick shields has been found experimentally to depend mainly on the highest energy photons in the beam, and, for a beam of any given minimum wavelength, to be influenced relatively little

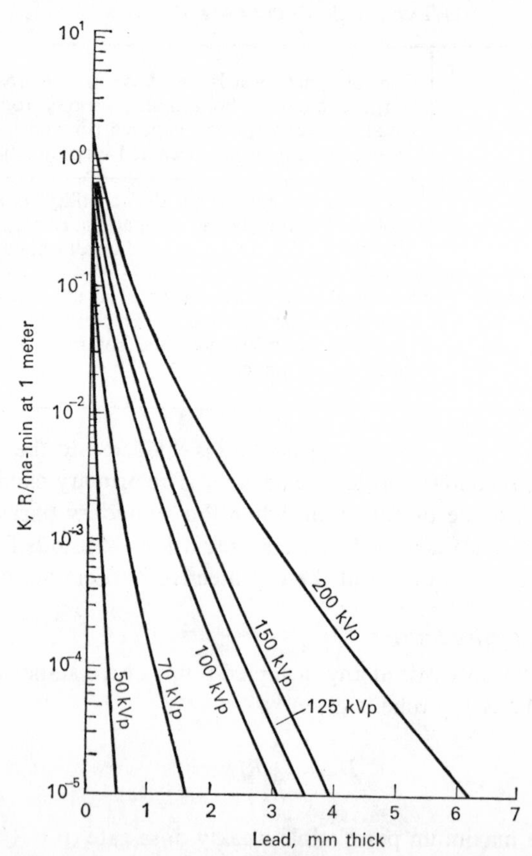

FIG. 10.11. Broad beam attenuation in lead of X-rays produced by potentials of 50–200 kV peak. The measurements were made with a 90° angle between the electron beam and the axis of the pulsed wave form X-ray beam. The 50, 70, 100, and 125 kVp X-rays were filtered with 0.5 mm aluminum; the 150 and 200 kVp X-rays were filtered with 3 mm aluminum. (From N.B.S. Handbook 76, *Medical X-ray Protection up to Three Million Volts*, Government Printing Office, Washington, 1961.)

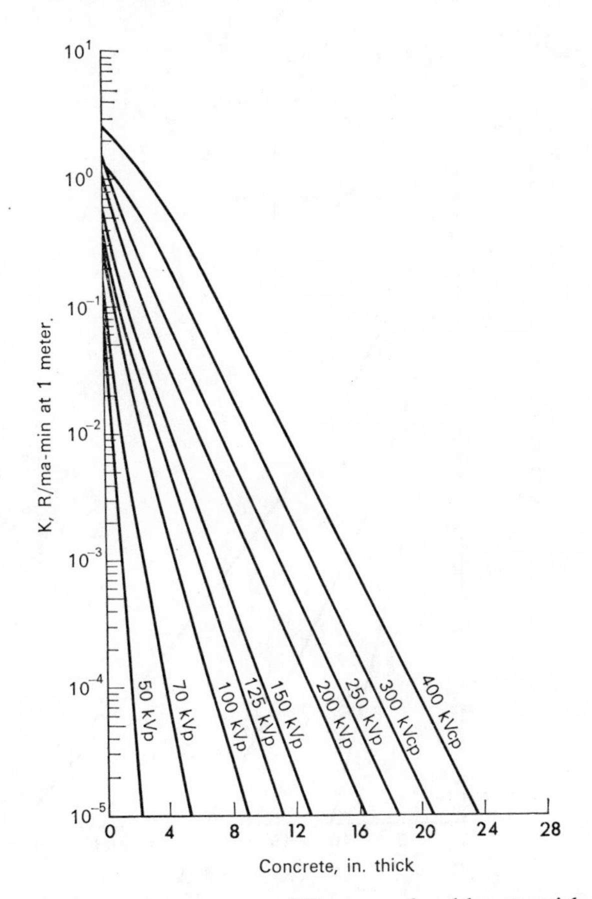

FIG. 10.12. Attenuation in concrete of X-rays produced by potentials of 50 to 400 kV. The measurements were made with an angle of 90° between the electron beam and the axis of the X-ray beam. The curves for 50–250 kVp are for a pulsed wave form. The filtrations were 1 mm Al for 50 kVp, 1.5 mm Al for 70 kVp, 2 mm Al for 100 kVp, and 3 mm Al for 125–250 kVp. The 300 and 400 kVcp were interpolated from data obtained with a constant potential generator and inherent filtration of approximately 3 mm Cu. (From N.B.S. Handbook 76, *Medical X-ray Protection up to Three Million Volts*, Government Printing Office, Washington, 1961.)

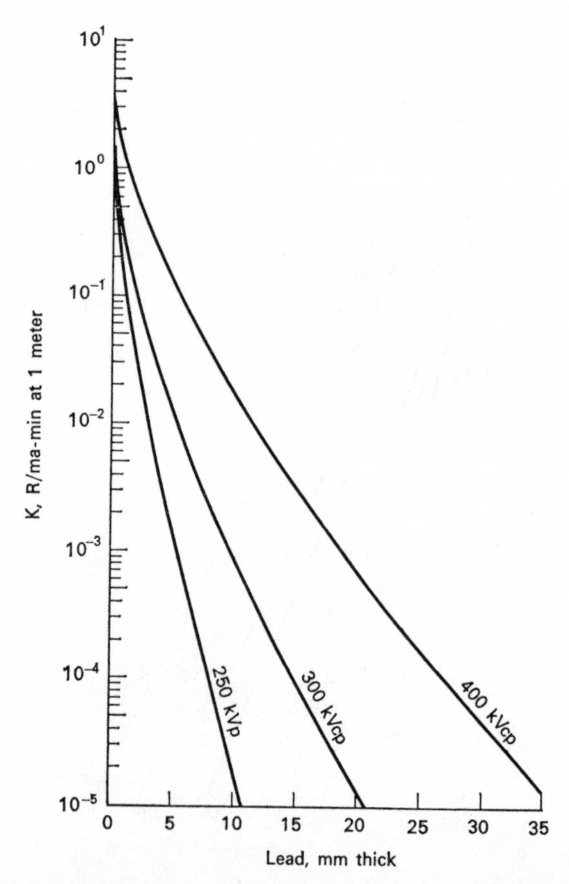

FIG. 10.13. Attenuation in lead of X-rays produced by potentials of 250–400 kV. The measurements were made with an angle of 90° between the axes of the electron beam and the X-ray beam. The 250 kVp curve is for a pulsed wave form and a filtration of 3 mm Al. The 300 and 400 kVcp curves were interpolated from data obtained with a constant potential generator and inherent filtration of approximately 3 mm Cu. (From N.B.S. Handbook 76, *Medical X-ray Protection up to Three Million Volts*, Government Printing Office, Washington, 1961.)

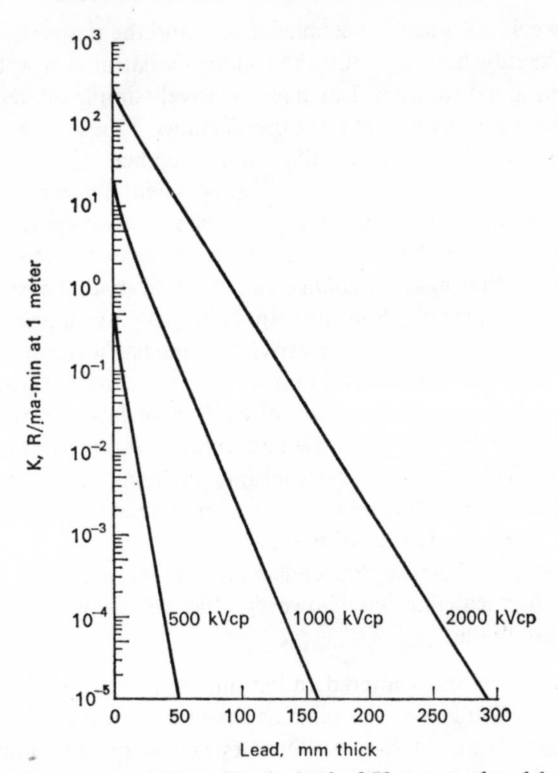

FIG. 10.14. Broad beam attenuation in lead of X-rays produced by constant potentials of 500 to 2000 kV. The measurements were made with a 0° angle between the electron beam and the axis of the X-ray beam, and with a constant potential generator. The 500 and 1000 kVcp curves were obtained with filtration of 2.8 mm tungsten, 2.8 mm Cu, 2.1 mm brass, and 18.7 mm water. For the 2000 kVcp curve, the inherent filtration was equivalent to 6.8 mm lead. (From N.B.S. Handbook 76, *Medical X-ray Protection up to Three Million Volts*, Government Printing Office, Washington, 1961.)

by the quality of the beam (that is, by the half-value layer for that beam). Accordingly, the X-ray transmission data shown in Figs. 10.11 through 10.14 may be used for the design of shielding against any X-rays within the range of the maximum potentials shown in the figures. To design the primary protective barrier, the value of K is computed from equation (10.12) and the required barrier thickness is read from the appropriate figure at the intersection of the ordinate, K, with the curve representing the X-ray energy. The use of the curves in the design of X-ray shielding is illustrated in Example 10.5, which follows the discussion of secondary protective barriers.

2. *Secondary protective barrier*

The requirements for the secondary protective barrier are determined by the operating conditions of the X-ray tube, including kilovoltage, milliamp minutes per week, distance to occupied areas, and the nature of the occupied areas. Since the tube housing limits the leakage radiation dose rate to a known and fixed value at a distance of 1 m, it is a relatively simple matter to determine the required barrier thickness at any other distance if the energy of the leakage radiation is known. The leakage radiation, having been filtered and hardened by passing through the lead tube housing, is essentially monochromatic; its half-value layer therefore depends only on the kilovoltage across the tube. Tables 10.2 and 10.3 list the required thicknesses, as the number of half-value layers, for protection against leakage radiation from diagnostic and therapeutic, and protective tube housings. In Table 10.4 are tabulated the thicknesses of lead and concrete that correspond to one half-value layer for various tube potentials. The required thickness of either concrete or lead is obtained by multiplying the required number of half-value layers from Tables 10.2 or 10.3 by the actual thickness of the half-value found under the appropriate kilo-voltage in Table 10.4. For structural shielding made of commercial building materials other than concrete, the concrete equivalent thickness may be calculated from the density of the material.

Scattered radiation intensity depends on a number of factors. For purposes of barrier design calculations, however, the following three simplifying assumptions are made:

1. The energy of the scattered radiation, for X-rays whose energy does not exceed 500 keV is the same as the energy in the primary beam.
2. A primary beam of X-rays whose energy is greater than 500 keV is degraded in energy to 500 keV after being scattered.

These simplifications are based on the fact that, for quantum energies of 500 keV or less, a photon loses relatively little energy in a scattering interaction, while higher energy X-rays that are scattered through 90° are degraded in energy to a quality approximately equivalent in energy distribution to X-rays generated by a potential of 500 kV. For X-ray energies up to 500 keV, therefore, the transmission curves for the respective kilovoltages are used, while for all X-rays of higher energy, the 500-kV transmission curves are used in the design of shielding against scattered X-rays. The third simplification is:

3. The dose rate of radiation scattered through 90°, at a distance of 1 m from the scatterer, is 0.1% of the incident dose rate at the scatterer.

Since it is assumed, for the purpose of estimating shielding thickness, that the quality of scattered high-energy X-rays is equivalent to that of a beam generated at 500 kV, the 500-kV transmission curve is used for all scattered supervoltage X-rays. However, the output of an X-ray machine increases as the kilovoltage increases. To account for a scattered dose rate only 0.1% as great as the incident dose rate, K in equation (10.9) is increased by a factor

TABLE 10.2. SECONDARY BARRIER REQUIREMENTS FOR LEAKAGE RADIATION FROM DIAGNOSTIC-TYPE PROTECTIVE TUBE HOUSINGS FOR CONTROLLED AREAS

Distance from target (ft)	Operating time, hours per week					
	2	5	10	15	25	40
	Number of half-value layers					
3	1.3	2.6	3.6	4.2	4.9	5.6
4	0.5	1.8	2.8	3.4	4.1	4.8
5	—	1.2	2.1	2.7	3.5	4.1
6	—	0.6	1.6	2.2	2.9	3.6
7	—	0.2	1.2	1.8	2.5	3.2
8	—	—	0.8	1.3	2.1	2.8
9	—	—	0.4	1.0	1.7	2.4
10	—	—	0.2	0.7	1.4	2.1
12	—	—	—	0.3	0.9	1.6

From N.B.S. Handbook 76, *Medical X-ray Protection up to Three Million Volts*, Government Printing Office, Washington, 1961.

TABLE 10.3. SECONDARY BARRIER REQUIREMENTS FOR LEAKAGE RADIATION FROM THERAPEUTIC-TYPE PROTECTIVE TUBE HOUSINGS FOR CONTROLLED AREAS

Distance from target (ft)	Operating time, hours per week					
	2	5	10	15	25	40
	Number of half-value layers					
3	4.6	5.9	6.9	7.5	8.2	8.9
4	3.8	5.1	6.1	6.7	7.4	8.1
5	3.2	4.5	5.5	6.1	6.8	7.4
6	2.6	3.9	4.9	5.5	6.3	7.0
7	2.2	3.5	4.5	5.1	5.8	6.5
8	1.8	3.1	4.1	4.7	5.4	6.1
9	1.5	2.8	3.7	4.3	5.1	5.8
10	1.2	2.5	3.5	4.0	4.8	5.5
12	0.6	1.9	2.9	3.5	4.3	4.9
15	0	1.3	2.3	2.9	3.6	4.3
20	—	0.5	1.5	2.0	2.8	3.5
30	—	—	0.3	0.9	1.6	2.3

From N.B.S. Handbook 76, *Medical X-ray Protection up to Three Million Volts*, Government Printing Office, Washington, 1961.

TABLE 10.4. HALF-VALUE LAYERS

	Approximate values obtained at high filtration for the indicated peak kilovolts across the tube under broad beam conditions											
						Half-value layer						
Attenuating material	50	70	100	125	150	250	250	300	400	500	1000	2000	3000
Lead, mm	0.05	0.18	0.24	0.27	0.3	0.5	0.8	1.3	2.2	3.6	8.0	12.0	15.0
Concrete, in.	0.2	0.5	0.7	0.8	0.9	1.0	1.1	1.2	1.3	1.4	1.8	2.45	2.95
Concrete, cm	0.51	1.27	1.8	2.0	2.3	2.5	2.8	3.0	3.3	3.6	4.6	6.2	7.5

From N.B.S. Handbook 76, *Medical X-ray Protection up to Three Million Volts*, Government Printing Office, Washington, 1961.

of 1000; to account for the increasing X-ray output with increasing kilovoltage, K is reduced by a factor f, that depends on the kilovoltage. Applying these corrections to equation (10.12), and using a value of 1 for the use factor U (since the scattered radiation is assumed to be isotropic), we have

$$K = \frac{1000 \times d^2 \times P}{f \times W \times T} \text{ R/mA} - \text{min/week} \qquad (10.13)$$

at a distance of 1 m from the scatterer. Values commonly assigned to f are:

kV	f
500 or less	1
1000	20
2000	120
3000	300

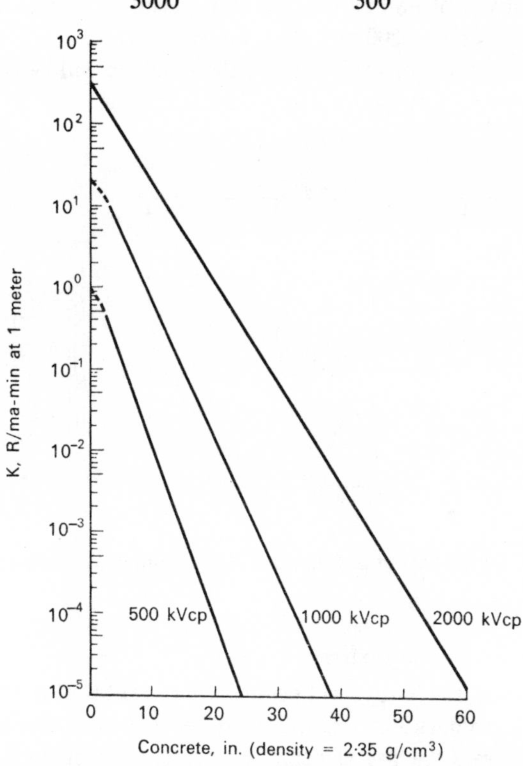

FIG. 10.15. Broad beam attenuation in concrete of X-rays produced by potentials of 500 to 2000 kVcp. The measurements were made with a 0° angle between the electron beam and the axis of the X-ray beam, and with a constant potential generator. The 500 and 1000 kVcp curves were obtained with filtration of 2.8 mm Cu, 2.1 mm brass, and 18.7 mm water. For the 2000 kVcp curve, the inherent filtration was equivalent to 6.8 mm lead. (From N.B.S. Handbook 76, *Medical X-ray Protection up to Three Million Volts*, Government Printing Office, Washington, 1961.)

The barrier thicknesses for leakage and scattered radiation are determined separately. If they are found to be about the same, then the thicker one may be increased by one half-value layer to obtain the required thickness. If one of the barriers is at least three half-value layers greater than the other, then it alone will suffice.

Example 10.5

The room shown in Fig. 10.16 will be used for diagnostic radiology. From the floor to the ceiling is 9 ft 7 in. The room above is another office that is not controlled by the radiologist. The X-ray room is on the ground floor of the building; there is no occupied space below. The floor and ceiling are made of concrete 5 in. thick, wall *A* is made of concrete 4 in. thick. Walls *B*, *C*, and *D* are made of hollow tile and thin plaster. The maximum machine ratings are 125 kVcp and 200 mA. As a fluoroscope, the machine will operate 15 hr/week at $3\frac{1}{2}$ mA. For radiography, the machine will be operated at 200 mA for 5 min/week. Compute the required shielding.

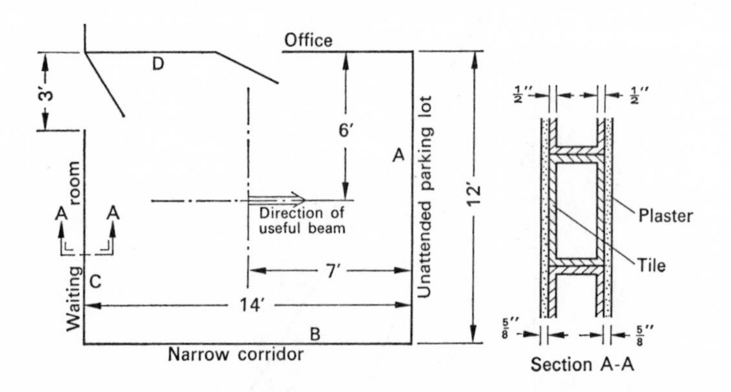

FIG. 10.16. Layout of diagnostic X-ray room of Example 10.5.

To compute the primary barrier:

For radiography, the X-rays will be directed only at wall *A*. This wall, therefore, must be a primary barrier, and all other walls are secondary barriers. During fluoroscopy, the useful beam always is intercepted by the viewing screen, which is shielded by leaded glass. No other primary barrier is needed. Considering the radiographic operation only, therefore, the workload is

$$W = 5 \text{ min/week} \times 200 \text{ mA} = 1000 \text{ mA} - \text{min/week},$$

the use factor, *U*, is equal to 1, and the occupancy factor, *T*, is equal to $\frac{1}{4}$ (for a parking lot). The distance from the target in the X-ray tube to the wall is

7 ft (or 7/3.28 m), and the maximum permissible weekly dose rate to an uncontrolled area is 0.01 R. Substituting these values into equation (10.12) we have

$$K = \frac{(7/3.28)^2 \times 0.01}{1000 \times 1 \times \frac{1}{4}} = 1.83 \times 10^{-4} \text{ R/ma} - \text{min at 1 m.}$$

From Fig. 10.12, using the curve for 125 kVp, we find that 8 in. of concrete are required. The wall is only 4 in. thick, therefore lead of thickness equivalent to 4 in. of concrete must be added to the wall. From Table 10.4, we find the half-value layer for 125-kVp X-rays to be 0.8 in. concrete or 0.27-mm lead. Since 4 in. concrete is equal to five half-value layers, the thickness of the additional lead is 5 times 0.27, or 1.35 mm. For fluoroscopy, the useful beam always is intercepted by the leaded glass of the viewing screen, and no other primary protective barrier is needed.

To compute the secondary barrier:

(a) *Ceiling: leakage radiation.* Assume the X-ray tube to be midway between the floor and the floor above, at a height of 5 ft above the floor. For design purposes, when the radiation originates below the floor, the dose rate at a height of 3 ft above the floor is considered as the criterion for safety. In this case, therefore, we have a total of 8 ft between the target in the X-ray machine and the point that we wish to protect. Since the X-ray machine will operate 15 hr/week, the required barrier thickness against leakage is, according to Table 10.2, 1.3 + 3.3 (because the office upstairs is an uncontrolled area) = 4.6 half-value layers. According to Table 10.4, the half-value layer for 125-kVp X-rays is 0.8 in. concrete; the required thickness of concrete is, therefore, 3.68 in. The floor is already 4 in. thick, and thus already exceeds the required thickness.

(b) *Ceiling: scattered radiation.* Equation (10.13) is used to calculate the required barrier thickness. The value for the workload is

$$W = \left[15 \frac{\text{hr}}{\text{wk}} \times 60 \frac{\text{min}}{\text{hr}} \times 3.5 \text{ mA}\right] + 5 \frac{\text{min}}{\text{wk}} \times 200 \text{ mA}$$
$$= 4150 \text{ mA} - \text{min/week.}$$

Using the following values for the other terms in equation (10.13): $f = 1$; $T = 1$; $d = 8$ ft (from the X-ray tube target at a height of 5 ft to 3 ft above the floor in the upstairs office); $P = 0.01$ R/week, we have

$$K = \frac{1000 \times (8/3.28)^2 \times 0.01}{1 \times 4150 \times 1} = 0.014 \text{ R/mA} - \text{min/week at 1 m.}$$

From Fig. 10.12, we read that 3.2 in. concrete are required. This value is slightly less than that needed to protect against the leakage radiation. We therefore add one half-value layer to the thicker barrier to give us 3.68 plus 0.8, or 4.48 in. concrete. Since the ceiling already is 5 in. thick, no additional shielding is necessary.

(c) *Wall B: leakage radiation.* From Table 10.2, we find the barrier thickness to be $2.2 + 3.3 = 5.5$ half-value layers. However, since the occupancy factor for a narrow corridor is 1/4, the allowable radiation dose rate may be increased by a factor of 4. This is accomplished by decreasing the barrier thickness by two half-value layers, giving 3.5 half-value layers as the required thickness. The wall is made of hollow tile and plaster, as shown in the diagram. The concrete equivalent of the wall is calculated as follows, using the densities of commerical building materials listed in Table 10.5:

$$(\text{density thickness})_{\text{wall}} = (\text{density thickness})_{\text{concrete}},$$

$$2.54\,\frac{\text{cm}}{\text{in}}\left[1.54\,\frac{\text{gm}}{\text{cm}^3} \times 2 \times \tfrac{5}{8}\text{ in.} + 1.93\,\frac{\text{gm}}{\text{cm}^3} \times 2 \times \tfrac{1}{2}\text{ in.}\right]$$

$$= 2.54\,\frac{\text{cm}}{\text{in.}} \times 2.35\,\frac{\text{g}}{\text{cm}^3} \times \chi\text{ in.}$$

$$\chi = 1.62\text{ in. concrete,}$$

or two half-value layers. We therefore need 1.5 additional half-value layers, or

$$1.5\text{ HVL} \times 0.27\text{ mm Pb/HVL} = 0.405\text{ mm lead.}$$

(d) *Wall B: scattered radiation.* Substituting the appropriate values into equation (10.13), we have

$$K = \frac{1000 \times (6/3.28)^2 \times 0.01}{1 \times 4150 \times 1/4} = 0.032\text{ R/mA} - \text{min/week at 1 m.}$$

From Fig. 10.11, we find that 0.6 mm lead, or two half-value layers, are required. However, the wall already is two half-value layers thick, and is therefore sufficient for protection against the scattered radiation. Since the barriers calculated for leakage and scattered radiation differ by less than three half-value layers, the required secondary barrier thickness is obtained by adding one half-value layer to the thicker barrier, giving $0.41 + 0.27 = 0.68$ mm lead that must be added to the plastered tile wall.

TABLE 10.5. DENSITIES OF COMMERCIAL BUILDING MATERIALS

Material	Density range g/cm³	Density of average sample g/cm³
Brick	1.6–2.5	1.9
Granite	2.60–2.70	2.63
Limestone	1.87–2.69	2.30
Marble	2.47–2.86	2.70
Sandplaster	—	1.54
Sandstone	1.90–2.69	2.20
Siliceous concrete	2.25–2.40	2.35
Tile	1.6–2.5	1.9

TABLE 10.6. COMMERCIAL LEAD SHEETS

Thickness mm	Thickness in.	Weight #/ft²
0.79	1/32	2
1.00	5/128	2½
1.19	3/64	3
1.58	1/16	4
1.98	5/64	5
2.38	3/32	6
3.17	1/8	8
4.76	3/16	12
6.35	1/4	16
8.50	1/3	20
10.1	2/5	24
12.7	1/2	30
16.9	2/3	40
25.4	1	60

Using the same type of reasoning as above, and applying the appropriate factors, we find the shielding requirements for the X-ray room to be those listed below:

Location	Additional lead, calculated mm	Additional lead, commercial mm
Wall *A*	1.35	1.58
Wall *B*	0.68	0.79
Wall *C*	0.22	0.79
Door in wall *C*	0.76	0.79
Wall *D*	0.95	1.00
Door in wall *D*	1.5	1.58
Floor	None	
Ceiling	None	

Commercially available lead sheet is listed in Table 10.6. In the example above, the commercial sheet of the next thickness greater than the calculated value would be used in the actual installation. When shielding the walls of an X-ray room, the lead should extend to a height of at least 7 ft.

Beta-ray shielding

Two factors must be considered when designing a shield against high-intensity radiation, viz. the beta-rays and the bremsstrahlung that are generated due to absorption in the source itself and in the shield. Because of these factors, the beta shield consists of a low-atomic-numbered substance (to minimize the production of bremsstrahlung) sufficiently thick to stop all the beta-rays, followed by a high-atomic-numbered material thick enough to attenuate the bremsstrahlung intensity to a safe level.

Example 10.6

Fifty milliliters of aqueous solution containing 10 Ci carrier-free ^{90}Sr in equilibrium with ^{90}Y, is to be stored in a laboratory. The health physicist requires the exposure rate at a distance of 1 m from the source to be no greater than 10 mR/hr. Design a source holder and shield to meet the specifications for radiation safety during storage.

The maximum and mean beta-ray energies of ^{90}Sr and ^{90}Y are:

	E_{max}, MeV	$\bar{E}$, MeV
^{90}Sr	0.54	0.19
^{90}Y	2.27	0.93

The beta shield must be sufficiently thick to stop the 2.27 MeV ^{90}Y betas. From Fig. 5.4, the range of a 2.27 MeV beta is found to be 1.1 g/cm². Let us use a bottle made of polyethylene, specific gravity 0.95, as the container for the radioactive solution. The wall thickness, therefore, must be

$$E = \frac{1.1 \text{ g/cm}^2}{0.95 \text{ g/cm}^3} = 1.06 \text{ cm}.$$

The bremsstrahlung are especially important, since ^{90}Sr and ^{90}Y are pure beta emitters, and thus there are ordinarily no gammas to be shielded. The radiation dose rate due to the bremsstrahlung at a distance of 1 m from the source may be estimated with the aid of equation (5.11),

$$f = 3.5 \times 10^{-4} Z \times E_{max}.$$

In this case, most of the beta-ray energy will be absorbed in the water. If we use for Z, in equation (5.11), the effective atomic number for water,

$$Z_w = \frac{2}{18} \times 1 + \frac{16}{18} \times 8 = 7.22,$$

and 2.27 MeV for E_{max}, we find f, the fraction of the ^{90}Y beta-ray energy that is converted into bremsstrahlung to be 5.73×10^{-3}. The rate at which energy is carried off by the ^{90}Y betas is

$$E_\beta = 10 \text{ Ci} \times 3.7 \times 10^{10} \frac{\text{dps}}{\text{Ci}} \times 0.93 \frac{\text{MeV}}{d} = 3.44 \times 10^{11} \frac{\text{MeV}}{\text{sec}}.$$

If the bremsstrahlung are considered to radiate from a virtual point in the center of the source, then the exposure dose rate at a distance of 1 m from this point is given by:

$$D = \frac{f \times E_\beta \text{ MeV/sec} \times 1.6 \times 10^{-6} \text{ erg/MeV} \times \mu_e \text{cm}^{-1} \times 3.6 \times 10^3 \text{ sec/hr} \times \times 10^3 \text{ mR/R}}{1.293 \times 10^{-3} \text{ g/cm}^3 \times 4\pi (100 \text{ cm})^2 \times 87.8 \text{ erg/g/R}}, \tag{10.14}$$

where μ_e is the energy absorption coefficient corresponding to the quantum energy of the bremsstrahlung. In calculating bremsstrahlung dose rate for purposes of radiation protection, the quantum energy is assumed to be equal to that of the maximum energy beta particle. Using the value $\mu = 3 \times 10^{-5}$ cm^{-1} (from Fig. 5.18), the bremsstrahlung dose rate due to the ^{90}Y beta rays is 23.9 mR/hr at 1 m. Similarly, the dose rate due to ^{90}Sr betas is 1.5 mR/hr.

If we have a shield of thickness t cm, then the dose rate attenuation from the bremsstrahling can be calculated from equation (5.19), using a value of 0.51 cm^{-1} for the total linear absorption coefficient for lead for the 2.27 MeV X-rays from ^{90}Y and 1.5 cm^{-1} for the 0.54 MeV X-rays from ^{90}Sr. Furthermore, the sum of the ^{90}Y and ^{90}Sr dose rates must not exceed 10 mR/hr. These considerations give the following set of simultaneous equations

$$I_Y = 23.9\, e^{-0.51t},$$

$$I_{Sr} = 1.5\, e^{-1.5t},$$

$$I_Y + I_{Sr} = 10,$$

which can be solved to give $t = 1.75$ cm.

Consideration of a build-up factor in this case is not necessary because it was assumed, in the calculation of the thickness of lead, that all the bremsstrahlung had a quantum energy equal to the maximum energy of the beta particle that gave rise to the X-ray. Since this quantum energy is in fact the upper energy limit of the bremsstrahlung, and since most of the bremsstrahlung are much lower in energy than this upper limit, the thickness calculated above overestimates the required thickness. This overestimate just about compensates for the build-up factor, thus making consideration of the build-up factor unnecessary. The shipping container, therefore, consists of a polyethylene bottle, whose wall thickness is 1.06 cm, placed into a lead container whose sides, top, and bottom are 1.75 cm thick. However, a shield of this construction would be unnecessarily heavy. The total amount of lead in the shield increases with the square of the diameter; thus, even a small decrease in shield diameter may result in a significant weight reduction. If weight were a problem, then such a reduction could be achieved by decreasing the thickness of the polyethylene bottle. If, instead of 1.06 cm, the thickness were reduced to $\frac{1}{2}$ cm, all of the ^{90}Sr betas, plus most of the ^{90}Y betas, would be stopped. Furthermore, those ^{90}Y betas that would get through would be greatly reduced in energy. The resulting bremsstrahlung from these betas, therefore, would contribute very little to the radiation dose rate outside the lead shield.

Neutron shielding

Shielding against neutrons is based on slowing down fast neutrons and absorbing thermal neutrons. In Chapter 5 it was seen that attenuation and absorption of neutrons is a complex series of events. Despite the complexity, however, the required shielding around a neutron source can be estimated by the use of removal cross sections. (For neutron energies up to 30 MeV, the removal cross section is about three-quarters of the total cross section.) In designing shielding against neutrons, it must be borne in mind that absorption of neutrons can lead to induced radioactivity and to the production of gamma radiation.

Example 10.7

Design a shield for a 5-Ci Pu–Be neutron source that emits 5×10^6 neutrons per sec, such that the dose rate at the outside surface of the shield will not exceed 2.5 mrem/hr. The mean energy of the neutrons produced in this source is 4 MeV.

Let us make the shield of water, and compute the minimum radius for the case of a spherical shield. Since we know that the capture of a neutron by hydrogen produces a 2.26 MeV gamma-ray, let us allow for the gamma-ray dose by designing the shield to give a maximum fast-neutron dose rate of 2 mrem/hr, which corresponds to a fast flux of 16 neutrons per cm^2/sec (Table 9.3). The total cross section for 4 MeV neutrons for hydrogen and oxygen are 1.9 and 1.7 barns, respectively. Since water contains 6.7×10^{22} hydrogen atoms and 3.35×10^{22} oxygen atoms per cm^3, the linear absorption coefficient of water is

$$\begin{aligned}\Sigma &= 3/4\,(1.9 \times 10^{-24}\ \text{cm}^2/\text{atom} \times 6.7 \times 10^{22}\ \text{atom/cm}^3 + 1.7 \\ &\quad \times 10^{-24}\ \text{cm}^2/\text{atom} \times 3.5 \times 10^{22}\ \text{atoms/cm}^3) \\ &= 0.138\ \text{cm}^{-1},\end{aligned}$$

which corresponds to a half-value layer of 5.01 cm. The Pu–Be may be considered as a point source of neutrons, with the neutron flux decreasing with increasing distance as a result of both inverse square dispersion and attenuation by the water. If S is the source strength in neutrons per second, T is the half-value layer in cm, n is the number of half-value layers, and B is the build-up factor, the fast-neutron flux, after passing through a thickness of nT cm, is

$$\phi = B \frac{\frac{S}{4\pi (nT)^2}}{2} \frac{\text{neutrons}}{\text{cm}^2\ \text{sec}}. \tag{10.15}$$

For radioactive neutron sources on the order of several curies, the shield thickness is relatively thin, and a significant dose build-up due to scattered neutrons results. For a hydrogenous shield at least 20 cm thick, the dose build-up factor is approximately 5. Using a value of 16 neutrons per cm²/sec for ϕ, 5 cm for T, and 5 for B, equation (10.15) may be solved for n to give about 7.8 half-value layers, which corresponds to a thickness of 39 cm water.

The thermal neutrons that would escape from the surface of a spherical water shield may be estimated with the aid of equation (5.52):

$$\phi_{th} = \frac{n_0}{2\,\pi\,RL}\, e^{-R/L}.$$

Since the shield radius calculated above is much greater than the fast diffusion length (which is equal to 5.75 cm), we may assume, for the purpose of this calculation, that essentially all the fast neutrons are thermalized, and that the thermal neutrons are diffusing outward from the center. Substituting the appropriate numbers into equation (5.52), we have

$$\phi_{th} = \frac{5 \times 10^6 \text{ neutrons/sec}}{2\,\pi \times 39 \text{ cm} \times 2.88 \text{ cm}}\, e^{-39/2.88} = 9 \times 10^{-3}$$

thermal neutrons per cm²/sec. This thermal neutron flux is so small relative to the maximum permissible thermal flux of 670 neutrons per cm²/sec, that it may, for most practical purposes, be ignored.

Capture of a thermal neutron by a hydrogen atom results in the emission of a 2.26 MeV gamma-ray. The water shield, therefore, acts as a distributed source of gamma-radiation. Since 16 neutrons per cm²/sec escape from the surface, the total number that escape from a sphere of radius 39 cm is 3×10^5 neutrons per sec, or approximately 5% of the source neutrons. The remaining 95% are absorbed in the water, thus giving a mean "specific activity" for 2.26 MeV photons of

$$\frac{4.8 \times 10^6 \text{ photons/sec}}{\frac{4}{3}\pi (39 \text{ cm})^3 \times 3.7 \times 10^7 \text{ photons/sec/mCi}} = 5 \times 10^{-7} \frac{\text{"mCi"}}{\text{cm}^3}.$$

The dose rate at the surface of a sphere containing a uniformly distributed gamma emitter is, from equations (6.27) and (6.32),

$$D = \tfrac{1}{2} \times C\Gamma \times \frac{4\pi}{\mu}(1 - e^{-\mu r}). \tag{10.16}$$

Using a value of 10^4 mrad/hr per mCi at 1 cm for Γ, 0.046 cm^{-1} for μ for 2.26 MeV photons in water, and 39 cm for the radius gives

$$D = \tfrac{1}{2} \times 5 \times 10^{-7} \times 10^4 \times \frac{4\pi}{4.6 \times 10^{-2}}(1 - e^{-0.046 \times 39})$$

$$= 0.56 \text{ mrad/hr.}$$

The dose rate at the surface of the shield due to both neutrons and gamma-rays is 2.56 mrem/hr, which is very close to the desired figure of 2.5. The gamma-ray dose rate could be reduced either by increasing the gamma-ray absorption coefficient of the water shield by dissolving a high-atomic-numbered substance, such as $BaCl_2$; or by reducing the rate of production of the gamma radiation. Of these possible alternatives, the simplest one is the reduction in the production of gamma radiation. This is easily accomplished merely by dissolving a boron compound in the water. Boron captures thermal neutrons with a capture cross section of 755 barns, according to the reaction $^{10}B + n^1 \rightarrow {}^7Li + \gamma$ (0.48 MeV). The 0.48 MeV gamma is emitted in 93% of the captures. Either sodium tetraborate (borax), $Na_2B_4O_7 . 10\,H_2O$, or boric acid, H_3BO_3, both of which are highly soluble in water and very inexpensive, may be considered for this application. If suppression of gamma radiation is the objective, then boric acid may be preferred over borax, since the sodium in the borax has a relatively high cross-section, 505 millibarns, for the $^{23}Na\,(n, \gamma)\,^{24}Na$ reaction. As a consequence of this reaction, a 6.96 MeV capture gamma is emitted and radioactive ^{24}Na, which emits one 1.39 MeV beta, one 1.37 MeV gamma, and one 2.75 MeV gamma per disintegration is produced.

The solubility of boric acid in water at room temperature is 63.2 g/l. The formula weight of H_3BO_3 is 61.84. The concentration of boron atoms in the saturated solution is

$$\frac{63.2 \text{ g/l} \times 10^{-3} \text{ l/ml} \times 6.03 \times 10^{23} \text{ molecules/mole} \times 1 \text{ atom B/molecule}}{61.84 \text{ g/mole}}$$

$$= 6.17 \times 10^{20} \frac{\text{atoms B}}{\text{ml}}.$$

If we consider the macroscopic cross sections for thermal neutron capture of the dissolved boron and of the hydrogen, we find that

$$\frac{\Sigma H}{\Sigma B} = \frac{0.13 \text{ cm}^{-1}}{0.42 \text{ cm}^{-1}} = 0.31.$$

The flux of 2.26 MeV hydrogen gamma-rays, and consequently the dose rate, will be reduced by this factor to $0.31 \times 0.56 = 0.17$ mR/hr. The dose rate due to the ^{10}B capture gammas, which is calculated from equation (10.16) using a photon "specific activity" of (0.69) $(0.93 \times 4.93 \times 10^{-7})$ "mCi per cm³", an absorption coefficient for 0.48-MeV photons in water of 0.097 cm^{-1}, and a value for Γ of 0.23×10^3 mrad cm^2/mCi hr, is found to be 0.046 mrad/hr. The total dose rate at the surface of the shield, therefore, is 2 mrem/hr fast neutrons plus 0.17 mR/hr due to the hydrogen capture gammas plus 0.05 mR/hr due to the boron capture gammas, or 2.22 mrem/hr if we saturate the water with boric acid.

Problems

1. A Po–Be neutron source emits 10^7 neutrons per second, of average energy 4 MeV. The source is to be stored in a paraffin shield of sufficient thickness to reduce the fast flux at the surface to 10 neutrons per cm^2/sec. Consider paraffin to be essentially CH_2 (for the purpose of this problem) and to have a density of 0.89 gm/cm^3.

(a) What is the minimum thickness of the paraffin shield?

(b) If the slowing down length is 6 cm and the thermal diffusion length is 3 cm, what will be the thermal neutron leakage flux at the surface of the shield?

(c) What is the gamma-ray dose rate, due to the hydrogen capture gammas, at the surface of the paraffin shield?

2. An X-ray therapy machine operates at 250 kVp and 20 mA. At a target to skin distance of 100 cm, the exposure rate is 20 R/min. The work load is 10,000 mA min/week. The X-ray tube is constrained to point vertically downward. At a distance of 4 m from the target is an uncontrolled waiting room. Calculate the thickness of lead to be added to the wall if the total thickness of the wall (which is made of hollow tile and plaster, density 2.35 g/cm^3) is 2 in.

3. A 2000-Ci ^{60}Co teletherapy unit is to be installed in an existing concrete room in the basement of a hospital so that the source is 4 m from the north and west walls—which are 30 in thick. Beyond the north wall is a fully occupied controlled room. Beyond the west wall is a public parking lot. The useful beam is to be directed towards the north wall for a maximum of 5 hr per week during radiation therapy. The beam will be directed at the west wall 1 hr per week. Considering only the radiation from the primary beam, how much additional shielding, if any, is required for each of the walls?

4. A radiochemist wants to carry a small vial containing 50 mCi ^{60}Co solution from one hood to another. If the estimated carrying time is 3 min, what would be the minimum length of the tongs used to carry the vial in order that his exposure not exceed 6 mR during the operation?

5. A viewing window for use with an isotope that emits 1 MeV gamma-rays is to be made from a saturated aqueous solution of KI in a rectangular battery jar. What will be the attenuation factor, assuming conditions of good geometry, if the solution thickness is 10 cm, and if the glass walls are equivalent in their attenuation property to 1 mm lead? A saturated solution of KI may be made by adding 30 g KI to 21 ml water to give 30 ml solution at 25°C. Total attenuation cross sections for 1 MeV gamma-rays for the elements in the solution are:

$$K = 4 \text{ barns} \qquad H = 0.2 \text{ barns}$$
$$I = 12 \text{ barns} \qquad O = 1.7 \text{ barns}$$

6. Lead foil consists of an alloy containing 87% Pb and 13% Sn, and 1% Cu. Its specific gravity is 10.4. If the mass attenuation coefficients for these three elements are 3.50, 1.17, and 0.325 cm^2/g respectively for X-rays whose wavelength is 0.098 Å, and if the specific gravities of the three elements are 11.3, 7.3, and 8.9 respectively,

(a) calculate the mass and linear absorption coefficients for lead foil,

(b) what thickness of lead foil would be required to attenuate the intensity of ^{57}Co gamma-rays by a factor of 25?

7. A hypodermic syringe that will be used in an experiment in which ^{90}Sr solution will be injected has a glass barrel whose wall is 1.5 mm thick. If the density of the glass is 2.5 g/cm^3, how many mm thick must we make a Lucite sleeve that will fit around the syringe if no beta particles are to come through the Lucite? The density of the Lucite is 1.2 g/cm^3.

8. A room in which 2000 Ci ^{137}Cs source will be exposed has the following layout

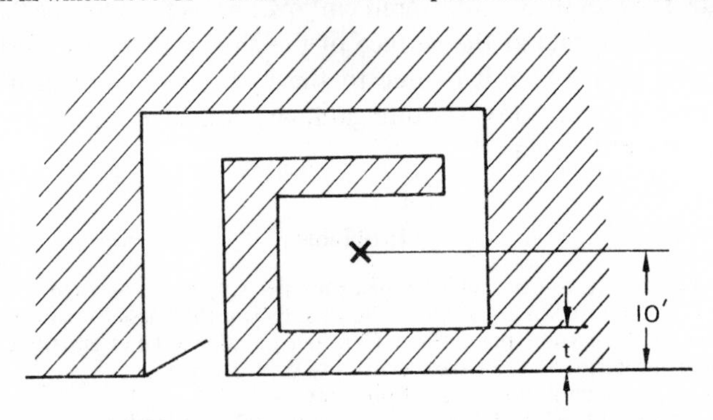

Calculate the thickness t of concrete so that the exposure rate at the outside surface of the wall does not exceed 2.5 mR/hr.

9. What minimum density thickness must a pair of gloves have to protect the hands from ^{32}P radiation?

10. When a radium source containing 50 mg Ra encapsulated in 0.5-mm-thick Pt is placed into a Pb storage container, the measured exposure rate at a distance of 1 m from the source is 21 mR/hr. If this same container is used for storing ^{137}Cs, how many mCi may be kept in it for a period of 4 hr without exceeding an exposure of 40 mR at a distance of 50 cm from the source?

11. What is the maximum working time in a mixed radiation field consisting of 20-mR/hr gamma, 4 mrad/hr fast neutrons and 5 mrad/hr thermal neutrons if a maximum dose of 300 mrem has been specified for the job?

Suggested References

1. PRICE, B. T., HORTON, C. C. and SPINNEY, K. T.: *Radiation Shielding*, Pergamon, New York, 1957.
2. BRAESTRUP, C. B. and WYCKOFF, H. O.: *Radiation Protection*, Charles C. Thomas, Springfield, 1958.
3. JAEGER, T.: *Principles of Radiation Protection Engineering* (translated by L. DRESNER), McGraw-Hill, New York, 1965.
4. The following Handbooks from the National Bureau of Standards, Government Printing Office, Washington, D.C.:
   (a) *Medical X-ray Protection up to Three Million Volts*, N.B.S. Handbook 76, Government Printing Office, Washington, 1961.
   (b) *Protection Against Neutron Radiation up to 30 Million Electron Volts*, N.B.S. Handbook 63, Government Printing Office, Washington, 1957.
   (c) *Shielding for High Energy Electron Accelerator Installations*, N.B.S. Handbook 97, Government Printing Office, Washington, 1964.
   (d) *Safety Standard for Non-Medical X-ray and Sealed Gamma-Ray Sources, Part I General*, N.B.S. Handbook 93, Government Printing Office, Washington, 1964.
   (e) *Protection Against Radiation from Sealed Gamma-Sources*, N.B.S. Handbook 73, Government Printing Office, Washington, 1960.

CHAPTER 11

INTERNAL RADIATION PROTECTION

Internal Radiation Hazard

Internal radiation results when the body is contaminated—either internally or externally—with a radioisotope. Accordingly, internal radiation protection is concerned with preventing or minimizing the deposition of radioactive substances on or in personnel. This is accomplished by a program designed to keep the contamination of the environment within acceptable limits, and at levels as low as possible. The last point—keeping environmental levels as low as possible—is especially important in the context of internal radiation protection. External radiation exposure is due to radiation originating in sources outside the body; there is no physical contact with the radiation source, and exposure ceases when one leaves the radiation area or the source is removed. Since external radiation may be measured with relative ease and accuracy, the potential or actual hazards may be estimated with a good deal of confidence. In the case of contamination, on the other hand, the radioactive material is deposited on or within the body. As a consequence, the contaminated person continues to suffer irradiation even after he leaves the area where he became contaminated. Furthermore, radioisotopes within the body may become systemically fixed. Their elimination then can be hastened—in those cases where possible—only with relative difficulty. Biological turnover rates, which may exhibit a marked variability, then become significant in determining the radiation absorbed dose from the internally deposited isotopes. For these reasons, it is relatively difficult to assess the hazard from internal emitters, and great emphasis is therefore placed on the prevention of contamination of personnel with radioactive materials.

Principle of Control

Radioactive substances, like other toxic agents, may gain entry into the body through three portals:

1. Inhalation—by breathing radioactive dust and gas.
2. Ingestion—by drinking contaminated water, eating contaminated food, or by tactilly transferring radioactivity to the mouth.
3. Absorption through the intact skin or through wounds.

Basically, therefore, protective measures to counter internal radiation are designed either to block the portals of entry into the body, or to interrupt the transmission of radioactivity from the source to man. This interruption can be effected either at the source—by enclosing and confining it, or by controlling the environment—by ventilation and good housekeeping, or at the man—by providing him with protective clothing and with respiratory protective devices. It should be noted that these control measures do not differ from those employed by the industrial hygienist in the protection of workers from the effects of non-radioactive noxious substances. However, the degree of control required for radiological safety almost always greatly exceeds the requirements for chemical safety. This point is made clear by the figures in Table 11.1, which compare the maximum allowable atmospheric concentrations of several non-radioactive toxic substances to the maximum concentrations recommended by the ICRP for radioactive forms of the same element.

TABLE 11.1. MAXIMUM ALLOWABLE CONCENTRATIONS OF SEVERAL SUBSTANCES BASED ON CHEMICAL AND RADIOLOGICAL TOXICITY

	Milligrams per meter^3		
	Non-radioactive	Radioactive	
Beryllium	0.002	^{7}Be	1.7×10^{-8}
Mercury	0.1	^{203}Hg	5×10^{-9}
Lead	0.2	^{210}Pb	1×10^{-9}
Arsenic	0.5	^{74}As	3×10^{-9}
Cadmium	0.1	^{115}Cd	4×10^{-10}
Zinc	5	^{65}Zn	1.2×10^{-8}

Control of the Source: Confinement

The simplest type of confinement and enclosure may be accomplished by limiting the handling of radioactive materials to well-defined, separated areas within a laboratory, and by the use of sub-isolating units, such as trays. For low-level work where there is no likelihood of atmospheric contamination with more than one maximum recommended body burden, this may be sufficient. If the possibility exists of the release to the atmosphere, either as a gas or an aerosol, of amounts of activity between 1 and 10 times the maximum recommended body burden, the usual practice is to use a ventilated hood. The purpose of the ventilated hood is to dilute and to sweep out the released radioactivity with the air that flows through the hood. It is thus essential to have a sufficient amount of air flowing through the hood at all times in order to accomplish this purpose. Constant air-flow may be maintained by a by-pass that opens as the face is closed. Openings along the bottom of the front-face frame facilitate the flow of air when the face is closed. Figure 11.1 shows a typical radiochemistry fume hood. The face velocity must be great enough to prevent contaminated air from flowing out of the face into the laboratory, but not great enough to produce turbulence around the edges that would allow contaminated air from the hood to spill out into the laboratory. It has been found that face velocities of 125–275 ft/min are required. To minimize the probability of contaminating the working environment with the exhaust from the hood, all the ductwork must be kept under a negative pressure. Any leakage in the ductwork will then be *into* the duct. This is most easily accomplished by locating the exhaust fan at the discharge end of the exhaust line, as shown below in Fig. 11.1.

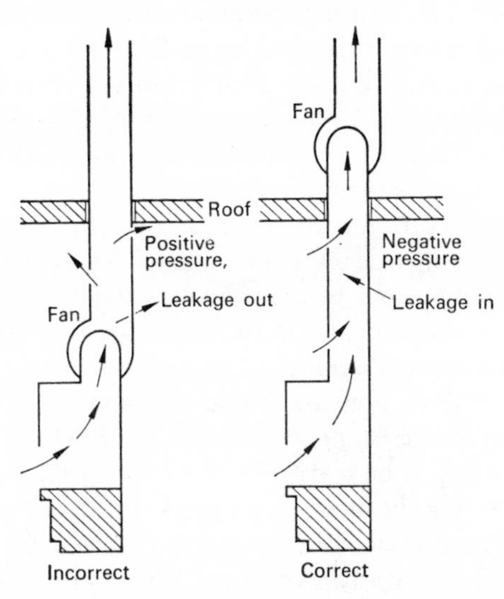

FIG. 11.1. Effect of fan location on direction of leakage in the ductwork. The fan should be close to the discharge end. Thereby creating a negative pressure in the ductwork and causing leakage in the ductwork to be into the duct.

If the hood is designed to remove only gases, vapors, or fumes, an air velocity in the ductwork of about 2000 ft/min is sufficient. However, since particulate matter tends to settle out, the air transport velocity must be on the order of 3500–4500 ft/min if particulate matter fallout is to be minimized. If the exhaust from the hood is of such a nature that it may create a radioactive pollution problem, then the effluent from the hood should be decontaminated by an appropriate air cleaning device. For this purpose, if the pollutant is a dust, either a rough filter alone or a rough filter, followed by a fire-resistant, high-efficiency filter is commonly employed. As used in this context, a high-efficiency filter is one that removes at least 99.995% of 0.3 μ diameter homogeneous particles of dioctyl phthalate (DOP). The filter should not offer a resistance greater than one inch water when air at 70°F and 29.9 in Hg flows through it at its rated capacity. A manometer, or other device, shoul be used to indicate when the filter is loaded and ready to be changed. A filte loaded with radioactive dust can easily become a source of contamination i adequate precautions to prevent the dust from falling off the filter during th changing operation are not taken. A simple way to minimize dispersal o loose dust when removing the filter is to spray the filter faces with an aeroso laquer before removing the filter, thereby trapping the radioactive dust in th filter. For this purpose, access ports upstream and downstream of the filte should be provided in the ductwork.

If the nature of an operation involving radioactivity is such that it must b completely enclosed, that is, if the operation is potentially capable of contam inating the working environment with more than 10 times the recommende maximum body burden, or when the large quantities of air required by a hoo are not available, then a glove box (Fig. 11.2) is used. It should be re-empha sized here that whereas the main function of a fume hood is to dilute an remove atmospheric contaminants, the main function of a glove box is t isolate the contaminant from the environment by confining it to the enclose volume. This is especially true if the contaminant is a particulate. Accordingly the air flow through the glove box may be very small—on the order of 25–5(ft^3/min. Air is usually admitted into the glove box through a high efficienc fiber-glass filter (to prevent discharge of radioactive dust into the room in cas of an accidental positive pressure inside the glove box) and is exhauste through a series of fire-resistant rough and high efficiency filters. Air-born particles small enough to be carried by this flow of air are thus transferre out of the glove box into the filter; larger particles fall out inside the glove bo and remain there until cleaned out. A negative pressure of at least 0.5 in water inside the glove box assures that any air leakage will be into the box Despite the negative pressure, however, it may be assumed that a small fraction, about 10^{-8}, of the activity inside the glove box will leak out during the course of normal use of the glove box. The laboratory should be prepared to handle such contamination, and the health physicist should be prepared to account for this activity in the design and operation of his surveillance program. For maximum safety, transfer of materials and apparatus into or out of the glove box always is done through an air lock. The viewing panel may be heat-resistant safety plate glass. Glove boxes are unshielded when used for handling radioisotopes that do not create high-level radiation levels. For radioisotopes that do create such high levels of radiation, shielding must be added. When handling a high energy beta-emitting radioisotope, it may be necessary to use extra thick gloves.

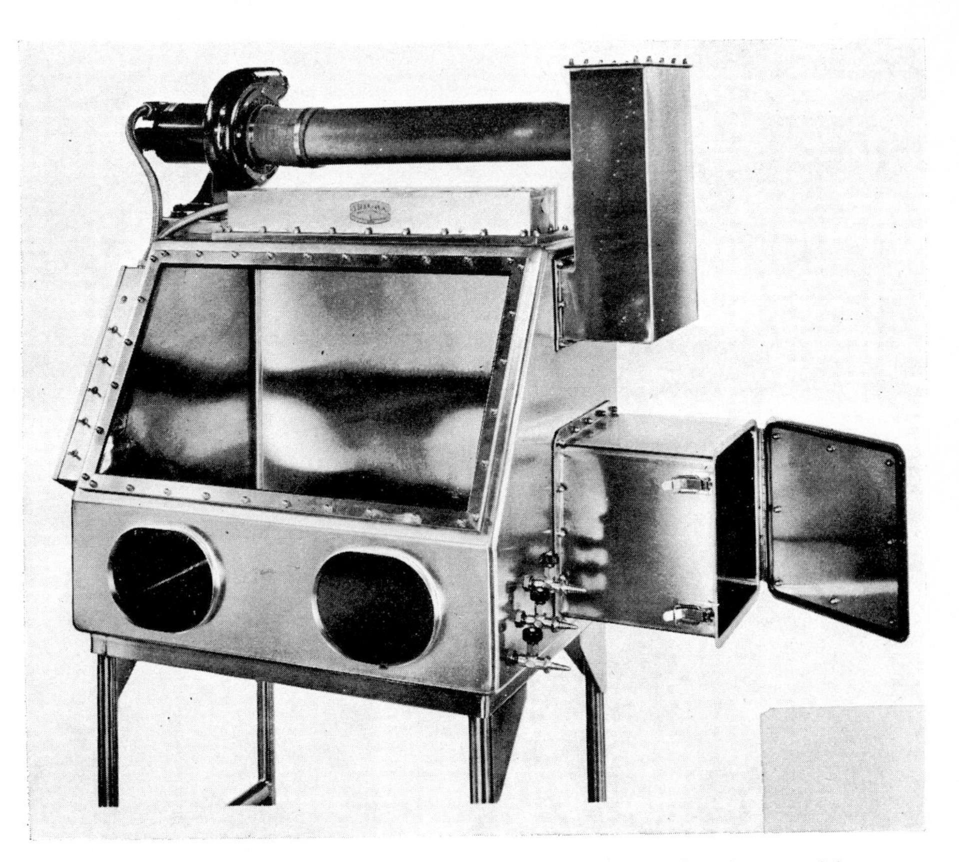

FIG. 11.2. Glove box for operations with low-intensity radioactive materials that might accidentally become dispersed into the environment if not handled in an enclosed volume. In use long rubber gloves fit over the port-flanges; material is transferred into and out of the glove box through the air lock. (Courtesy S. Blickman, Inc.)

Environmental Control

Environmental control of hazards from radioactive contamination begins with the proper design of the buildings, rooms, or physical facilities in which radioisotopes will be used; and continues with the proper design of the procedures and processes in which radioactivity will be employed. Since a finite probability exists that an accidental breakdown of a mechanical device or a human failure will occur despite the best efforts to prevent such a breakdown, the course of action to be taken in the event of an emergency must be known before the emergency occurs. In the design of the physical facilities, attention must be paid to the decontaminability of working surfaces, floors, and walls; to plumbing and means for monitoring or storing radioactive waste, both liquid and solid; to means for incineration of radioactive waste; to isotope storage facilities; to change rooms and showers; and to ventilation and the direction of air flow: office to corridor to area where radioisotopes are handled to exhaust through an air-cleaning system that will assure radiological safety outside the building. Strict control, including monitoring of all persons, materials, and equipment leaving the radiation area, must be maintained over the area where radioisotopes are being used or stored in order to prevent the spread of contamination outside the radiation area. The degree to which each of these control measures is implemented depends, of course, on the types and amounts of isotopes handled, and on the consequences of an accidental release of radioactivity to the environment.

In order to maintain control over internal radiation hazards, good housekeeping and good ventilation must be practiced. In regard to ventilation requirements, several important facts should be emphasized. The first is that

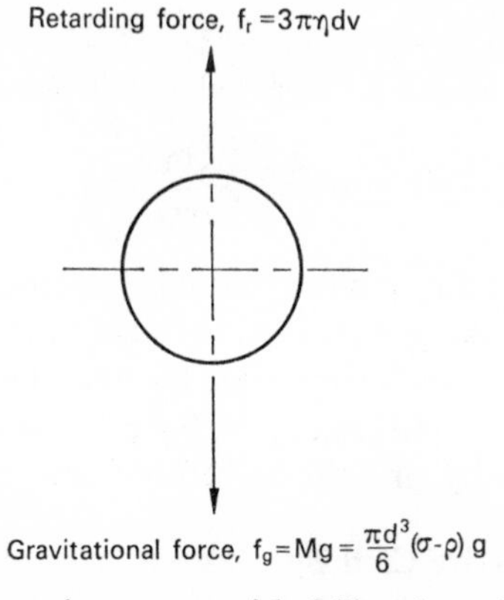

FIG. 11.3. The forces acting on a particle falling through air. η = viscosity of air = 185 μpoise at room temperature (1 poise = 1 dyne sec/cm^2); d = particle diameter, cm; v = velocity of fall, cm/sec; M = particle mass, g; g = acceleration due to gravity, 980 cm/sec^2 or 980 dynes/g; σ = particle density, g/cm^3; ρ = air density, g/cm^3.

fine particles under the influence of gravity do not, for practical purposes move independently of the air in which they are suspended. Such particle behave effectively as if they are weightless, and can be assumed to remai suspended indefinitely in the air. *Control of airborne dust particles thus i reduced to a matter of air flow control.* A particle released into the atmospher (with no horizontal motion) is acted upon by two forces—the downwar force of gravity and the upward retarding force due to the resistance to fre fall offered by the air (Fig. 11.3). When the retarding force is equal to th gravitational force, there is no unbalanced force to accelerate the fallin particle, and a constant velocity, called the *terminal velocity,* is attained b the falling particle. Equating f_r to f_g and solving for the terminal velocity, w have

$$f_r = 3\pi\eta dv = \frac{\pi d^3}{6}(\sigma - \rho)g = f_g \qquad \text{(11.1A)}$$

$$v_t = v = \frac{d^2(\sigma - \rho)g}{18\eta}. \qquad \text{(11.1B)}$$

Example 11.1

Calculate the terminal velocities of spherical particles of U_3O_8 whos diameters are 1 μ and 20 μ. The density of U_3O_8 is 8.30 g/cm³.

Substituting into equation (11.1B), we have for the 1 μ particle,

$$v_t = \frac{(10^{-4}\ \text{cm}^2)\ (8.30 - 1.29 \times 10^{-3})\ \text{g/cm}^3 \times 980\ \text{dynes/g}}{18 \times 1.85 \times 10^{-4}\ \text{dyne sec/cm}^2}$$

$$= 0.00244\ \frac{\text{cm}}{\text{sec}},$$

and for the 20 μ particle, the terminal velocity is calculated as

$$v_t = 0.976\ \frac{\text{cm}}{\text{sec}}.$$

The difference between the terminal settling velocities of the 1 μ and 20 diameter particles is striking. However, even the settling velocity of the 20 particle, which is 2.3 ft/min, is very much less than the ambient air velocitie of about 25 ft/min in occupied space. A 20 μ particle could thus be carried relatively long distance by air currents before falling out.

Control of Man: Protective Clothing

The philosophy of radiation protection advocates the restriction of radia tion exposure to levels as far below the maximum recommended doses a practically possible. Since it is extremely difficult to maintain absolute radic logical asepsis when working with unsealed sources, and since the possibilit of an accidental spill or release to the environment of radioactivity always exists, it is customary to require isotope workers to wear protective clothing whose use is restricted to the radioactivity area. Such protective clothing may include laboratory coats, coveralls, caps, gloves, and shoes or shoe covers. Protective clothing is always assumed to be contaminated and, therefore, is removed when the worker leaves the radioactivity area. To be most effective, the protective clothing should be so designed that the worker can remove it easily and without transferring contamination from the clothing either to himself or to his environment. He should always be monitored before leaving the radioactivity area.

Protective clothing must, by its very nature, become contaminated; its main function is to intercept radioactivity that would otherwise contaminate the worker's skin or the clothing that he wears outside the radioactivity area. The degree of allowable contamination in the protective clothing varies with the type of work that the wearer does. For this reason, the maximum degree of contamination permitted on protective clothing is determined by the individual installation.

For most isotope laboratories, laundering protective clothing presents no special problem. In most instances, ordinary laundering procedures, repeated more than once if necessary, suffice. Sodium hexametaphosphate, or a complexing agent, such as sodium ethylene-diamine-tatraacetic acid (Na EDTA), may be added to the pre-wash rinse to facilitate the removal of radionuclides. After laundering, the protective clothing should be monitored to ascertain that it has, in fact, been decontaminated to some previously determined limit. If a piece of protective clothing is unusually or very severely contaminated, it may be simpler and cheaper to dispose of the item rather than to try to decontaminate it. In most cases of industrial or medical use of radioisotopes, the concentration of radioactivity in the wash water is low enough to be discharged directly into the sanitary sewer system. However, before being discharged, the wash water must be checked to verify that it can be safely discharged. If the activity level is too high, then the wash water must be treated as radioactive waste.

Control of Man: Respiratory Protection

When a man is likely to be exposed to a high concentration of air-borne radioactivity, respiratory protection is required. The exact type of respiratory protection depends on the nature of the air-borne contaminant. It must be emphasized that *respiratory protective devices may be used only for those hazards for which they are designed.*

Respiratory protective devices for radiological applications may be classified into two main categories:

1. Filter type respirators that are suitable only for dusts. *These respirators are not designed to provide protection against radioactive gases.* When

using the half-face respirator, the wearer must try to eliminate possible leaks around the face piece. Half-face respirators are considered suitable for air-borne dust concentrations up to ten (10) times the recommended maximum atmospheric concentration; respirators with full face masks are considered suitable up to fifty (50) times the recommended maximum atmospheric concentration.

2. Supplied air masks that may be used either against dusts or gases or both. In this category we have two subdivisions: (a) air line hoods, which utilize uncontaminated air, under positive (with respect to the atmosphere) pressure supplied from a remote source, and (b) self-contained breathing apparatus, in which breathing air is supplied either from a bottle carried by the man, or from a cannister containing oxygen generating chemicals. The advantage of the supplied air device is that the pressure in the breathing zone is higher than atmospheric pressure. As a consequence, all leakage is from the inside out. When using a supplied air device, it is imperative to know the time limitation on the supply of air.

A third type of respiratory protective device is the gas-mask. In this device, contaminated air is cleared by chemicals in a cannister through which the air passes. Because of the specific action of the chemical agents on the contaminant, different cannisters must be used for different gases. For this reason, as well as for the fact that air may leak into the face-piece of a gas-mask, gas-masks are not recommended for use against radioactive gas.

Surface Contamination Limits

Contamination of personnel and/or equipment may occur either from normal operations or as a result of the breakdown of protective measures. An exact quantitative definition of contamination that would be applicable in all situations cannot be given. Generally, contamination means the presence of undesirable radioactivity—undesirable either in the context of health or for technical reasons, such as increased background, interference with tracer studies, etc. In this discussion, only the health aspects of contamination are considered.

Surface contamination falls into two categories, fixed and loose. In the case of fixed contamination, the radioactivity cannot be transmitted to personnel, and the hazard, consequently, is that of external radiation. For fixed contamination, therefore, the degree of acceptable contamination is directly related to the external radiation dose rate. Setting a maximum limit for fixed surface contamination thus becomes a relatively simple matter. The hazard from loose surface contamination arises mainly from the possibility of tactile transmission of the radioactive contaminant to the mouth or to the skin, or of resuspending the contaminant and then inhaling it. It follows that the degree of hazard from surface contamination is strongly dependent on the degree to which the contaminant is fixed to the surface.

Dealing with loose surface contamination limits is not as straightforward as dealing with contamination of air and water. In the case of air and water contamination, safety standards can be easily set—at least in theory—on the basis of recommended maximum absorbed doses (Dose = 5 (N-18), etc.). Using these criteria, we can calculate maximum permissible body burdens for each of the radioisotopes. From the calculated body burden we go one step further from the basic radiation safety criteria, and compute maximum concentrations in air and water which, if continuously inhaled or ingested, would result in a body burden less than the calculated maximum. For the case of surface contamination, we go one more step away from the basic criteria; we try to estimate the surface contamination which, if it should be dispersed into the environment, would result in concentrations that may lead to an excessive body burden. Thus, specification of limits for loose surface contamination is three steps removed from the basic safety requirements.

From the foregoing discussion, it is clear that maximum limits for surface contamination cannot be fixed in the same sense as limits for the concentration of radionuclides in air and water. Nevertheless, it is useful to compute a number that may serve as a guide in the evaluation of the hazard to workers from surface contamination, and to assist the health physicist in deciding whether or not to require the use of special protective measures for workers in contaminated areas.

On the basis of per-unit-quantity of radioactivity, inhalation is considered the most serious route of exposure. Surface contamination, therefore, is usually limited by the inhalation hazard that may arise from resuspension of the contaminant. The quantitative relationship between the concentration of loose surface contamination and consequent atmospheric concentration above the contaminated surface due to stirring up the surface is called the *resuspension factor*, f_r, and is defined by

$$f_r = \frac{\text{atmospheric concentration, } \mu\text{Ci/cm}^3}{\text{surface concentration } \mu\text{Ci/cm}^2}. \qquad (11.2)$$

Experimental investigation of the resuspension of loose surface contamination shows that the resuspension factor varies from about 10^{-4} to 10^{-8}, depending on the conditions under which the studies were conducted. A value of 10^{-6} is reasonable for the purpose of estimating the hazard from surface contamination.

Example 11.2

Estimate the maximum surface contamination of ^{90}Sr dust that may be allowed before taking special safety measures to protect personnel against a contamination hazard.

The maximum atmospheric concentration of ^{90}Sr recommended by the ICRP is 2×10^{-10} μCi/cm^3. Using a value of 10^{-6} cm^{-1} for the resuspension factor in equation (11.2), we have

$$10^{-6}\ \text{cm}^{-1} = \frac{2 \times 10^{-10}\ \mu\text{Ci/cm}^3}{\text{surface concentration}}$$

$$\therefore \text{surface concentration} = 2 \times 10^{-4}\ \mu\text{Ci/cm}^2.$$

It should be emphasized that a figure for loose surface contamination calculated by the method of Example 11.2 is intended only as a guide. In any particular case, the health physicist may, at his discretion, and depending on the nature of the operation, the degree of ventilation, and other relevant factors, insist on more or less stringent requirements for surface contamination before requiring the use of protective devices for the worker.

TABLE 11.2. UNITED KINGDOM ATOMIC ENERGY AUTHORITY MAXIMUM PERMISSIBLE LEVELS OF SURFACE CONTAMINATION, μCi/cm^2 [(a)]

Type of surface	Principal alpha emitters[(b)]	Low toxicity alpha emitters[(c)]	Beta emitters
Inactive and low activity areas	10^{-5}	10^{-4}	10^{-4}
Active areas	10^{-4}	10^{-3}	10^{-3}
Personal clothing	10^{-5}	10^{-4}	10^{-4}
Authority clothing not normally worn in active areas	10^{-4}	10^{-3}	10^{-3}
Skin	10^{-5}	10^{-5}	10^{-4}

The contamination of surfaces by radioactive materials may give rise to external radiation or to intake of radioactive materials by persons. The control of surface contamination is, therefore, an essential part of the safe handling of radioactive materials. Surface contamination should be controlled to the lowest practicable levels, and in any case, within the maximum permissible levels specified above, unless relaxations (e.g. for firmly fixed contamination or low toxicity contaminants) have been permitted on the advice of the health physicist. The requirements of the maximum permissible doses from external radiation must be observed.

(a) Averaging is permitted over inanimate areas of up to 300 cm^2 or, for floors, walls, and ceilings, 1000 cm^2. Averaging is permitted over 100 cm^2 for skin; for the hands, over the whole area of one hand, nominally 300 cm^2.

(b) All alpha emitters other than those listed in note (c).

(c) Uranium isotopes.
Enriched and depleted uranium.
Natural uranium.
Natural thorium.
Short-lived radionuclides, such as radon daughters.
Thorium-232.
Thorium-228 and thorium-230 when diluted to a specific activity of the same order as that of natural uranium and natural thorium.

(From the U.K.A.E.A. Health and Safety Code, *Maximum Permissible Doses from Inhaled and Ingested Radioactive Materials*, Authority Code No. E.1.2, Issue No. 1, London, June 1961.)

Various laboratories and nuclear installations have set their own limits for contamination of personnel, equipment, and protective clothing. Tables 11.2 and 11.3 are given to illustrate some of the contamination standards maintained by several large users of radioisotopes.

TABLE 11.3. U.S.S.R. SURFACE CONTAMINATION LIMITS

Object of contamination	Contamination from 150 cm² in 1 min			
	Alpha particles		Beta particles	
	Before cleaning	After cleaning	Before cleaning	After cleaning
Hands	75	bg	5000	bg
Special linens and towels	75	bg	5000	bg
Cotton special work clothes	500	100	25,000	5000
"Pellicular" clothing	500	200	25,000	10,000
Gloves, outside	500	100	25,000	5000
Special shoes, outside	500	200	25,000	5000
Work surfaces and equipment	500	200	25,000	5000

Note. No contamination of the body is permitted.

(From *Sanitary Regulations for Work with Radioactive Substances and Sources of Ionizing Radiation*, Ministry of Health, U.S.S.R., Moscow, 1960.)

Waste Disposal

Proper collection and disposal of radioactive waste is an inherent part of contamination control and internal radiation protection. In one sense, we cannot dispose of radioactive waste. All other types (non-radioactive) of toxic wastes can be treated either chemically, physically, or biologically in such a manner as to reduce their toxicity. In the case of radioactive wastes, on the other hand, nothing can be done to decrease the radioactivity, and hence, the inherent toxicity of the waste. The only means of ultimate disposal is time—to allow for the decay of the radioactivity.

Radioactive wastes originate from any operation in which radioisotopes are used or produced. For purposes of management and treatment, wastes may be classified as high, intermediate, and low level. Low-level wastes are defined as those that must be diluted by a factor of no more than 10^3 before discharge into the environment, for intermediate levels, $10^3 < DF \leq 10^5$, and for high-level waste, the required dilution factor would be greater than 10^5. High-level wastes are associated with the inventory of fission products in the burned-up fuel of nuclear reactors and with the chemical and metallurgical processes involved in the separation of the fission products from the unspent uranium or plutonium in the burned-up fuel.

Intermediate- and low-level wastes are associated with operations on irradiated fuel as well as with other operations involving radioisotopes. The amount of waste produced in power reactors is astronomical. A 500 megawatt reactor, whose fuel reprocessing cycle is 180 days, will have a fission product inventory of about 4×10^8 Ci, including about 3×10^5 Ci ^{90}Sr. It has been estimated that, by the year 2000, about 2.5×10^6 megawatts of power will be produced in nuclear reactors, thereby producing enormous quantities of waste materials. The ^{90}Sr alone will amount to about 1.5×10^9 Ci every 180 days. If we consider the atmosphere up to 10,000 ft, where almost all the atmospheric mixing occurs, the volume of air is about 1.5×10^{18} m^3. If the ^{90}Sr were released to the air, the resulting world-wide atmospheric concentration of this isotope alone would be 1×10^{-9} μCi/cm^3, which is about 333 times greater than the recommended maximum atmospheric concentration. This estimate, it should be pointed out, is due to operation for only one-half year. Release of these quantities of radioactive waste into the sea is not much better than release to the atmosphere. The volume of the sea is about the same as that of the atmosphere below 10,000 ft. Concentrations of activity in the sea water, too, would be high. Furthermore, because of the tremendous harvest of sea food from the ocean and the known ability of aquatic organisms to concentrate certain isotopes, indiscriminate release into the sea of high-level radioactive wastes could create serious health hazards for man. At the present time, the quantities of waste produced by the nuclear energy industry are relatively small and quite manageable in regard to both health and safety and economics. The high-level waste products are simply treated to minimize their volume, and then stored in underground containers. Quite clearly, however, one of the limiting factors in the full exploitation of nuclear energy may very well be the public health problems associated with the radioactive waste.

The range of radioactivity in waste is very large; for liquids, for example, high-level wastes contain on the order of hundreds to thousands of curies per gallon, while low-level wastes are considered as having activity concentrations down to the order of a microcurie per gallon. Because of this wide range of activity, several basically different methods are used in the management of radioactive waste. For large amounts of activity, the general principle is to concentrate and confine the waste, while for small amounts, the waste is usually diluted and dispersed. The exact manner in which each of these principles is applied depends on whether the waste is liquid, solid, or gaseous, and on the level of activity.

High-level Liquid Wastes

High-level wastes originate mainly as highly acidic solutions from the chemical processing of burned-up fuel, at specific activities on the order of hundreds of curies per gallon. These liquids are reduced in volume by evaporation to about one-tenth and then are stored in underground tanks. The problem of storing high-level liquid waste is complicated by the fact that the rate of heat production due to the radioactivity is high. One thousand curies, for example, assuming a mean decay energy of 1 MeV per disintegration, generates 3.7 watts of power as heat. Provision, therefore, must be made to transfer this heat. In one scheme for reducing the volume of the radioactivity, this heat is used to boil the liquid waste until the solution is sufficiently concentrated for transfer to the storage tank. The condensate from this evaporation process is treated as intermediate-level waste. A storage tank for high-level waste must be designed for strength, corrosion resistance, heat removal, and monitoring for leak detection. Typical tanks, of capacity on the order of 10^5–10^6 gal, are steel-lined, reinforced concrete, equipped with cooling coils and external condensers to remove the heat from the heat transfer fluid.

Although storage in tanks has so far been practical, the potential hazards from leakage during the required long-term retention time, which is on the order of hundreds of years, are relatively serious. Much experimental work has, therefore, been directed towards the conversion of liquid waste into a solid from which the radioactivity cannot be leached. Incorporating the activity into ceramics, clays, and glass seem to be the most promising technics. After conversion to a solid form, the high level activity could then be stored in salt beds or other suitable geological formations without serious risk of releasing the radioactivity to the environment. Salt beds, because of their unique characteristics, seem to offer an almost ideal media for disposal of the solid waste. Salt beds are not associated with groundwater; the salt is relatively plastic, which results in rapid closure of fractures; and salt beds are extensive in area, underlying about 400,000 square miles in the United States.

Intermediate- and Low-level Liquid Wastes

Intermediate- and low-level wastes can, under certain conditions, be discharged into the sea or into the ground; or they may be treated either chemically, physically, or biologically in order to separate the radioactive solutes from the non-radioactive solvent.

The tremendous volume of the ocean seems to make it an ideal medium for the dilute and disperse technic for management of low and intermediate levels of waste. Furthermore, since sea water already contains a significant amount of radioactivity, mostly in the form of ^{40}K (it is estimated that the total activity content of the oceans is on the order of one-half million megacuries) addition of relatively small quantities of activity in the form of low and intermediate levels of waste would thus seem to add very little to the total activity of the oceans. However, because of the uncertainties regarding the diverse physical, chemical, and biological processes that govern the distribution of radioisotopes in the sea and the transmission of the radioisotopes through the food chain to man, it is very difficult to specify maximum amounts of radioactivity that may be discharged into the seas. According to the recommendations of the International Atomic Energy Agency (*Radioactive Waste Disposal into the Sea*, 1961), wastes of low and intermediate activity may be disposed of into the sea under controlled and specified conditions; providing that accurate records of all waste disposal activities are kept, and that the oceans be under continuous radiological surveillance in order to ascertain that radiologic safety is being maintained.

Disposal of low- and intermediate-level wastes into the ground may be practiced if the hydrologic factors, the ion-exchange properties of the soil, and the population density are favorable. This method of disposal can be called the "delay and decay" method because the slow movement of the radioactivity through the ground affords the activity sufficient time to decay by a significant degree. Characteristics favorable to ground disposal include a deep-water table, good ion-exchange properties of the soil in order to extract relatively large fractions of radionuclides from the waste as the waste percolates through the ground, few bodies of surface water in order to maximize the time of underground flow, a large volume of underground water flow in order to maximize dilution, and a very low population density in the area of the ground disposal site. In the practice of ground disposal, a wood-lined pit, called a "crib", of appropriate capacity, is built in the ground, and is filled with gravel. The liquid waste is pumped into the crib, and slowly percolates out into the ground. When dispersing radioactive wastes into the ground, it is important to follow the course of the underground activity and to monitor the surface waters. In a suitable environment, such as is found at Hanford, ground disposal may be practised safely.

Chemical processes for the decontamination of intermediate- and low-level liquid wastes include the standard methods of water treatment and ion-exchange methods. Hydroxide flocs, which are produced by adding alum or ferric salts to the liquid wastes, and then increasing the pH until aluminum or ferric hydroxide is precipitated, are useful for removing cations other than those of the alkali metals and alkaline earths. This treatment is especially effective for removing alpha emitters; it is not very effective for removing ^{90}Sr. Removal of about 95% of ^{90}Sr may be effected with a calcium phosphate floc under highly basic conditions (pH $\sim$ 11.5). Radiostrontium can also be effectively removed by lime-soda softening the water. The degree of removal of ^{90}Sr is proportional to the degree of softening, since the $^{90}Sr\,CO_3$ is precipitated with $CaCO_3$. Under certain conditions, liquid wastes may be decontaminated by ion-exchange methods. However, since non-radioactive ions are also absorbed on the ion exchanger, the effectiveness of this method depends on the relative concentrations of radioactive and non-radioactive ions. Better than 99% reduction in radioactivity can be achieved under optimum conditions.

Decontamination of water by biological means may also be used. However, biological removal of radionuclides is less effective than chemical treatment.

Its main use, consequently, is for those cases where organic matter must be destroyed, as in sewage treatment or where high concentrations of organic complexing agents makes ordinary chemical treatment difficult.

For non-volatile radioactivity, evaporation is an effective means for decontaminating water. However, because evaporation requires removal of the solvent or the suspending medium, and since this component of the liquid waste usually accounts for more than 95% of the total pre-treated volume, evaporation is a relatively expensive method for the treatment of liquid waste. Evaporation is usually reserved for those cases where a very high degree of decontamination is required. By means of evaporation, decontamination factors on the order of 10^4–10^6 may be obtained at vapor mass velocities ranging from about 20–3000 kg/m^2/hr. The separated radioactivity, now in a relatively small volume, is processed further, as described in the next section, for disposal.

After removal of the bulk of the radioactivity from the suspending liquid, the decontaminated water can often be discharged into the storm sewer. Very low-level wastes, such as those produced in a laboratory handling tracer amounts of radioactivity, also are usually discharged into the sewer. Such discharges cannot be done indiscriminately, but are subject to regulatory control. The rule that is usually adopted for discharge into the public sewer system is that the quantity released per day, when diluted by the average daily quantity of flow into the sewer from the institution, does not cause the mean concentration of the sewage to exceed the recommended maximum concentration for drinking water.

A convenient way to immobilize the radioactivity in aqueous waste is to convert the liquid into concrete, then to package the solidified waste in an impermeable container and either to bury it on land or dump it into the sea. Several methods are used in packaging the radioactive concrete. The simplest is to pour a concrete bottom in a 55-gal drum several inches thick, then, after it hardens, to pour the radioactive mix to within several inches of the top. After the radioactive concrete hardens, it is topped with more concrete, then the drum is sealed with its cover. For higher-level waste, the radioactive concrete is poured into a metal container (whose maximum size is a 30-gal drum). This sealed container is then placed into a concrete-bottomed 55-gal drum, concrete is poured around and on top of the smaller radioactive container, and the drum is sealed. If the drum is to be transported elsewhere for burial (on land or sea), it must be properly marked, according to the regulations; and the radiation level must not exceed 200 mR/hr on the surface or 10 mR/hr at 1 m from the surface.

Airborne Wastes

Airborne radioactivity may be either gaseous or particulate. Gases may arise from neutron activation of cooling air in a reactor and from gaseous fission products, and from radiochemical reactions in which a gaseous product is produced. Particulates may be due to a large variety of processes, ranging from condensate droplets formed during the treatment of high-level liquid wastes to dusts from incinerators in which inflammable solids are burned. Hazards from air-borne wastes are best controlled at the source of the waste by limiting the production of air-borne wastes. If air-borne wastes are produced, the air must be sufficiently decontaminated that it may be safely diluted and discharged into the atmosphere. If the levels of the air-borne radioactivity are sufficiently low, the waste may be diluted and dispersed into the environment without further treatment.

Gases are usually difficult to remove. For small quantities of iodine and the noble gases, adsorption on activated charcoal may be used. Most of the radioactive gaseous wastes of the atomic energy industry are very short lived. Accordingly, these gases may be compressed and stored in tanks until they decay. Some of the methods used against radioactive gases are summarized in Table 11.4. In many instances, the most expedient method for dealing with radioactive gas is to discharge it to the atmosphere from a high stack, and thus to dilute the radioactivity to an acceptable level.

Particulate matter may be removed from gases by a variety of different devices, listed in Table 11.5, whose operating principles may be based on gravitational, inertial, electrostatic, thermal, or sonic forces; on physicochemical effects, or on filtration or barrier effects. The collection efficiencies of the different devices vary over a wide range. In considering an air-cleaning device for radioactive dusts, it should be borne in mind that the collection efficiency given by the manufacturer of air-cleaning devices for non-radioactive dusts is usually based on *mass* collection. Since the mass of a particle is proportional to the cube of its diameter, a single 10-micron particle is equivalent to 1000 one-micron particles. Reference to Table 11.1 shows that the maximum allowable concentration of non-radioactive particles is on the order of a million times or more greater than the allowable concentration for radioactive particles. Air-cleaning devices that are designed to remove much mass from the air, and are thus designated as high-efficiency collectors, may nevertheless be inadequate for respirable radioactive dusts. When this is the case, the final air-cleaning device usually is a high-efficiency filter that is designed for radioactive dusts. The performance of some high-efficiency filters is given in Table 11.6.

The extremely rigorous filtration requirements for radioactive dusts makes it desirable to specify the performance of a filter in a more meaningful way than "collection efficiency". Rather than designate the effectiveness of filters by filtration efficiency, in which there appears to be only a small difference between 99.99% and 99.995% (the former passes twice as many particles as the latter, 10 per 100,000 versus 5 per 100,000), we often used the *decontamination factor* as the figure of merit for a filter. The decontamination factor, d*f*, for

TABLE 11.4. TREATMENT METHODS FOR RADIOACTIVE GAS

Treatment	Gas	Efficiency %	Velocity fpm	Pressure drop, in. of water	Comments
Detention chamber	Noble gases	100	0	0	Use to hold up relatively small volumes
Spray tower	Halogens, HF	70–99	50	0.1–1.0	Precleaning or final cleaning for iodine removal
Packed tower	Radioiodine	95–99	50–200	1–10	Heated Berl Saddles coated with silver nitrate
Adsorbent beds	Iodine and noble gases	99.95	168	2.8	Activated charcoal or molecular sieves: may be used to decay xenon. May be refrigerated
Limestone beds	Halogens, HF	94–99.9	30	1–3	Experimental only. Some hood applications
Liquifaction column	Noble gases	99.9	—	—	Used to recover small amounts
Stripping column		90–95	—	—	Pilot studies only
Refrigerated carbon catalyst and carbon pellets	Xenon, krypton	99.9	—	—	Liquid nitrogen used for refrigerant. Gases recovered by desorption

From L. Silverman, Economic aspects of air and gas cleaning for nuclear energy processes, *Disposal of Radioactive Wastes*, Vol. 1, p. 147, I.A.E.A., Vienna, 1960.

TABLE 11.5. BASIC CHARACTERISTICS OF AIR CLEANING EQUIPMENT

Type of equipment	Particle size range mass median μ	Efficiency for size in col. 2 %	Velocity fpm	Pressure loss, in. of water	Current application in U.S. atomic energy programs
Simple settling chambers	> 50	60–80	25–75	0.2–0.5	Rarely used except for chips and recovery operations
Cyclones, large diameter	> 5	40–85	2000–3500 (entry)	0.5–2.5	Precleaners in mining, ore handling and machining operations
Cyclones, small diameter	> 5	40–95	2500–3500 (entry)	2–4.5	Same as above
Mechanical centrifugal collectors	> 5	20–85	2500–4000	—	Same as large cyclone application
Baffle chambers	> 5	10–40	1000–1500	0.5–1.0	Incorporated in chip traps for metal turning
Spray washers	> 5	20–40	200–500	0.1–0.2	Rarely used, occasionally as cooling for hot gases
Wet filters	Gases and 0.1–25 μ mists	90–99	100	1–5	Used in laboratory hoods and chemical separation operations
Packed towers	Gases and soluble particles > 5	90	200–500	1–10	Gas absorption and precleaning for acid mists
Cyclone scrubber	> 5	40–85	2000–3500 (entry)	1–5	Pyrophoric materials in machining and casting operations, mining, and ore handling. Roughing for incinerators
Inertial scrubbers, power-driven	8–10	90–95	—	3 to 5 HP/1000 cfm	Pyrophoric materials in machining and casting operations, mining and ore handling
Venturi scrubber	> 1	99 for H_2SO_4 mist. SiO_2, oil smoke, etc. 60–70	12,000 24,000 at throat	6–30	Incorporated in air cleaning train of incinerators
Viscous air conditioning filters	10–25	70–85	300–500	0.03–0.15	General ventilation air

Dry spun-glass filters	5	85–90	30–35	0.1–0.3	General ventilation air. Precleaning from chemical and metallurgical hoods
Packed beds of graded glass fibres 1–20 μ. 40 in. deep	<1	99.90–99.99	20	10–30	Dissolver off-gas cleaning
High-efficiency cellulose-asbestos filters	<1	99.95–99.98	5 through media. 250 at face	1.0–2.0	Final cleaning for hoods, glove boxes, reactor air and incinerators
All-glass web filters	<1	99.95–99.99	5 through media. 250 at face	1.0–2.0	Same as above
Conventional fabric filters	>1	90–99.9	3–5	5–7	Dust and fumes in feed materials production
Reverse-jet fabric	>1	90–99.9	15–50	2–5	Same as above
Single-stage electrostatic precipitator	<1	90–99	200–400	0.25–0.75	Final clean-up for chemical and metallurgical hoods. Uranium machining
Two-stage electrostatic precipitator	<1–5	85–99	200–400	0.25–0.50	Not widely used for decontamination

From L. Silverman, Economic aspects of air and gas cleaning for nuclear energy processes, *Disposal of Radioactive Wastes*, Vol. 1, pp. 139–79, I.A.E.A., Vienna, 1960.

TABLE 11.6. PERFORMANCE OF HIGH-EFFICIENCY FILTERS
(At normal air temperatures and standard density air)

Medium	Test aerosol		Air velocity		Resistance, in. of water	Efficiency %	Method	Remarks
	Name	Size μ (homogeneous except*)	fpm	cm/sec				
CC-6 Cellulose-asbestos paper	Methylene blue	—	4	2	0.8	99.9871	Discoloration	
	Dioctyl phthalate (DOP)	0.3	5	2.5	0.67	99.85	Penetrometer	
	Atmospheric dust	0.5*	5	2.5	0.67	99.9+	Count	
	Duralumin	0.18	500	250	100	97.7	Count	Note excessive velocity causes greater penetration of fine size
	Duralumin	0.18	2	1	0.28	99.7	Count	Reduced velocity improves performance
	Potassium permanganate ($KMnO_4$)	0.02	20	10	2.7	93.0	Count	Size for maximum penetration
AEC No. 1 Cellulose-asbestos	DOP	0.3	5	2.6	0.7	99.78	Penetrometer	
	Duralumin	0.18	2	1	0.28	92.9	Count	
	Duralumin	0.18	40	20	5.6	99.6	Count	
	Atmospheric dust	0.58	5	2.5	0.7	99.98	Count	
	$KMnO_4$	0.01	4	2	0.56	91.0	Count	
		0.02						Size for maximum penetration

All-glass superfine fibres—Hurlbut-MSA1106B	DOP	0.3	5	2.5	1.05	99.999	Penetrometer	
	Atmospheric dust	0.5*	5	2.5	1.05	99.9+	Count	
	$KMnO_4$	0.015	20	10	4.4	93.0	Count	Size for maximum penetration
Sand	Cell ventilation gases	1	3–5	1.5–2.5	4.5–5.5	99.5–99.8	Radioactivity	69 in. deep: graded sizes from 2¼-50 mesh. Will not withstand high moisture conditions
Composite glass wool	Process off-gases	1	20	10	4.0	99.0		Composition given in reference
	Methylene blue	0.6 MMD	20	10	4.0	99.99	Gravimetric	Same
Compressed glass fibres	Atmospheric dust	0.5	5.25	2.6	0.69	99.997	Count	0.02 in. thick 50% 1.3 μ and 50% 3.0 μ fibers
Resin wool	Atmospheric dust	0.5	14	7	0.3	99.6	Discoloration	These filters are known to decrease in performance when exposed to ionizing radiation
Glass	Uranium oxide	0.12	2.3–7.8	1.2–3.9	0.30–1.23	95.5–99.5	Gravimetric	Special glass formulation developed by A.D. Little. Aluminum separators and furnace cement seals

From L. Silverman, Economic aspects of air and gas cleaning for nuclear energy processes, *Disposal of Radioactive Wastes*, Vol. 1, pp. 139–79, I.A.E.A., Vienna, 1960.

a filter whose efficiency is E per cent, is defined as

$$df = \frac{100}{100 - E}. \tag{11.3}$$

The filter whose efficiency is 99.99% and, thus, passes 10 particles per 100,000 has a decontamination factor of 10,000, while the filter of 99.995% efficiency, which passes 5 particles per 100,000, has a decontamination factor of 20,000.

Following filtration, the remaining radioactive particulates are discharged into the atmosphere for dispersion of the non-filterable low levels of activity. If the particles are small, i.e. $<1\ \mu$, the particulate terminal settling velocity is very low, and the particles may be considered as part of the gas in regard to their diffusion into the atmosphere and transport with the gases that issue forth from the exhaust stack.

Meteorological considerations

When a contaminant is discharged from a chimney, it is assumed that the contaminant will be carried downwind, while at the same time it diffuses laterally and vertically. The two main consequences of this dispersion in the atmosphere are dilution of the contaminant, and its eventual return to the breathing zone at ground level. Of particular interest in evaluating the safety of discharge into the air is the relationships between the rate of discharge and the ground level concentrations—both in the breathing zone and on the ground (as fallout)—of the discharged radioactivity. The ground-level distribution of the discharged radioactivity depends on a number of factors, including atmospheric stability, wind velocity, type of terrain and the nature of the boundary layer of air (the air layer immediately over the ground for a distance of several hundred feet) and height of the chimney. It is thus very difficult to predict precisely the pattern of ground-level distribution, although reasonable estimates may be made from one of several different sets of atmospheric diffusion equations.

Atmospheric stability depends on the temperature gradient of the air (Fig. 11.4). Meteorologists refer to the temperature gradient of the atmosphere as the *lapse rate*. A parcel of air that is rising expands as a result of the decreasing atmospheric pressure. If no heat is gained or lost by this parcel of air, the expansion will be adiabatic, and the temperature of the air parcel will drop. For dry air, this adiabatic cooling results in a temperature decrease of 5.4°F per 1000 ft of ascent; for average moist air, the lapse rate is 3.5°F per 1000 ft. If the temperature gradient of the atmosphere is less than adiabatic, but still negative, we have a *stable lapse rate*. In this case, a rising parcel of air cools faster than the surrounding atmosphere. It, therefore, is denser than the air in which it is immersed, and tends to sink. A sinking parcel of air is warmer than the surrounding air, and thus is less dense, which results in a tendency to rise. A stable lapse rate, therefore, tends to restrict the width of

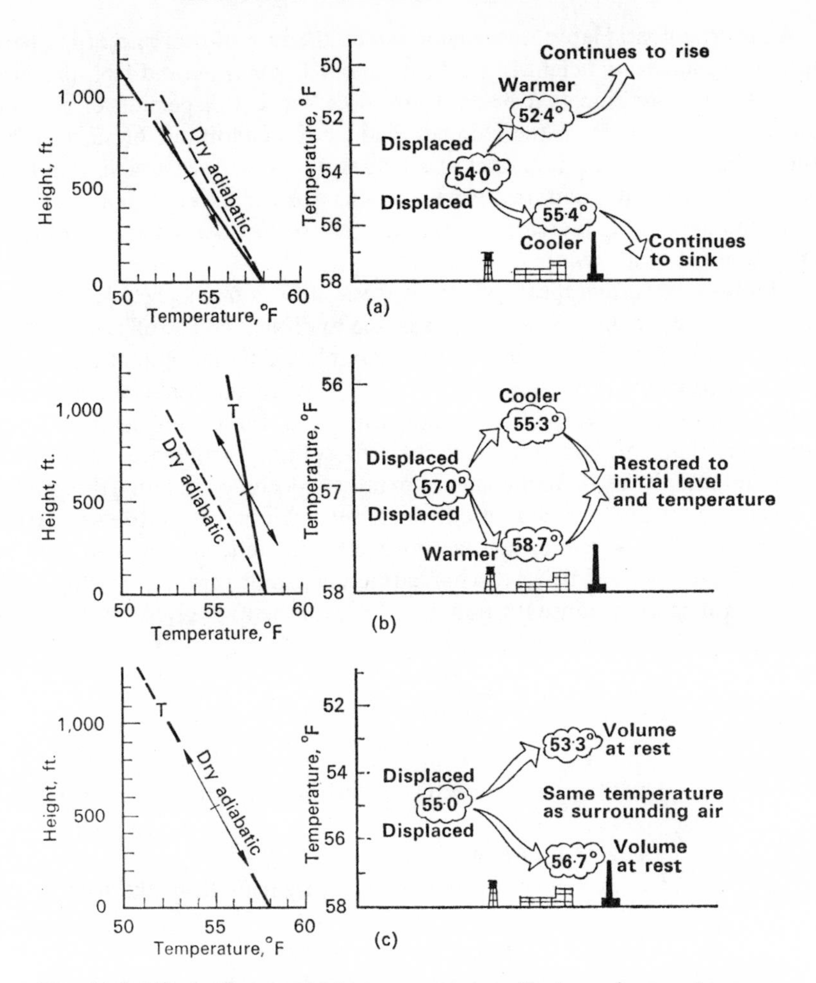

FIG. 11.4. Effect of atmospheric temperature gradient—or lapse rate—on a displaced volume of air. (a) Unstable lapse rate. (b) Stable lapse rate. (c) Neutral lapse rate. (From *Meteorology and Atomic Energy*, U.S.A.E.C., Washington, D.C., 1955.)

the plume in the vertical direction, thereby decreasing the dilution effect of the atmosphere.

If the lapse rate is positive, that is, if the air temperature increases with increasing height, then the super-stable condition known as an *inversion* occurs (since the temperature gradient is "inverted"). The rising effluent from the chimney becomes much denser than the surrounding air as it cools adiabatically, and thus sinks. The overall effect of an inversion is to trap the effluent from the chimney, and to prevent its ascent to higher altitudes.

A superadiabatic lapse rate, one in which the rate of decrease of temperature with increasing height is greater than 5.4°F per thousand feet, produces an unstable condition that helps to promote vertical dispersion of the contaminated effluent from the chimney. Under the conditions of such an unstable lapse rate, a rising parcel of air does not cool fast enough, due to its adiabatic expansion, and, therefore, remains warmer and less dense than the surrounding air. It, therefore, continues to rise. By the same reasoning, a falling parcel of air continues to fall.

Although we often speak of atmospheric diffusion, the fact is that very little atmospheric dispersion of gases is due to diffusion. The effects of turbulence usually are so great as to mask completely molecular diffusion. For this reason, estimates of the dispersion of gases in the atmosphere are not based on the classical diffusion equation, but rather on semi-empirical equations that consider the effects of atmospheric turbulence. One of the more commonly used mathematical expressions for estimating the ground level concentration of a gaseous effluent from a chimney is Sutton's equation. For conditions where we have a single elevated (effective height $\geq$25 m) "point" source of gas being discharged at a constant rate, the average (over 3 min) ground level concentration of the contaminant is given by

$$\bar{\chi}(x, y) = \frac{2Q}{\pi C^2 \mu x^{2-n}} e^{-[(1/C^2 x^{2-n})(y^2+h^2)]}, \tag{11.4}$$

where
$\bar{\chi}$ = volumetric concentration of the contaminant, mCi per m³, etc.,
Q = emission rate, mCi/sec, etc.,
x, y = coordinates of point of measurement from the foot of the chimney, meters,
μ = mean wind speed, meters per second,
C = virtual diffusion coefficients in lateral and vertical directions (given in Table 11.6),
n = dimensionless parameter determined by the atmospheric stability (given in Table 11.6),
h = effective chimney height, meters.

If the effluent gas has a significant exit velocity, or if it is at a high temperature, it will rise to a level higher than the chimney. The effective chimney height, therefore, is the sum of the actual chimney height, a factor due to the exit velocity, and a temperature factor:

$$h = h_a + d\left(\frac{v}{\mu}\right)^{1.4}\left(1 + \frac{\Delta T}{T}\right), \tag{11.5}$$

where h_a = actual height of chimney, meters,
d = chimney outlet diameter, meters,
v = exit velocity of gas, meters/second,
μ = mean wind speed, meters/second,
ΔT = difference between ambient and gas temperatures, °C,
T = absolute temperature of gas, °K.

The standard deviation of the concentration predicted by Sutton's equation is a factor of 5. That is, 68 times out of 100, the true concentration can be expected to lie between $\bar{\chi}/5$ and $5\ \bar{\chi}$, while 96 times out of 100, the true concentrations can be expected to lie between $\bar{\chi}/10$ and $10\ \bar{\chi}$.

TABLE 11.7. DIFFUSION (C) AND STABILITY (n) PARAMETERS FOR SUTTON'S EQUATION

Lapse rate	n	C^2 Chimney height, meters			
		25	50	75	100
Superadiabatic	0.20	0.043	0.030	0.024	0.015
Stable	0.25	0.014	0.010	0.008	0.005
Moderate inversion	0.33	0.006	0.004	0.003	0.002
Large inversion	0.50	0.004	0.003	0.002	0.001

Example 11.3

The ^{41}A effluent from an air-cooled reactor is 1000 μCi/sec. The effluent is discharged through a chimney 75 m high, the mean wind velocity is 2 m/sec, and the lapse rate is stable. The temperature of the gas is 87°C, the ambient temperature is 27°C, the effluent velocity is 10 m/sec, and the diameter of the chimney orifice is 2 m.

(a) What is the ground level concentration of ^{41}A at a distance of 200 meters from the chimney, on the center line of the plume?

The effective height of the chimney, as calculated from equation (11.5), is 98 m. Table 11.7 gives the stability parameter, n, as 0.25, and the diffusion parameter, C^2, as 0.005. Substituting the appropriate numerical values into equation (11.4), we have

$$\bar{\chi}(200,0) = \frac{2 \times 10^3}{\pi \times 0.005 \times 2 \times (200)^{2-0\cdot25}} e^{-[[1/0.005\ (200)^{2-0.25}]\ (0+98^2)]}$$

$$= 5.9\ e^{-178}$$

$$\approx 0.$$

The calculated result shows that at a distance of 200 m, the plume has not yet spread sufficiently to reach the ground.

(b) At what distance from the chimney will the ground-level concentration be maximum?

On the centerline of the plume, $y = 0$, and equation (11.4) becomes

$$\bar{\chi}(x, 0) = \frac{2Q}{\pi\, C^2\, \mu\, x^{2-n}}\, e^{-(h^2/C^2\, x^{2-n})} \tag{11.6}$$

and the point of maximum ground level concentration occurs where $d\bar{\chi}/dx = 0$. Differentiating equation (11.6) with respect to x, and solving for this distance, we have

$$x_m = \left(\frac{h^2}{C^2}\right)^{1/2-n}. \tag{11.7}$$

Substituting into equation (11.7) yields

$$x_m = \left(\frac{98^2}{0.005}\right)^{1/2-0.25} = 3800 \text{ meters.}$$

(c) What is the maximum ground level concentration?

Substituting the value for x from equation (11.7) into equation (11.6) shows the maximum concentration to be

$$\bar{\chi}_{\max} = \frac{2Q}{\pi\, \mu\, h^2\, e}, \tag{11.8}$$

which yields, when the appropriate numbers are inserted,

$$\bar{\chi}_{\max} = \frac{2 \times 10^3}{\pi \times 2 \times (98)^2 \times 2.718} = 1.2 \times 10^{-2}\, \frac{\text{mCi}}{\text{m}^3}.$$

In this calculation, no allowance was made for the fact that about 30 min travel time are required for the plume to travel 3800 m. Since ^{41}A has a half-life of 110 min, about 10% of it will decay during the transit time. Generally, if the transit time from the chimney to the point in question is long relative to the half-life of the radioisotope, the emission rate, Q, in equations (11.4), (11.6), and (11.8) is multiplied by the decay factor, $e^{-\lambda t}$, where t is the transit time.

If a chimney emits particulate matter of significant settling velocity, then depletion of the radioactivity in the plume due to fallout of the particles must be considered in estimating the downwind concentration. The effect of the fallout is to tilt the axis of the plume downward through an angle $\theta = \tan^{-1} v_t/\mu$ from the horizontal, where v_t is the terminal settling velocity and μ is the mean wind speed. The elevation of the plume, z', at any downwind distance x thus is

$$z' = h - x \tan\theta, \tag{11.9}$$

or

$$z' = h - \frac{x v_t}{\mu}, \tag{11.10}$$

and Sutton's equation for estimating the ground level concentration, equation (11.4), is modified to give

$$\bar{\chi}(x, y) = \frac{Q}{\pi C^2 \mu x^{2-n}} e^{-[y^2+(z')^2/C^2 x^{2-n}]}. \tag{11.11}$$

The rate of ground deposition of particles at point (x, y) is found by multiplying the ground level concentration, equation (11.11), by the terminal deposition velocity, v_g, of the particulate matter:

$$w \frac{\text{units}}{\text{m}^2\,\text{sec}} = \bar{\chi} \frac{\text{units}}{\text{m}^3} \times v_g \frac{\text{m}}{\text{sec}}. \tag{11.12}$$

Deposition velocity is determined mainly by the micro-meteorological conditions near the surface, and thus cannot be calculated with any reasonable degree of accuracy. Experimentally, values ranging from about 0.2–2.5 cm/sec have been obtained. For dust from nuclear weapons, the average deposition velocity is about 1 cm/sec. (A. C. Chamberlain, *Aspects of Travel and Deposition of Aerosol and Vapour Clouds*, U.K.A.E.A., Report HP/R 1261, 1955.)

Solid Wastes

Not very much can be done to solid radioactive waste in order to reduce its radioactivity content. The chief type of processing is reduction of volume prior to storage, or burial in land, or dumping into the sea. Long-term storage is usually restricted to high-level solids, and may be accomplished by use of an underground concrete vault. Properly packaged waste may be buried, that is, placed directly into the ground and covered with earth in special areas that are designated by the appropriate regulative agencies as radioactive waste-disposal burial grounds. The exact type of packaging depends on the quantity and the half-life of the activity. High-level wastes may by encased in concrete filled steel drums, while for low-level or short-lived wastes it may be sufficient to place the activity into polyethylene bags, and then to seal the bags into steel cans or drums. A good-quality, painted (inside and outside) steel drum may be expected to maintain its integrity for at least 10 years; concrete casing may be expected to retain its integrity for 100 years.

When dumping solid radioactive waste into the sea, consideration is given to the density of the package (in order to assure sinking to the ocean bottom), corrosion resistance, and sufficient strength to resist the great hydrostatic pressures at the bottom of the sea. Concrete encased waste in a 55 gal drum is considered to be satisfactory. A good drum has a lifetime in sea water of about 10 years; concrete may be expected to have a lifetime in the ocean of about 30 years. Clearly, the concrete serves more than a weighting function; it is, in fact, the chief containment vessel. Accordingly, it should be of good quality and low porosity. A package for sea disposal should be free of voids for two reasons: (1) an empty space inside the package may make the package subject to collapse from the hydrostatic pressure of the ocean, and (2) it decreases the overall density of the package. For disposal into the deep sea, the specific gravity should not be less than 1.2 (75 lb/ft^3, or 10 lb/gal). If the waste is to be dumped on the continental shelf, the specific gravity of the container and its contents should not be less than 1.5 (12.5 lb/gal). Furthermore, the loaded container should contain no voids if collapse of the container due to the tremendous pressure of the water is to be prevented.

Relatively large quantities of combustible low-level radioactive waste, such as paper towels, protective clothing, rags, animal excreta and carcasses, etc., as well as non-combustible low-level wastes, such as broken glassware, hypodermic needles, etc., are produced in many radioisotope laboratories and nuclear medicine clinics. Often, the non-combustible and combustible waste is accumulated in different containers. The non-burnable waste is packaged for disposal into the sea or into the ground, while the combustible waste is burned. Incineration may either concentrate the activity by burning away the substrate in which activity is held, if the activity is non-volatile, or it may disperse the activity with the effluent from the chimney if the activity is volatile or if the contaminated waste is transformed physically into particulates. In the instance where the activity is concentrated in the ashes, we have a case of volume reduction; the ashes still must be picked up and packaged for disposal. (The collection of ashes from an incinerator in which radioactivity had been burned should be done under the supervision of a health physicist. Appropriate respiratory protective equipment should be available if necessary.) If the activity may go up the stack to be dispersed in the atmosphere, the rate of incineration of radioactive waste should be controlled in order to limit the activity in the stack effluent to acceptable levels.

Example 11.4

Dead rats that had been injected with a ^{14}C-tagged compound are to be burned in an incinerator. The carbon is expected to go up the stack as CO_2, but some of it may also be discharged as carbon particulates. About 250 lb of dry non-radioactive waste are burned per 8-hr day in this incinerator. How much activity may be incinerated if the concentration at the top of the chimney is not to exceed 10 times the maximum allowable concentration of 10^{-6} μCi/cm^3? (It should be noted that this does not violate the ICRP recommendation of 1/30 the continuous occupational value for large population groups, since the concentration referred to is at a point where it can be inhaled by the large population group. The top of a chimney does not meet this requirement. The higher discharge concentration in this case is based on dilution by the atmosphere to a safe level before the activity can be inhaled.)

At least 3.5 lb of air per pound waste must be supplied to the incinerator.

Since one cubic foot of air weighs 0.075 lb, the amount of air used during the day is

$$250 \frac{\#\text{ waste}}{\text{day}} \times \frac{3.5 \#\text{ air}/\#\text{ waste}}{0.075 \#\text{ air/ft}^3} \times 2.83 \times 10^4 \text{ cm}^3/\text{ft}^3 = 3.3 \times 10^8 \text{ cm}^3/\text{day}.$$

To attain a maximum concentration of 10×10^{-6} μCi/cm^3, we cannot incinerate more than

$$3.3 \times 10^8 \frac{\text{cm}^3}{\text{day}} \times 1 \times 10^{-5} \frac{\mu\text{Ci}}{\text{cm}^3} = 3300 \frac{\mu\text{Ci}}{\text{day}}.$$

Furthermore, since N.B.S. Handbook 53 lists 4 μCi/g carbon as the maximum specific activity of any carbon particles that may be discharged from the stack, we will adhere to this limit. If the waste that is incinerated is assumed to be 25% carbon by weight, then the amount of carbon incinerated is

$$250 \frac{\text{lb waste}}{\text{day}} \times 0.25 \frac{\text{lb C}}{\text{lb waste}} \times 454 \frac{\text{g}}{\text{lb}} = 2.83 \times 10^4 \frac{\text{g C}}{\text{day}}.$$

The mean specific activity, therefore, of the ^{14}C, if 3300 μCi are burned per day, is

$$\frac{3.3 \times 10^3 \mu\text{Ci/day}}{2.83 \times 10^4 \text{ g C/day}} = 0.117 \frac{\mu\text{Ci}}{\text{g C}}.$$

This is much less than the recommended maximum of 4 μCi/g. Let us now consider the rat. It consists of 18% carbon by weight. The maximum activity, A_m, in a rat that could be incinerated, if there were no isotope dilution, is:

$$A_m = 4 \times 0.18 \times W \frac{\mu\text{Ci}}{\text{rat}}.$$

For a 300-g animal, this corresponds to 216 μCi per rat. However, because of the very large dilution by the waste, we may allow the daily maximum incinerated activity of 3300 μCi to be distributed over any number of rats.

In certain other instances, the radioactivity could be converted into a gas, and then discharged to the atmosphere. An illustration of how this may be accomplished within the limits prescribed by our radiation safety criteria is shown in the following example.

Example 11.5

Ten millicuries ^{14}C waste, in the form of 1 g $BaCO_3$, will be disposed of by changing the chemical form of the carbon to $^{14}CO_2$, and then discharging the radioactive gas to the atmosphere. The chemical manipulations will be carried out in a fume hood whose face opening is 6 ft wide and 3 ft high, and whose face velocity is 100 ft/min. The $^{14}CO_2$ will escape to the atmosphere via the exhaust stack from the hood. The chemical conversion from the carbonate to the gas will be accomplished by the addition of 1 N HCl. What is the maximum rate at which acid may be added to the $BaCO_3$ if the maximum permissible atmospheric concentration of 10^{-6} μCi/cm³ air is not to be exceeded at the discharge end of the exhaust stack?

The conversion of the carbonate to CO_2 proceeds according to the reaction

$$BaCO_3 + 2HCl \rightarrow BaCl_2 + CO_2.$$

Since the formula weight of $BaCO_3$ is 197.4 (the additional weight due to the ^{14}C is very small, and may be neglected), 1 g $BaCO_3$ is

$$\frac{1 \text{ g}}{197.4 \text{ g/mole}} = 0.00506 \text{ moles.}$$

To convert all the $BaCO_3$ to CO_2, $2 \times 0.00506 = 0.01012$ moles HCl are needed. Since 1 N HCl contains 1 mole acid per liter, the required amount of acid will be contained in 0.01012 liters, or 10.12 ml HCl. According to the chemical equation, 1 mole $BaCO_3$ reacted with 2 moles acid yields 1 mole CO_2. Therefore, if the reaction goes to completion, 0.00506 moles CO_2 will be produced. This gas will occupy a volume, under standard conditions of temperature and pressure, of

$$5.06 \times 10^{-3} \text{ moles} \times 22.4 \frac{\text{liters}}{\text{mole}} = 0.1155 \text{ liters.}$$

The MPC (maximum permissible concentration) for ^{14}C is 10^{-6} μCi/cm³, or 10^{-3} μCi per liter air. The specific activity of the $^{14}CO_2$ produced in this reaction is

$$\frac{10\text{mCi} \times 10^3 \ \mu\text{Ci/mCi}}{1.155 \times 10^{-1} \text{ liters}} = 8.65 \times 10^4 \frac{\mu\text{Ci}}{\text{liter}}.$$

The amount of additional air that must be added to the radioactive CO_2 in order that the resulting mixture have a concentration of 10^{-3} μCi/liter may be calculated as follows:

$$\frac{1 \times 10^4 \ \mu\text{Ci}}{(0.1155 + X) \text{ liters}} = \frac{10^{-3} \ \mu\text{Ci}}{1 \text{ liter}},$$

$$X = 10^7 - 0.1155,$$

$$X \approx 10^7 \text{ liters air.}$$

The volume of air that flows through the discharge stack is

$$Q = \text{face area} \times \text{velocity}$$

$$= 3 \text{ ft} \times 6 \text{ ft} \times 100 \frac{\text{ft}}{\text{min}} \times 28.32 \frac{\text{liters}}{\text{ft}^3}$$

$$= 5.1 \times 10^4 \frac{\text{liters}}{\text{minute}}.$$

If the conversion of the $BaCO_3$ to CO_2 proceeds at a uniform rate of speed, it must take at least

$$\frac{10^7 \text{ liters}}{5.1 \times 10^4 \text{ liters/min}} = 196 \text{ minutes.}$$

The 1 N HCl must, therefore, flow into the gas generator at a rate not exceeding

$$\frac{10.12 \text{ ml}}{196 \text{ min}} = 5.17 \times 10^{-2} \frac{\text{ml}}{\text{min}},$$

or about 20 min/ml.

Assessment of Hazard

A realistic assessment of a hazard from an internally deposited radioisotope requires more consideration than merely comparing an environmental concentration with a recommended maximum allowable concentration. The maximum allowable concentrations, it should be re-emphasized, are based on continuous exposure for 50 years. In recognition of this fact, the NCRP permits exposure at atmospheric concentrations up to 1200 times the recommended maximum based on 50-year exposure if the actual exposure does not exceed 1 hr, and provided that this exposure would not result in a 13-week dose that exceeds the basic safety criterion. Furthermore, the recommended maxima do not—in most instances—consider the chemical form of the radioisotope, nor the influence of the chemical form on the metabolic properties of the isotope, nor the consequent effect on the absorbed dose from exposure to the isotope. If the metabolic properties are known, they may be used to assess the hazard from an internal emitter, as shown below in the following example:

Example 11.6

^{14}CO will be produced in a pilot study in which excess H_2SO_4 will react with $H{-}^{14}CO{-}ONa$, whose specific activity is 20 mCi/millimole, to produce ^{14}CO. The threshold limit value (TLV) for CO gas, based on its chemical effects, is 50 parts per million for occupational exposure. The recommended maximum atmospheric concentration, based on radiological considerations, for occupational exposure to ^{14}C is 4×10^{-6} $\mu Ci/cm^3$ air.

(a) Will the industrial hygiene control that limits CO to 50 ppm be sufficient to meet the handbook requirements of radiological safety?

To find the molar concentration of CO in the atmosphere that corresponds to 50 ppm:

$$50 \text{ ppm} = \frac{50 \text{ moles CO}}{10^6 \text{ moles atmosphere}} = 5 \times 10^{-5} \frac{\text{moles CO}}{\text{mole atmosphere}}.$$

Since there is one carbon atom per molecule of sodium formate, and also one carbon atom per CO molecule, the specific activity of the ^{14}CO will also be 20 mCi/mm; and the concentration of radioactivity corresponding to 50 ppm is

$$\frac{5 \times 10^{-5} \text{ moles CO/mole atm} \times 2 \times 10^{1} \text{ curies/mole CO} \times 10^{6}\ \mu\text{Ci/curie}}{2.24 \times 10^{4} \text{ cm}^3\text{/mole atm}} =$$

$$= 4.46 \times 10^{-2} \frac{\mu\text{Ci}}{\text{cm}^3 \text{ atm}}.$$

Use of industrial hygiene criteria would, in this case, lead to an atmospheric concentration of ^{14}C of $4.46 \times 10^{-2} \div 4 \times 10^{-6} = 11{,}000$ times the maximum recommended concentration for continuous exposure, and about 10 times the concentration for a 1 hr exposure.

(b) The industrial hygienist, believing that control of the ^{14}CO according to the chemical TLV is sufficient for the radiological hazard, allows a chemical engineer to be exposed to 50 ppm of the ^{14}CO for a period of 2 hr. What is the dose, in rads, to the chemical engineer as a result of his exposure?

In order to calculate the absorbed dose, certain facts must be known about the physiological behavior of CO. When CO is inhaled, it diffuses across the capillary bed in the lungs, and dissolves in the blood. It then is absorbed by the erythrocytes and combines with the hemoglobin to form carboxyhemoglobin. Since carboxyhemoglobin is incapable of transporting oxygen, the inhalation of CO leads to cellular anoxia, which in turn may lead to unconsciousness or to death—depending on the amount of CO that is absorbed into the blood. The maximum amount of an inhaled gas that can be absorbed, which is called the *saturation value*, depends on the partial pressure of the gas in the atmosphere. The saturation value for CO, S_∞, as per cent hemoglobin tied up as carboxy hemoglobin, is given by

$$S_\infty = \frac{210 \times \text{p CO}}{210 \times \text{p CO} + \text{p O}_2} \times 100, \tag{11.13}$$

where p CO is the percent CO in the air, and p O_2 is the percent oxygen in the alveolar air (p O_2 is usually equal to 15). One hundred percent saturation corresponds to 20 ml CO per 100 ml blood. Rate of absorption, in most cases, follows first order kinetics; that is, the fractional approach to saturation per unit time remains constant. Thus, if 1% of the saturation value is absorbed in 1 min after beginning inhaling the gas, 1% of the remaining 99% will be absorbed during the second minute, then 1% of the 98.01% left and so on. Since saturation is approached assymptotically, we usually refer to the *half saturation time* to designate the rate of absorption of an inhaled gas. The numerical value for the half saturation time is independent of the atmospheric concentration of the gas (except for very high concentrations). For CO, the half saturation time is about 47 min. The absorption of CO is analogous to the buildup of a radioactive daughter as it approaches secular equilibrium, and is described by a similar equation:

$$S = S_\infty (1 - e^{-0.693/T \times t_i}), \tag{11.14}$$

where S_∞ is the saturation value corresponding to a particular atmospheric concentration of the CO, S is the percent of the hemoglobin bound with CO, T is the half saturation time, and t_i is the inhalation time.

For an atmospheric concentration of 50 ppm (50×10^{-6} parts CO per part air), which corresponds to 50×10^{-4}%, or 0.005%, the hemoglobin saturation value is calculated from equation (11.13) to be

$$S_\infty = \frac{210 \times 0.005}{210 \times 0.005 + 15} \times 100 = 6.55\%.$$

After the 2-hr exposure, the percentage of the worker's hemoglobin that is bound with CO is calculated from equation (11.14) to be

$$S = 6.55\,(1 - e^{-0.693/47 \times 120}) = 5.45\%.$$

The blood volume of the standard man is 7.7% of his weight, or 5.4 liters for a 70-kg man; it, therefore, can hold 1/5 of 5.4 liters, or 1080 ml CO or oxygen. Since 5.45% of this capacity is tied up with CO, the quantity of CO in the man's body is

$$0.0545 \times 1080 \text{ ml} = 59 \text{ ml } ^{14}\text{CO at NTP}.$$

Since the specific activity of the ^{14}CO is 20 mCi/millimole, the body burden following two hours of inhalation is:

$$59 \text{ ml} \times \frac{1 \text{ millimole}}{22.4 \text{ ml}} \times 20 \frac{\text{mCi}}{\text{millimole}} = 52.7 \text{ mCi}.$$

Assuming the blood, and hence the ^{14}C to be uniformly distributed throughout the body of the 70 kg man, the dose rate due to this body burden of ^{14}C is, from equation (6.12):

$$D = \frac{52.7 \text{ mCi} \times 3.7 \times 10^7 \text{ dps/mCi} \times 5 \times 10^{-2} \text{ MeV/dis} \times 1.6 \times 10^{-6} \text{ erg/MeV} \times 3.6 \times 10^3 \text{ sec/hr}}{7 \times 10^4 \text{ g} \times 10^2 \text{ erg/g/rad}}$$

$$= 8.03 \times 10^{-2} \text{ rad/hr}.$$

If inhalation had continued until the hemoglobin saturation value was reached, the body burden would have reached

$$\frac{6.55}{5.45} \times 52.7 = 63.5 \text{ mCi},$$

and the dose rate would have been proportionally increased to D_∞, the maximum possible value, under the conditions of the exposure, of 9.65×10^{-2}

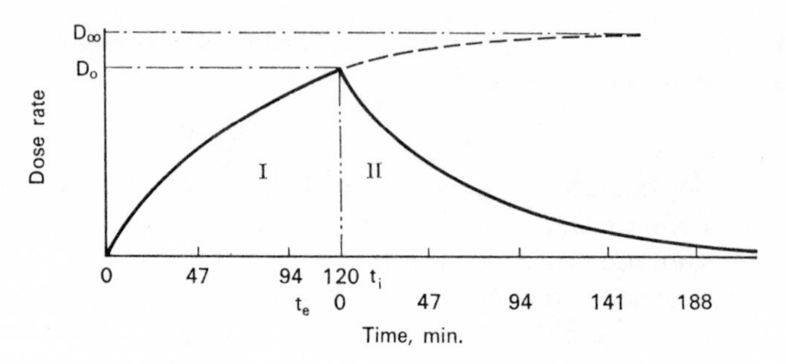

FIG. 11.5. Variation of dose rate with time after beginning of inhalation of ^{14}CO at a concentration of 50 ppm and 4.46×10^{-7} $\mu Ci/cm^3$. Region I under the curve represents the period of inhalation, region II represents the period of exhalation. D_o, the dose rate at the end of the period of inhalation, is 8.03×10^{-2} rad/hr; D_∞, the dose rate due to the saturation amount of ^{14}CO, is 9.65×10^{-2} rad/hr.

rad/hr. In this case, the body burden, and hence the dose rate, varied with time, as shown graphically in Fig. 11.5. The instantaneous dose rate during the period of inhalation (period I in Fig. 11.5) is given by

$$DR = D_\infty (1 - e^{-kt_i}), \tag{11.15}$$

where k is the carboxyhemoglobin dissociation constant, $\frac{0.693}{T}$ and t_i is the time of inhalation. The total dose during the period of inhalation is

$$D = D_\infty \int_0^{t_i} (1 - e^{-kt_i}) dt_i, \tag{11.16}$$

which, when integrated, yields

$$D = D_\infty \left[t_i + \frac{1}{k} (e^{-kt_i} - 1) \right]. \tag{11.17}$$

The total dose during the time period t_e (period II in Fig. 11.5) after termination of inhalation, under conditions of decreasing body burden, is given by equation (6.22) as

$$D = \frac{D_0}{\lambda_E} (1 - e^{-\lambda_E t_e}), \tag{6.22}$$

where D_0 is the instantaneous dose rate at time $t = 0$, that is, the time when inhalation ceased, and λ_E is the same as k in equation (11.15). Substituting 9.65×10^{-2} rad/hr for D_∞, 2 hr for t_i, and 0.885 per hour for k into equation (11.17), the radiation absorbed dose during buildup is found to be 0.103 rad. When 8.03×10^{-2} rad/hr is substituted for D_0, 0.885 per hour for λ_E, and ∞ for t in equation (6.21), the radiation-absorbed dose from the ^{14}C during the period of elimination is found to be 0.091 rad. The total dose due to the 2-hr inhalation of the ^{14}CO is the sum of these two doses 0.194 rad, or 194 mrad. More than 99% of this dose is absorbed within 7 effective hours after beginning inhalation of the radioactive carbon monoxide.

Although the worker was exposed to a concentration of radiocarbon very much higher than the maximum recommended value, the consequent radiation absorbed dose—which is the basic criterion for radiation safety—is very much less than the 3000 mrem maximum that is allowed during a period of 13 consecutive weeks.

The concept of a maximum credible accident is useful in advance planning for the purpose of minimizing radiation dose in the event of an accident, or for designing safety limitations into an experiment. Consider the following example.

Example 11.7

An engineer wishes to use tritiated water in an experimental study of a closed pressurized system. The system's capacity is 3 liters water which will be kept at a temperature of 300°F. The experiment will be done in a ventilated laboratory whose dimensions are 10′ × 10′ × 10′. The maximum credible accident is one in which the system will rupture, and the entire 3 liters of tritiated water will be sprayed into the room. The laboratory ventilation rate is 346 ft^3/min. (The laboratory has its own exhaust line and stack, so that there is no possibility of spreading the tritium, in the event of an accident, to other laboratories in the building.) What is the maximum amount of tritiated water, as H^3OH, that may be in the system, assuming a maximum credible accident, if the engineer is not to inhale more tritium than that which would deliver a dose of 3 rad over a period of 13 weeks. In the event of such an accident, it is estimated that the engineer might remain in the laboratory for as long as 2 min.

For tritium:

(a) the critical organ is the total body (weight = 70 kg);
(b) the biological half-time is 12 days;
(c) the radiological half-life is 12.3 years,
 (i) the effective half-life, from equation (6.19) is 12 days,
 (ii) the effective elimination constant, from equation (6.17), is 0.0578 day^{-1};
(d) pure beta emitter, average beta energy = 0.006 MeV;
(e) all the inhaled tritium is assumed to be absorbed.

The initial dose rate, D_0, that will result in a total dose of 3 rad over a period of 13 weeks (91 days) is calculated from equation (6.22):

$$3 = \frac{D_0}{0.0578 \text{ day}^{-1}}(1 - e^{-0.0578 \times 91}),$$

$$D_0 = 0.174 \text{ rad/day}.$$

The body burden, Q mCi, that will deliver this initial dose rate is calculated from equation (6.12) to be

$$0.174\ \frac{\text{rad}}{\text{day}} = \frac{Q\ \text{mCi} \times 3.7 \times 10^{7}\ \text{dps/mCi} \times 6 \times 10^{-3}\ \text{MeV/d} \times 1.6 \times \times 10^{-6}\ \text{erg/MeV} \times 8.64 \times 10^{4}\ \text{sec/day}}{7 \times 10^{4}\ \text{g} \times 10^{2}\ \text{erg/g/rad}},$$

$$Q = 39.7\ \text{mCi}$$

If the entire 3 liters of water were vaporized, the density of the steam would be

$$\frac{3\ \text{kg} \times 2.24\ \text{lb/kg}}{10^{3}\ \text{ft}^{3}} = 6.72 \times 10^{-3}\ \frac{\text{lb}}{\text{ft}^{3}}.$$

For the density of water vapor to be this great, the temperature must be 132°F. This is an unreasonably high ambient temperature. If we assume an ambient temperature of 100°F, then the saturated water vapor density is 2.852×10^{-3} lb/ft³, and the amount of water in the air is

$$\frac{2.852}{6.72} \times 3\ \text{kg} = 1.275\ \text{kg}.$$

Included in this quantity of water vapor must be an amount of tritium such that the worker inhales enough during 2 min to deposit 39.7 mCi.

Assume that the worker's breathing rate is 20 respirations per minute, and that the tidal volume is 0.5 liters. If there were no ventilation, and the concentration of the tritium had remained constant, the atmospheric concentration C mCi/liter that would lead to the required body burden is

$$C\ \frac{\text{mCi}}{\text{liter}} \times \text{Respiration rate, } \frac{\text{liter}}{\text{min}} \times \text{Exposure time, min} = Q\ \text{mCi}.$$

However, the atmospheric concentration does not remain constant; the ventilation system changes one-half the air of the laboratory in $1\frac{1}{2}$ min, which corresponds to a rate of 0.3465 per minute. The atmospheric concentration of tritium in the laboratory, C, at any time t after release, assuming instantaneous release and distribution to give an initial concentration of C_0 is given by

$$C = C_0\, e^{-kt}, \tag{11.18}$$

where k is the turnover rate of the air. The total amount of inhaled tritium, during any exposure at a mean respiration rate of RR, is

$$Q = RR \times C_0 \int_0^t e^{-kt}\, dt, \tag{11.19}$$

which yields, upon integration,

$$Q = RR \times \frac{C_0}{k}(1 - e^{-kt}). \tag{11.20}$$

In the case under consideration, the respiration rate is 0.5 liters per inspiration $\times$ 20 inspirations per minute = 10 liters per minute. Substituting into equation (11.20) to solve for C_0, we have

$$39.7 \text{ mCi} = 10 \frac{\text{liter}}{\text{min}} \times \frac{C_0 \text{ mCi/liter}}{0.3465 \text{ min}^{-1}}(1 - e^{-0.3465 \times 2}),$$

$$C_0 = 2.75 \frac{\text{mCi}}{\text{liter}}.$$

Since the volume of the room was 1000 ft^3, or 28,300 liters, and recalling that only 1.275 kg of the 3 kg of water will be in the vapor state, the maximum amount of tritium that may be in the tritiated water is:

$$\frac{3}{1.275} \times 2.75 \frac{\text{mCi}}{\text{liter vapour}} \times 2.83 \times 10^4 \text{ liter vapour} = 183.5 \times 10^3 \text{ mCi},$$

or 183.5 Ci. The specific activity of tritium, calculated from equation (4.29), is 9.8×10^3 Ci/g. The weight of 183.5 Ci, therefore, is 18.75×10^{-3} g. The molecular weight of H^3OH is 20, and the amount of tritiated water containing 18.75×10^{-3} g tritium is

$$\frac{3}{20} \times W = 18.75 \times 10^{-3}$$

$$= 0.125 \text{ g tritiated water.}$$

According to these calculations, the use of no more than this amount of tritiated water would insure against overexposure in the event of the maximum credible accident. If this is not enough activity for the purpose of the experiment, then additional precautions would have to be taken, such as enclosure of the process or increased ventilation.

Problems

1. A health physicist finds that a radiochemist was inhaling $Ba^{35}SO_4$ particulates that were leaking out of a faulty glove box. The radiochemist had been inhaling the dust, whose mean radioactivity concentration was 9×10^{-5} μCi/cm^3, for a period of 2 hours. Using the two-compartment ICRP lung model, calculate the absorbed dose to the lung during the 13-week period and during the 1-year period immediately following inhalation.

2. A tank, of volume 100 liters, contained ^{85}Kr gas at a pressure of 10.0 kg/cm^2. The specific activity of the krypton is 20 Ci/g. The tank is in an unventilated storage room, at a temperature of 27°C, whose dimensions are $3 \times 3 \times 2$ m. As a result of a very small leak, the gas leaked out until the pressure in the tank was 9.9 kg/cm^2. A man unknowingly then spent 1 hr in the storage room. Assume the half saturation time for krypton solution in the body fluids to be 3 min. Henry's law constant for Kr in water at body temperature is 2.13×10^7. Calculate (a) the immersion dose, (b) the internal dose due to the inhaled krypton. The partition ratio of Kr in water to Kr in fat is 1:10.

3. If the man in problem 2 turned on a small ventilation fan of capacity 100 ft^3/min as he entered the room, calculate his immersion and inhalation doses.

4. An accidental discharge of ^{89}Sr into a reservoir resulted in a contamination level of 10^{-3} μCi/cm^3 of water.

(a) Using the *basic* radiological health criterion of the ICRP, would this water be acceptable for drinking purposes for the general public if the turn-over half time of the water in the reservoir is 30 days?

(b) If the water were ingested continuously; what maximum body burden would be reached?

(c) How long after ingestion started would this maximum occur?

(d) What would be the absorbed dose during the first 13 weeks of ingestion?

(e) What would be the absorbed dose during the first year?

(f) What would be the absorbed dose during 50 years following the start of ingestion?

5. Nickel carbonyl $Ni(CO)_4$ has a maximum permissible atmospheric concentration of 1 part per billion (ppb) based on its chemical toxicity. A chemist is going to use this compound tagged with ^{63}Ni. The specific activity of the nickel is 6.75 mCi/g. The industrial hygienist is planning to limit the atmospheric concentration of $Ni(CO)_4$ in the lab to 0.5 ppb. Will this restriction meet the requirement for the radioactivity maximum allowable concentration of 6×10^{-8} μCi/ml?

6. Chlorine-36 tagged chloroform, $CHCl_3$, whose specific activity is 100 μCi/millimole, is to be used under such conditions that 100 mg/hr may be lost by evaporation. The experiment is to be done in a laboratory of dimensions $15 \times 10 \times 8$ ft. The lab. is ventilated at a rate of 100 ft^3/min.

(a) Do any special measures have to be taken in order to control the atmospheric concentration of the ^{36}Cl to 10% of its MPC (MPC $= 4 \times 10^{-7}$ μCi/cm^3)?

(b) To what concentration of chloroform, in parts per million, does the radiological MPC correspond for this compound? Compare this concentration to the chemical MPC for chloroform.

7. For the purpose of estimating hazards from toxic vapors or gases of high molecular weight, it is sometimes *incorrectly* assumed that settling of the vapor is determined by the specific gravity of the pure vapor, which is defined as

$$\frac{\text{Molecular weight of the pure vapor}}{\text{“Molecular weight” of air}}$$

instead of the *correct* specific gravity given by

$$\frac{\text{“Molecular weight” of air and vapor mixture}}{\text{“Molecular weight” of air}}.$$

(a) If the vapor pressure of benzene (benzol), C_6H_6, is 160 mm Hg at 20°C, calculate the correct specific gravity of a saturated air mixture of benzene vapors, and compare it to the specific gravity of the pure vapor.

(b) If the chemical MPC for benzene is 25 ppm by volume, calculate the specific gravity of an air-benzene mixture of this concentration.

(c) What is the maximum specific activity of ^{14}C-tagged benzene in order that one-half the radiological MPC for ^{14}C (MPC $= 4 \times 10^{-6}$ μCi/cm^3) not be exceeded by a benzene concentration of 12.5 ppm?

8. Iodine-131 is to be continuously released to the environment through a chimney whose effective height is 100 ft, and whose discharge rate is 1000 ft^3/min. The average wind speed is 2 m.p.h. and the lapse rate is stable.

(a) At what maximum rate may the radioiodine be discharged if the maximum downwind ground level concentration is not to exceed 10% of the occupational MPC of 3×10^{-9} μCi/cm^3.

(b) How far from the chimney will this maximum occur?

9. Inhalation exposure is often described as the product of atmospheric concentration and time, as in units of mCi sec/m^3. Using the ICRP assumptions that 23% of inhaled iodine is deposited in the thyroid, and that the thyroid weighs 20 g, calculate the rad dose corresponding to an acute exposure of 1 (mCi sec)/m^3 of (a) ^{131}I, (b) ^{133}I. (c) Assuming that the other 73% of the inhaled iodine is absorbed into the blood and is bound to the protein, calculate the total body doses due to the protein-bound iodine.

10. A possible way for a small isotope user to treat his radioactive waste prior to disposal into the sanitary sewer is to isotopically dilute the radioisotope to a specific activity such that continuous ingestion of the contaminated water would result in a dose to the critical organ that does not exceed the maximum permissible. For the case of ^{75}Se, calculate the maximum specific activity to meet this criterion. According to the ICRP, the critical organ for Se is the kidneys, weight 300 g. The maximum permissible amount in the kidneys is 90 μCi, and this activity is 4% of the total body burden. Four per cent of the ingested Se is deposited in the kidney.

11. A graphite-moderated reactor is cooled by passing 1,500,000 lb air per hour through the core. The mean temperature in the core is 300°C, and the thermal neutron flux is 5×10^{12} neutrons/cm^2/sec. If the air spends an average of 10 sec in the reactor core, what is the rate of production of ^{41}A? If the chimney through which the air is discharged is 320 ft high and has an orifice diameter of 7 ft; and the temperature of the effluent air is 170°C, while the ambient temperature is 30°C; and if the mean wind velocity is 2 m/sec, at what distance from the chimney will the ground level concentration of ^{41}A be a maximum? What will be the value of this maximum concentration (in μCi/cm^3). How does this figure compare to the MPC for ^{41}A?

Suggested References

1. *Design Guide for a Radioisotope Laboratory*, American Institute of Chemical Engineers, New York, 1964.
2. Manov, G. G. and Bizzel, O.: *Design of Radioisotope Laboratories for Low and Intermediate Levels of Activity*, A.S.T.M. Special Technics Publication 159, American Society for Testing Materials, New York, 1953.
3. Hawkins, M. B.: The design of laboratories for the safe handling of radioisotopes, in *Laboratory Design*, edited by H. S. Coleman, Reinhold Publishing Co., New York, 1951.
4. *Meteorology and Atomic Energy*, U.S. Atomic Energy Comm., Washington, 1955.
5. *Handbook on Aerosols*, U.S. Atomic Energy Comm., Washington, 1950.
6. Friedlander, S. K., Silverman, L., Drinker, P. and First, M. W.: *Handbook on Air Cleaning*, U.S. Govt. Printing Office, Washington, 1952.
7. Drinker, P. and Hatch, T. F.: *Industrial Dust*, McGraw-Hill, New York, 1954.
8. Green, H. L. and Lane, W. R.: *Particulate Clouds: Dusts, Smokes, and Mists*, D. Van Nostrand, Princeton, 1964.
9. Straub, C. P.: *Low-level Radioactive Wastes, Their Handling, Treatment, and Disposals*, U.S. Govt. Printing Office, Washington, 1964.
10. Glueckauf, E.: *Atomic Energy Waste*, Butterworths, London, 1961.
11. *Disposal of Radioactive Wastes*, vols. I and II, I.A.E.A., Vienna, 1960.

CHAPTER 12

CRITICALITY

Criticality Hazard

Of all the potential radiation hazards with which the health physicist deals, that of an accidental criticality is among the most serious. Criticality may be defined as the attainment of physical conditions such that a fissile material will sustain a chain reaction. During this chain reaction, the nuclei of the fissile material (material whose atoms can be made to fission) splits, thereby liberating tremendous amounts of energy in the form of radiation and producing large quantities of radioactive fission products. In a nuclear reactor, criticality is attained under conditions that are very rigorously controlled in regard to safety and power level. If criticality is accidentally attained outside a reactor during the processing or handling of nuclear fuel, the consequences to personnel and equipment are very grave. The utmost in care and controls, both technical and administrative, must be exercised in the handling, use, or transport of fissile materials if death or serious injury or property damage due to a criticality accident is to be avoided. Generally, such efforts to prevent criticality accidents are called *criticality control* or *nuclear safety.*

Nuclear Fission

The liquid drop model pictures the nucleus as a sphere inside of which the nucleons are in constant motion. As a consequence of this motion, the sphere may become distorted, and, under certain conditions, may become highly deformed and split into several parts: two smaller nuclei, which are called fission fragments, and several neutrons.

For fission to be possible, the following mass–energy relationship must hold:

$$E_f = (M - m_1 - m_2 - m_n)\, c^2, \tag{12.1}$$

where E_f is the energy released during fission, M is the mass of the fissioned nucleus, m_1 and m_2 are the masses of the fission fragments, and m_n is the mass of the neutrons. This condition can be met only by those isotopes whose atomic number and atomic mass number are such that $Z^2/A \geq 15$. However, although many isotopes at the upper end of the periodic table meet this requirement, spontaneous nuclear fission is an extremely unlikely event. Table 12.1 lists the spontaneous fission rates for several isotopes that are theoretically fissionable. Although the likelihood of spontaneous fission is almost infinitesimally small, spontaneous fission nevertheless is extremely important in criticality control, since it can, under the proper conditions, initiate an *accidental criticality*, or an uncontrolled nuclear chain reaction. For an isotope where $Z^2/A \geq 49$, the nucleus is unstable towards fission, and the isotope would undergo instantaneous spontaneous fission if it should be produced. All the naturally occurring fissionable isotopes are highly stable against spontaneous fission.

TABLE 12.1.
SPONTANEOUS FISSION RATES

Isotope	Fissions/g/sec
^{232}Th	4.1×10^{-5}
^{233}U	$<1.9 \times 10^{-4}$
^{234}U	3.5×10^{-3}
^{235}U	3.1×10^{-4}
^{236}U	2.8×10^{-3}
^{238}U	7.0×10^{-3}

From Studier and Huizenga, *Phys. Rev.* **96**, 546 (1954).

All fissionable isotopes require a certain amount of activation energy in order to cause them to fission. This is due to a potential barrier that must be exceeded before the nucleus splits. For example, let us consider the case where a nucleus of atomic number Z splits into two equal fission fragments of atomic number $Z/2$ and nuclear radius r. In order to part into two distinct fragments as a result of coulombic repulsion, these two fission fragments must be separated by a minimum distance of $2r$. At this separation, the potential energy in the system is at its maximum value of

$$E_m = \frac{(Ze/2)^2}{2r},$$

and decreases as the distance between the fission fragments increases. At distances less than $2r$, nuclear forces become operative, and the potential energy in the system again decreases, as shown in Fig. 12.1. In order for fission to occur, E_f must be equal to or greater than E_m. Thus if sufficient energy is added to the system to exceed the height of the potential barrier, the fission can be initiated. From Fig. 12.1 it can be seen that an isotope whose nuclear potential well is represented by the dotted line requires less activation energy than one whose nuclear potential well is described by the solid line. This difference in activation energy explains why some isotopes, such as ^{235}U and ^{239}Pu are more easily fissionable than others, such as ^{238}U and ^{240}Pu.

Activation energy for nuclear fission may be obtained from the binding energy released when a neutron is absorbed by a fissile nucleus. It was pointed out in Chapter 4 that even–even nuclei are most stable; that is, they have more binding energy per nucleon than isotopes with an odd number of nucleons. Because of this fact, the addition of a neutron to a nucleus containing an odd number of nucleons, thus producing a nucleus with an even

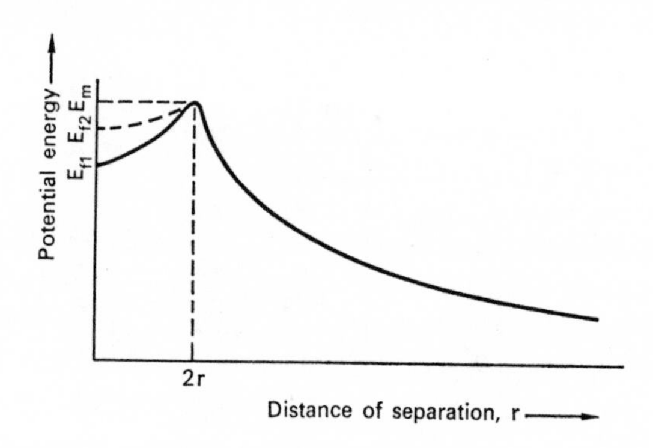

FIG. 12.1. Potential energy in a system of two equal fission fragments, each of atomic number $Z/2$ and radius r.

number of nucleons, releases more energy than the addition of a neutron to a nucleus containing an even number of nucleons. Nuclei with odd numbers of nucleons are consequently more easily fissioned than those with an even number of nucleons. For example, ^{235}U, which fissions after capturing a thermal neutron, is first transformed into an even–even nucleus by the captured thermal neutron, and then fissions:

$$^{235}_{92}U + ^{1}_{0}n \rightarrow ^{236}_{92}U \rightarrow \text{fission},$$

while ^{238}U, which can also capture a thermal neutron, is transformed into an even–odd nucleus, and rids itself of its excitation energy by emitting a gamma-ray:

$$^{238}_{92}U + ^{1}_{0}n \rightarrow ^{239}_{92}U \rightarrow \gamma.$$

The reason for the two different modes of de-excitation of the compound nucleus, nuclear fission and gamma emission, is shown in the following calculation, in which the energy of binding a neutron to ^{235}U and to ^{238}U is calculated. From equation (3.20), the nuclear mass for ^{235}U is calculated, then the mass of a neutron is added. We once more use equation (3.20), this time to calculate the mass of ^{236}U, then find the binding energy by subtracting the

^{236}U mass from the sum of the masses of ^{235}U and a neutron. This procedure gives

mass of ^{235}U	=	235.11240
mass of neutron	=	1.00893
sum of masses		236.12133
−mass of ^{236}U	=	236.11401
mass defect	=	0.00732 amu,

or

$$0.00732 \text{ amu} \times 931 \frac{\text{MeV}}{\text{amu}} = 6.81 \text{ MeV}.$$

In a similar manner, the binding energy of a neutron captured by a ^{238}U nucleus is calculated as 5.31 MeV. Thus, we see that more energy, in the amount of 1.5 MeV, is liberated when a ^{235}U nucleus binds a thermal neutron than when ^{238}U binds such a neutron. In the case of ^{235}U, this additional energy is sufficient to cause fission. Uranium-238 can also be made to fission. This occurs if the ^{238}U nucleus captures a fast neutron. The kinetic energy of the neutron plus the binding energy is sufficient to cause nuclear fission. Experimentally, it has been found that a neutron must have at least 1.1 MeV of kinetic energy in order to induce fission in ^{238}U. Although ^{238}U can undergo "fast fission", the probability of such a reaction is very low in comparison to the probability of thermal fission in ^{235}U. The cross section for fast fission of ^{238}U is 0.29 barns, while for thermal fission of ^{235}U, the cross section is 588 barns. This great difference in fission cross section is one of the chief reasons for the popularity of ^{235}U fueled thermal neutron reactors.

When an atom fissions, it splits into two fission fragments plus several neutrons (the mean number of neutrons per fission of ^{235}U is 2.5) plus gamma-rays—according to the conservation equation

$$^{236}_{92}\text{U} \rightarrow {}^{A_1}_{Z_1}F + {}^{A_2}_{Z_2}F + \nu \, {}^{1}_{0}n + Q. \tag{12.2}$$

The value of Q in equation (12.2), which may be calculated from the mass balance for any particular pair of fission fragments, is about 200 MeV. An approximate distribution of this energy is

Fission fragments, kinetic energy	167 MeV
Neutron kinetic energy	6
Fission gamma-rays	6
Radioactive decay	
Beta particle	5
Gamma-rays	5
Neutrinos	11
	200

Most of this energy is dissipated as heat within the critical assembly. In a power reactor, this heat energy is converted into electrical energy. Using a mean value of 190 MeV heat energy per fission, the rate of fission to generate one watt of power is calculated as follows:

$$1\ \text{W} = X\,\frac{\text{fiss}}{\text{sec}} \times 190\,\frac{\text{MeV}}{\text{fiss}} \times 1.6 \times 10^{-6}\,\frac{\text{erg}}{\text{MeV}} \times 1\,\frac{\text{J}}{10^{7}\ \text{erg}} \times 1\,\frac{\text{W}}{\text{J/sec}}$$

$$X = 3.3 \times 10^{10}\,\frac{\text{fissions}}{\text{second}}.$$

The amount of ^{235}U burned up to produce 1 megawatt day of heat energy is

$$\frac{3.3 \times 10^{10}\ \text{fiss/sec W} \times 10^{6}\ \text{W/MW} \times 8.64 \times 10^{4}\ \text{sec/day}}{6.03 \times 10^{23}\ \text{atoms/mole/235 g/mole}} = 1.125\,\frac{\text{g fissioned}}{\text{MWd}}.$$

Fission Products

The atomic numbers of the fission fragments range from 30 (^{72}Zn) to 64 (^{158}Gd). All fission fragments are radioactive, and decay, usually in chains of several members in length, to form *fission products.* Two such chains of fission products that are of special interest to health physicists are

$$(1)\quad {}^{90}_{36}\text{Kr} \xrightarrow[33\ \text{S}]{\beta} {}^{90}_{37}\text{Rb} \xrightarrow[2.74\ \text{m}]{\beta} {}^{90}_{38}\text{Sr} \xrightarrow[20\ \text{y}]{\beta} {}^{90}_{39}\text{Y} \xrightarrow[64.2\ \text{h}]{\beta} {}^{90}_{40}\text{Zr (stable)},$$

and

$$(2)\quad {}^{137}_{53}\text{I} \xrightarrow[22\ \text{S}]{\beta} {}^{137}_{54}\text{Xe} \xrightarrow[3.9\ \text{m}]{\beta} {}^{137}_{55}\text{Cs} \xrightarrow[27\ \text{y}]{\beta} {}^{137}_{56}\beta\text{a (stable)}.$$

The fission yield for the various fission products are shown in Fig. 12.2. After production, each species of fission product decays according to its unique disintegration rate. Although the total activity at any time after production, therefore, is the sum of the exponential decay curves for each fission product, the collective activity, between about 10 sec and 1000 hr after fission, decreases as $(\text{time})^{-1.2}$. The quantity of radioactivity at time T days after fission is given by

$$A = 1.03 \times 10^{-16}\, T^{-1.2}\,\frac{\text{curies}}{\text{fission}}, \tag{12.3}$$

where T is the time in days after fission.

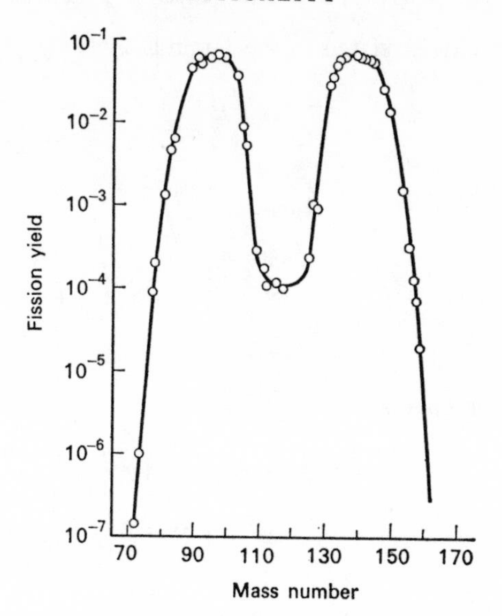

FIG. 12.2. Fission product yield from thermal fission of ^{235}U.

Example 12.1

A certain criticality accident was due to the accumulation, in a 225-gal tank, of 3 kg ^{239}Pu dissolved in an organic solvent (tributyl phosphate). When a stirrer began to operate, changes in the density and geometry of the system caused it to become critical, and, in a single burst of about 10 msec, about 10^{17} nuclei fissioned.

(a) How much energy, in kilowatt hours, was released?

$$E = 10^{17} \text{ fiss} \times 190 \frac{\text{MeV}}{\text{fiss}} \times 1.6 \times 10^{-6} \frac{\text{erg}}{\text{MeV}} \times 1 \frac{\text{J}}{10^7 \text{ erg}} \times 1 \frac{\text{W sec}}{\text{J}} \times$$

$$\times 1 \frac{\text{kWh}}{3.6 \times 10^6 \text{ W sec}} = 0.845 \text{ kWh}.$$

(b) What was the mean power level during criticality?

$$\frac{0.845 \text{ kWh} \times 3.6 \times 10^3 \text{ sec/hr}}{1 \times 10^{-3} \text{ sec}} = 304 \text{ megawatts}.$$

(c) What was the fission product inventory 1 min, 1 hr, 1 day after the accident?

From equation (12.3), we find the activity at 1 min ($1/1440 = 6.95 \times 10^{-4}$ days) to be

$$A = 1.03 \times 10^{-16} (6.95 \times 10^{-4})^{-1.2} \frac{\text{Ci}}{\text{fiss}} \times 10^{17} \text{ fiss} = 6.34 \times 10^4 \text{ Ci},$$

after 1 hr, the activity is related to the 1 min activity by

$$\frac{A_1}{A_2} = \frac{(T_1)^{-1.2}}{(T_2)^{-1.2}}$$

$$A_2 = A_1 \left(\frac{T_1}{T_2}\right)^{1.2} \tag{12.4}$$

$$= 6.34 \times 10^4 \left(\frac{1}{60}\right)^{1.2}$$

$$= 465 \text{ Ci},$$

and after 1 day, we have left

$$A_2 = 6.34 \times 10^4 \left(\frac{1}{1440}\right)^{1.2}$$

$$= 10.1 \text{ Ci}.$$

Criticality

Criticality is attained when at least one of the several neutrons that are emitted in a fission process causes a second nucleus to fission. However, although a chain reaction occurs whenever fissionable material is irradiated with neutrons, not all systems of fissionable material can go critical. If more neutrons are lost by escape from the system or by non-fission absorption in impurities or "poisons" than are produced in fission, then the chain reaction is not self sustaining, and dies out. In this case, the assembly of fissionable material is called *sub-critical.* If we have a sustained chain reaction, and if the rate of fission neutron production exceeds the rate of loss, the assembly is called *super-critical.* When exactly one neutron per fission is available for initiating another fission, the system is called *critical.*

Multiplication Factor: The Four Factor Formula

The measure of criticality in a given system is expressed by the effective multiplication factor, k_{eff}, which is defined as

$$k_{eff} = \frac{N_{f+1}}{N_f}, \tag{12.5}$$

where N_{f+1} is the number of neutrons produced in generation $f+1$ by the number of neutrons N_f of the previous generation. Neutrons of a given generation remain in that generation until they either cause fission or are lost through non-fission processes. When k_{eff} is less than 1.0000, the system is sub-critical; when k_{eff} exceeds 1.0000, the system is super-critical, and when k_{eff} is exactly equal to 1, the system is critical.

Criticality, or the value of k_{eff}, depends on the supply of neutrons of proper energy to initiate fission, and on the availability of fissile atoms. The numerical value of k_{eff}, in turn, depends on two conditions: the composition and the physical arrangement of the fissile assembly, and the size of the fissile assembly. Neutrons are lost from the assembly by several non-fission processes, including leakage from the fissile assembly. If the assembly is infinitely large, then no neutrons are lost through leakage, and the multiplication factor is known as the infinite multiplication factor, k_∞. The relationship between k_{eff} and k_∞ is

$$k_{eff} = L \times k_\infty, \tag{12.6}$$

where L is the non-leakage probability.

The value of k_∞ is determined by four factors whose mutual interrelationship can be demonstrated by following a group of n fission neutrons through a life-cycle. If η is the mean number of neutrons emitted per *absorption in uranium*, then capture of n thermal neutrons will result in $n\,\eta$ fission neutrons. The mean number of neutrons emitted per fission, ν, depends on the fuel. For ^{235}U, $\nu = 2.5$; and for ^{239}Pu, $\nu = 3.0$. However, since both ^{235}U and ^{239}Pu also absorb thermal neutrons without fissioning, not every absorption leads to fission, and hence the mean number of fission neutrons per absorption by the fissile material (the fuel) must be less than ν. For pure ^{235}U, $\eta = 2.1$, while for any other degree of enrichment, $\eta < 2.1$ because of non-fission absorption by ^{238}U. The exact value of η depends on the degree of enrichment of the uranium.

Example 12.2

Calculate η for natural uranium

$$\eta = \frac{\Sigma_f}{\Sigma_a} \times \nu = \frac{N_5\,\sigma_{f5}}{N_5\,\sigma_{a5} + N_8\,\sigma_{a8}} \times \nu = \frac{\sigma_{f5}}{\sigma_{a5} + (N_8/N_5)\,\sigma_{a8}} \times \nu \tag{12.7}$$

where
Σ_f = macroscopic fission cross section,
Σ_a = macroscopic absorption cross section,
N_5 = ^{235}U atoms per cm^3,
N_8 = ^{238}U atoms per cm^3,
σ_{f5} = fission cross section for ^{235}U = 549 barns,
σ_{a5} = absorption cross section for ^{235}U = 650 barns,
σ_{a8} = absorption cross section for ^{238}U = 2.8 barns,
ν = average number of neutrons per fission of ^{235}U = 2.5.

For natural uranium,

$$\frac{N_8}{N_5} = 139,$$

and

$$\eta = \frac{549}{650 + (139 \times 2.8)} \times 2.5 = 1.32.$$

Uranium-238 has a small cross section (0.29 barn) for fission by fast neutrons. Before becoming thermalized, therefore, some of the fast neutrons will be captured by ^{238}U, and will cause "fast fission". If we define the *fast fission factor*, ϵ, as

$$\epsilon = \frac{\text{Total number of fission neutrons}}{\text{Number of thermal fission neutrons}}, \tag{12.8}$$

then the capture of n thermal neutrons in the fuel will produce $n\ \eta\epsilon$ fission neutrons. The value for ϵ depends on the ratio of moderator to fuel, on the ratio of inelastic scattering cross section to fission cross section, and on the geometrical relationship between uranium and moderator. The maximum value of ϵ is 1.29, in the case of unmoderated pure uranium metal. In the case of a homogeneous fuel assembly, such as a solution of fuel, ϵ is very close to 1.

While the fast neutrons are being slowed down, they may be captured by ^{238}U without producing fission; resonances for such capture occur between 200 eV and 5 eV. The probability that a neutron will escape this resonance capture is called the *resonance escape probability*, p, and is defined as the fraction of the fast, fission-produced neutrons that finally become thermalized. The value of p depends on the ratio of moderator to fuel. For a very high ratio, p approaches 1, whereas for a very low ratio, p is very small. For pure unmoderated natural uranium p is 0, which means that natural uranium cannot become critical under any conditions if it is not moderated. The resonance escape probability is given by

$$p = \exp\left[-\frac{N_8}{\xi\Sigma_s}\int(\sigma_a)_{\text{eff}}\frac{dE}{E}\right], \tag{12.9}$$

where N_8 = number of ^{238}U atoms per cm^3,
ξ = average logarithmic energy decrement, as defined by equation (5.48),
Σ_s = macroscopic scattering cross section for the moderator–uranium mixture,
$(\sigma_a)_{\text{eff}}$ = effective absorption cross section for the moderator–uranium mixture.

Experimentally, it has been found that, for a homogeneous mixture of uranium in a moderator, the effective resonance integral can be approximated by

$$\int(\sigma_a)_{\text{eff}}\frac{dE}{E} = 3.9\left(\frac{\Sigma_s}{N_8}\right)^{0.415} \text{barns}, \tag{12.10}$$

for cases where $\Sigma_s/N_8 \leq 1000$ barns. When Σ_s/N_8 increases beyond 1000 barns, the value of the effective resonance integral increases to a limit of 240 barns. Some values for the integral are given in Table 12.2.

TABLE 12.2. EFFECTIVE RESONANCE INTEGRAL FOR SEVERAL VALUES OF Σ_s/N_0

Σ_s/N_o, barns	$\int (\sigma_a)_{eff} \frac{dE}{E}$, barns
8.2	9.3
50	20
100	26
300	42
500	51
1,000	69
2,000	90
3,000	101
10,000	125
∞	240

For a heterogeneous assembly consisting of natural uranium fuel rods,

$$\int (\sigma_a)_{eff} \frac{dE}{E} = 9.25 + 24.7 \frac{S}{M} \text{ barns}, \tag{12.11}$$

S = surface area of the uranium, cm², M = weight of uranium, grams.

From the original n thermal neutrons, we thus have $n \eta \epsilon p$ thermal neutrons. Not all of these thermal neutrons produce fission; some are absorbed by non-fuel atoms, and some are absorbed by ^{235}U without producing fission. (Only 84% of the thermal neutrons absorbed by ^{235}U cause fission.) The fraction of the total number of thermalized neutrons absorbed by the fuel (including all the uranium) is called the *thermal utilization factor*, f. The total number of new neutrons thus produced by the original n thermal neutrons is $n\eta\epsilon pf$. From the definition of k in equation (12.5), we obtain the *four-factor formula* for criticality in an infinitely large system:

$$k_\infty = \frac{N_{f+1}}{N_f} = \frac{n\eta\epsilon pf}{n} = \eta\epsilon pf. \tag{12.12}$$

One of the four factors, η, depends only on the fuel. The other factors, ϵ, p, and f, depend on the composition and physical arrangement of the fuel: ϵ varies from a maximum of 1.29 for unmoderated uranium to almost 1 for a homogeneous dispersion of fuel in a moderator; p is on the order of 0.8 to almost 1 (for pure ^{235}U fuel, $p = 1$; for high degrees of enrichment, $p \approx 1$). The thermal utilization factor can be calculated from its definition

$$f = \frac{\Sigma_{aU}}{\Sigma_{aU} + \Sigma_{aM} + \Sigma_{ap}} = \frac{\sigma_{a5} N_5 + \sigma_{a8} N_8}{\sigma_{a5} N_5 + \sigma_{a8} N_8 + \sigma_{aM} N_M + (\Sigma \sigma_{ai} N_i)_p}, \quad (12.13)$$

Σ_{aU} = macroscopic absorption cross section of uranium,
Σ_{aM} = macroscopic absorption cross section of the moderator,
Σ_{ap} = macroscopic absorption cross section of other substances in the fuel assembly,
σ_{ai} = absorption cross section of the ith element in the "other substances",
N_i = atoms/cm^3 of the ith element in the other substances,
σ_{a5} = absorption cross section of ^{235}U = 650 barns,
σ_{a8} = absorption cross section of ^{238}U = 2.8 barns,
N_5, N_8 = atoms/cm^3 of ^{235}U and ^{238}U respectively.

Example 12.3

Calculate f for a solution of 925-g uranium as uranyl sulfate, UO_2SO_4, in 14 liters water. The uranium is enriched to 93%.

The problem may be solved by application of equation (12.13). However, it can be slightly simplified since, in a homogeneous mixture, the number of atoms per cm^3 is directly proportional to the molar concentration of the various substances. Therefore, equation (12.13) can be rewritten as

$$f = \frac{\sigma_{a5} M_5 + \sigma_{a8} M_8}{\sigma_{a5} M_5 + \sigma_{a8} M_8 + \sigma_{aH_2O} M_{H_2O} + \sigma_{aO_2SO_4} M_{O_2SO_4}}, \quad (12.14)$$

where M represents the number of moles of the respective substances in the solution. In 925 g of 93% enriched uranium, we have

$$\frac{0.93 \text{ g } ^{235}\text{U/g U} \times 925 \text{ g U}}{235 \text{ g } ^{235}\text{U/mole}} = 3.66 \text{ moles } ^{235}\text{U}$$

and

$$\frac{0.07 \times 925}{238} = 0.27 \text{ moles } ^{238}\text{U}$$

for a total of 3.93 moles uranium as UO_2SO_4.

From a table of absorption coefficients, we find that

$$\sigma_{aH} = 0.332 \text{ barns} \qquad \sigma_{a5} = 650$$
$$\sigma_{aO} = 0.0002 \qquad \sigma_{a8} = 2.8$$
$$\sigma_{aS} = 0.49$$

which leads to the following:

Material	σ_a, barns	Moles	$\sigma_a \times$ moles
Uranium-235	650	3.66	2379.00
Uranium-238	2.8	0.27	0.76
Water	0.664	778	516.59
O_2SO_4	0.491	3.93	1.93

Substituting into equation (12.14), we have

$$f = \frac{2379 + 0.76}{2379 + 0.76 + 516.59 + 1.93} = 0.82.$$

Note that in this case we used absorption cross sections for 2200 m/sec neutrons, which leads to a conservative result when making criticality calculations for nuclear safety.

The effective multiplication constant is useful for estimating, for purposes of nuclear safety, the minimum concentration of a fissile material in a solution of moderator that could become critical.

Example 12.4

Estimate the minimum concentration of 93% enriched uranyl sulfate, UO_2SO_4, in water, in order that the solution form a critical mass.

The minimum concentration will occur when the size (or volume) of the solution is infinitely large. In that case, in order to attain criticality,

$$k_{\text{eff}} = \eta\epsilon p f = 1.$$

The value for η is calculated from equation (12.7), using the molar ratio of 0.27 to 3.66 for $^{238}N/^{235}N$:

$$\eta = \frac{549}{650 \times 0.27/3.66 \times 2.8} \times 2.5 = 2.11$$

To calculate f, using equation (12.14), let us arrange the data as shown in the following table, letting M represent the number of moles of UO_2SO_4 that is to be added to 1 mole of water, to give the minimum concentration that will result in $k_\infty = 1$. In Example 12.3, we calculated an atomic ratio of $^{235}U/U$ of 3.66/3.93. Using this value gives:

Material	σ_a, barns	Moles	$\sigma_a \times$ moles
Uranium-235	650	$\frac{3.66}{3.93}M$	605 M
Uranium-238	2.8	$\frac{0.27}{3.93}M$	0.192 M
Water	0.664	1	0.664
O_2SO_4	0.491	M	0.491 M

From the definition of f, equation (12.14), we have:

$$f = \frac{605\ M + 0.192\ M}{605\ M + 0.192\ M + 0.664 + 0.491\ M} = \frac{605.2\ M}{605.7\ M + 0.664}.$$

Making the very reasonable assumption that ϵ and p both are equal to 1, the four factor formula, equation (12.12) because:

$$k_\infty = \eta\,\epsilon\,p\,f = 1$$

$$1 = 2.11 \times 1 \times 1 \times \frac{605.2\ M}{605.7\ M + 0.664}$$

$$M = 9.85 \times 10^{-4}\ \text{moles},$$

and, since the molecular weight of the 93%-enriched uranium is $0.93 \times 235 + 0.07 \times 238 = 235.2$, we need

$$235.2\ \frac{\text{g U}}{\text{mole}} \times 9.85 \times 10^{-4}\ \text{moles} = 0.232\ \frac{\text{g U}}{\text{mole water}}.$$

Since 1 mole water = 18 g $\approx$ 18 ml, 0.232 g U/mole water corresponds to:

$$\frac{0.232\ \text{g U}}{18\ \text{ml}\ H_2O} = \frac{x\ \text{g U}}{1000\ \text{ml}\ H_2O}$$

$$x = 12.9\ \text{g U/liter water},$$

or

$$\frac{363\ \text{g}\ UO_2SO_4}{235.2\ \text{g U}} \times 12.9\ \frac{\text{g U}}{\text{liter water}} = \frac{20\ \text{g}\ UO_2SO_4}{\text{liter water}}.$$

Since the above calculation was based on an infinitely large volume, a concentration less than 12.9 g 93%-enriched uranium as UO_2SO_4 can never go critical.

Nuclear Reactor

In a nuclear reactor, these various factors are combined to produce a controlled, sustained chain reaction. The core of a nuclear reactor consists of fuel (^{235}U or ^{239}Pu), a moderator to thermalize the neutrons, a coolant to remove the heat, and control rods to control the chain reaction. In the case of a uranium-fueled reactor, the uranium is usually enriched in the ^{235}U isotope, since natural uranium contains only about 0.7% ^{235}U. Generally, the greater the degree of enrichment, the smaller is the size of the reactor. The control rod is made of a metal, such as cadmium, hafnium, or boron steel, that has a very high thermal neutron capture cross section. When the control rod is fully inserted into the reactor core, the multiplication constant is less than one because of the loss of neutrons to the absorber. As the control rod is withdrawn, neutrons that would have been captured by the control rod are now free to initiate a fission reaction. At a certain point of withdrawal, the multiplication constant becomes equal to 1 and the reactor is critical. Should the control rod be kept at that point, then there would be no further increase in the power level of the reactor. To increase the power level of the reactor, the rod is withdrawn, so that the multiplication constant k is greater than 1. The power level then increases when the desired power level is attained, the control rod is *reinserted* until k is decreased to exactly 1.00000. The reactor then continues to operate at that power level. To decrease the power level, the control rod is inserted into the core, causing k to become less then 1.0000, and the rate of nuclear fission to decrease. When the desired new power level is attained, the control rod is withdrawn to make k exactly equal to 1.0000 once more, and the reactor continues to operate at the reduced power level.

Reactivity and Reactor Control

The increase in the multiplication factor above 1 is called excess reactivity, and is defined by

$$\Delta k = k - 1. \tag{12.15}$$

For every n neutrons in one generation, we have $n\Delta k$ additional neutrons in the succeeding generation. If the lifetime of a neutron generation is l sec, the time rate of change of neutrons is

$$\frac{dn}{dt} = \frac{n\Delta k}{l}, \tag{12.16}$$

which yields, when integrated from n_0 to n,

$$\frac{n}{n_0} = e^{\Delta k/l \cdot t}. \tag{12.17}$$

If we define the *reactor period*, T, as the time during which the neutrons (and consequently the power level) would increase by a factor of e (an e-fold increase), then in equation (12.17),

$$\frac{\Delta k}{l} = \frac{1}{T},$$

and the reactor period, or e-folding time, is

$$T = \frac{l}{\Delta k}. \tag{12.18}$$

The mean lifetime, from birth of a neutron until its absorption in pure ^{235}U, is about 0.001 sec. Consider the case where the excess reactivity is 0.1 %, that is, $\Delta k = 0.001$. The reactor period is

$$T = \frac{0.001}{0.001} = 1 \text{ sec},$$

and the power level would increase by a factor of e, or 2.718 each second. If Δk were increased to 0.5 %, then

$$T = \frac{0.001}{0.005} = 0.2 \text{ sec},$$

and the power level increase in 1 sec would be

$$\frac{n}{n_0} = e^{t/T} = e^{1/0.2} = 150.$$

Such rapid increases in the power level as calculated in the example above would make it extremely difficult, if not impossible, to control a reactor. Fortunately, the calculation above is not applicable to a real reactor, because, although the mean lifetime of a single neutron, from birth until absorption, is about 0.001 sec, the mean lifetime of a whole generation of a large number of neutrons is much greater than 0.001 sec. This increased mean generation time is due to the fact that 0.6407 % of fission neutrons are delayed, that is, they are emitted as long as 80.39 sec after fission. Six distinct groups of delayed neutrons are observed, each group having its own mean delay time. Table 12.3 lists these groups, together with their mean delay (or generation) time.

TABLE 12.3. DELAYED NEUTRONS FROM THE FISSION OF ^{235}U

Group i	Yield, % n_i	Mean generation time, sec T_i	Yield × mean time $n_i \times T_i$
1	0.0267	0.33	0.009
2	0.0737	0.88	0.065
3	0.2526	3.31	0.836
4	0.1255	8.97	1.125
5	0.1401	32.78	4.592
6	0.0211	80.39	1.688
$\Sigma n_i = 0.6407$		$\Sigma n_i T_i = 8.315$	

The mean generation time for all the fission neutrons of a given generation is

$$\bar{T} = \frac{\sum_{i=0}^{6} n_i T_i}{\sum_{i=0}^{6} n_i} = \frac{8.315 + 99.359 \times 10^{-3}}{100} = 0.084 \text{ sec.}$$

The group $i = 0$ is the group of *prompt neutrons*, whose yield is 99.359%, and whose mean generation time is 0.001 sec.

If Δk is equal to or greater than 0.006407, the reactor is said to be in the *prompt critical* condition, since the chain reaction can be sustained by the prompt neutrons alone. If Δk is less than 0.006407, the reactor is in the *delayed critical* condition, because the delayed neutrons are essential to sustaining the chain reaction. In the delayed critical condition, the reactor period is sufficiently long to allow the power level to be easily controlled.

Example 12.5

Compute the reactor period and the increase in power level in 1 sec, for the case where the mean neutron generation time is 0.084 sec, for excess reactivity of 0.1% and 0.5%.

For $\Delta k = 0.001$,

$$T = \frac{l}{\Delta k} = \frac{0.084}{0.001} = 84 \text{ sec,}$$

and

$$\frac{n}{n_0} = e^{t/T} = e^{1/84} = 1.012;$$

for

$$\Delta k = 0.005,$$

$$T = \frac{l}{\Delta k} = \frac{0.084}{0.005} = 16.8 \text{ sec,}$$

and

$$\frac{n}{n_0} = e^{t/T} = e^{1/16.8} = 1.06.$$

Excess reactivity is measured in units of *dollars* and *cents* ($1 = 100¢) and in *inhours*. One dollar's worth of reactivity is that amount of excess reactivity that will cause the reactor to go prompt critical. One inhour (*in*verse hour) is that amount of excess reactivity that results in a reactor period of one hour. Two inhours of reactivity give a reactor period of one-half hour, etc.

Fission Product Inventory

As a reactor continues to operate, fission products are produced at a rate proportional to the power level, and the activity per fission is given by equation (12.3). If a nuclear reactor operates at a power level of P watts for a time dt, then the fission product activity at a time T days after shut down is

$$dA = 1.03 \times 10^{-16}\, T^{-1.2} \frac{\text{Ci}}{\text{fiss}} \times (3.3 \times 10^{10} \times 8.64 \times 10^{4}) \frac{\text{fiss}}{\text{day}}\ \text{watt} \times P\ \text{watts} \times dt\ \text{days} \tag{12.19}$$

or, combining the numerical constants,

$$dA = 0.293\, PT^{-1.2}\, dt. \tag{12.20}$$

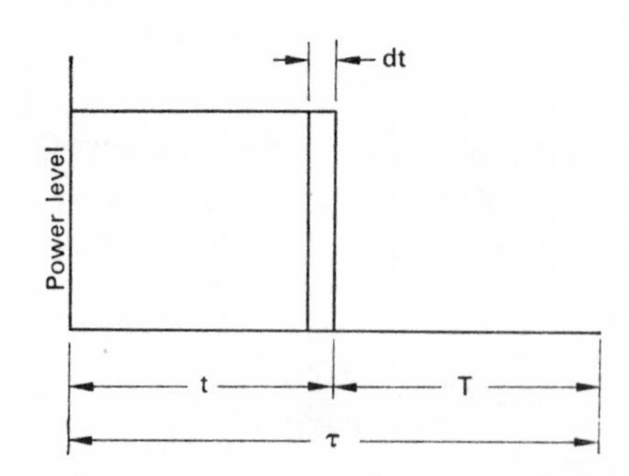

FIG. 12.3. Curve showing relationship between reactor operating time at a constant power level and post operation cooling time, that is used in the derivation of equation (12.23).

Now referring to Fig. 12.3,

if t = reactor operating time, days,
T = cooling time after reactor shut down, days,
τ = total time = $t + T$,

then $T = \tau - t$, and equation (12.20) may be written as

$$dA = 0.293\, P\, (\tau - t)^{-1.2}\, dt, \tag{12.21}$$

$$A = 0.293\, P \int_{0}^{t} (\tau - t)^{-1.2}\, dt, \tag{12.22}$$

$$A = 1.46\, P\, [(\tau - t)^{-0.2} - \tau^{-0.2}]\ \text{Ci}. \tag{12.23}$$

Example 12.6

A power reactor operates at a thermal power level of 500 megawatts for 200 days. It is then shut down for refuelling. What is the fission product inventory (a) a day after shut down and (b) 10 days after shut down?

Substituting into equation (12.23),

$$\text{(a)}\quad A = 1.46 \times 5 \times 10^8 [(201 - 200)^{-0.2} - (201)^{-0.2}]$$
$$= 1.46 \times 5 \times 10^8 [1 - 0.346] = 4.8 \times 10^8 \text{ Ci.}$$

$$\text{(b)}\quad A = 1.46 \times 5 \times 10^8 [(210 - 200)^{-0.2} - (210)^{-0.2}]$$
$$= 1.46 \times 5 \times 10^8 [0.631 - 0.343] = 2.1 \times 10^8 \text{ Ci.}$$

Tables 12.4 and 12.5 list the fission product activity resulting from operation at a power level of 1 megawatt; Fig. 12.4 graphically illustrates the fission product activity–time relationship.

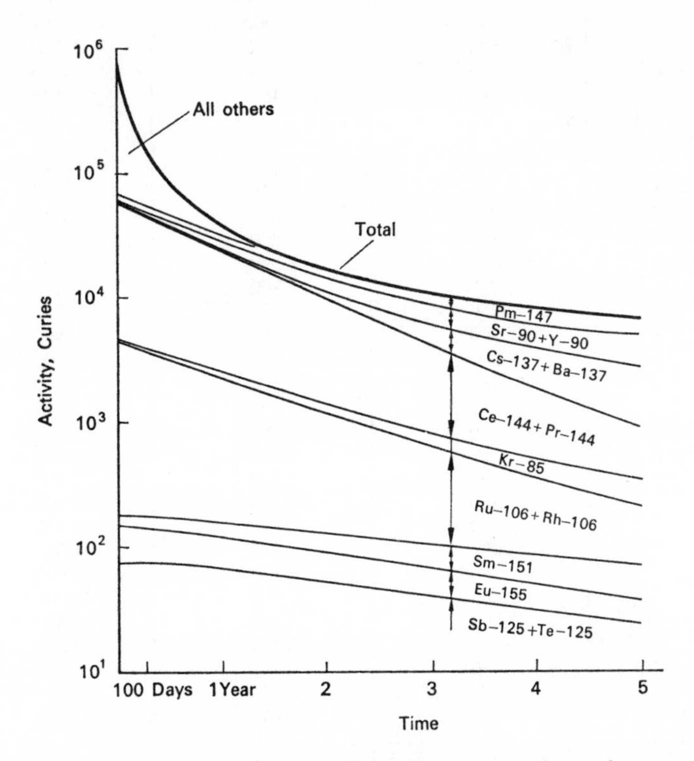

FIG. 12.4. Activity–time relationship of certain fission products after removal from a reactor that had been operating at a power level of 1 Mw for 1 year. (From *Radiological Health Handbook*, 1960.)

TABLE 12.4. PRODUCTION OF IMPORTANT FISSION PRODUCTS IN A REACTOR

Fission product	Activity (curies)[a] after selected periods of continuous operation of a reactor at a power level of 1000 kW		
	100 days	1 year	5 years
Kr-85	53	191	818
Rb-86	0.25	0.26	0.26
Sr-89	28,200	38,200	38,500
Sr-90	402	1,430	6,700
Y-90[b]	402	1,430	6,700
Y-91	34,800	48,900	49,500
Zr-95	32,900	49,200	50,300
Nb-95(90 H)[b]	446	687	704
Nb-95(35 D)[b]	20,900	48,200	50,500
Ru-103	25,100	30,900	31,000
Rh-103[b]	25,100	30,900	31,000
Ru-106	753	2,180	4,220
Rh-106[b]	753	2,180	4,220
Ag-111	151	151	151
Cd-115	4.8	5.9	5.9
Sn-117	83	84	84
Sn-119	<24	<24	<100
Sn-123	4	9	10
Sn-125	100	101	101
Sb-125[b]	12	43	139
Te-125[b]	5	34	136
Sb-127	787	787	787
Te-127(90 D)[b]	146	260	277
Te-127(9.3 H)[b]	808	922	939
Te-129(32 D)	1,410	1,590	1,590
Te-129(70 M)[b]	1,410	1,590	1,590
I-131	25,200	25,200	25,200
Xe-131[b]	250	252	252
Te-132	36,900	36,900	36,900
I-132[b]	36,900	36,900	36,900
Xe-133	55,300	55,300	55,300
Cs-136	52	52	52
Cs-137	300	1,080	5,170
Ba-137[b]	285	1,030	4,910
Ba-140	51,500	51,700	51,700
La-140[b]	51,300	51,700	51,700
Ce-141	43,000	47,800	47,800
Pr-143	45,000	45,300	45,300
Ce-144	9,860	26,700	44,000
Pr-144[b]	9,860	26,700	44,000
Nd-147	21,800	21,800	21,800
Pm-147[b]	1,290	4,900	16,000
Sm-151	9	37	175
Eu-155	23	74	207
Eu-156	108	109	109
Total	563,691	693,573	767,547

[a] Calculated using fission product yields. [b] Daughter product.
From *Radiological Health Handbook*, 1960.

TABLE 12.5. ACTIVITY OF FISSION PRODUCTS IN CURIES AT SPECIFIED TIMES (T) AFTER REMOVAL FROM A REACTOR THAT HAS OPERATED AT 1000 kW ENERGY FOR 1 YEAR

Fission product	$T = 0$[(a)]	$T = 100$ days	$T = 1$ year	$T = 5$ years
Kr-85	191	187	177	132
Rb-86	0.26	—	—	—
Sr-89	38,200	10,300	321	—
Sr-90	1,430	1,420	1,380	1,200
Y-90[(b)]	1,430	1,420	1,380	1,200
Y-91	48,900	14,500	577	—
Zr-95	49,200	17,000	1,000	—
Nb-95(90 H)[(b)]	687	152	15	—
Nb-95(35 D)[(b)]	48,200	28,700	2,140	—
Ru-103	30,900	5,920	74	—
Rh-103[(b)]	30,900	5,920	74	—
Ru-106	2,180	1,800	1,090	68
Rh-106[(b)]	2,180	1,800	1,090	68
Ag-111	151	—	—	—
Cd-115	5.9	1.2	—	—
Sn-117	84	0.7	—	—
Sn-119	<64	<48	<23	<0.4
Sn-123	9	5	1	—
Sn-125	101	0.1	—	—
Sb-125[(b)]	43	41	34	12
Te-125[(b)]	34	39	36	13
Sb-127	787	—	—	—
Te-127(90 D)[(b)]	260	123	16	—
Te-127(9.3 H)[(b)]	922	124	16	—
Te-129(32 D)	1,590	182	0.6	—
Te-129(70 M)[(b)]	1,590	182	0.6	—
I-131	25,200	4	—	—
Xe-131[(b)]	252	1.5	—	—
Te-132	36,900	—	—	—
I-132[(b)]	36,900	—	—	—
Xe-133	55,300	—	—	—
Cs-136	52	0.3	—	—
Cs-137	1,080	1,070	1,060	970
Ba-137[(b)]	1,030	1,020	1,010	920
Ba-140	51,700	230	—	—
La-140[(b)]	51,700	265	—	—
Ce-141	47,800	4,740	10	—
Pr-143	45,300	288	—	—
Ce-144	26,700	20,800	10,700	268
Pr-144[(b)]	26,700	20,800	10,700	268
Nd-147	21,800	40	—	—
Pm-147[(b)]	4,900	4,800	3,950	1,360
Sm-151	37	37	36.6	35
Eu-155	74	67	52	13
Eu-156	109	1.2	—	—
Total	693,573	144,130	36,964	6,527

[(a)] Calculated using fission product yields. [(b)] Daughter product.

From *Radiological Health Handbook*, 1960.

Criticality Control

The occurrence of an accidental criticality depends on the following factors:

1. Quantity of fissile material.
2. Geometry of the fissile assembly.
3. Presence or absence of a moderator.
4. Presence or absence of a neutron reflector.
5. Presence or absence of a strong neutron absorber (Poison).
6. Concentration of fissile material, if the fissile material is in solution.
7. Interaction between two or more assemblies or arrays of fissile materials, each one of which is sub-critical by itself. Consideration of this possibility is important in the transport and storage of fissile materials.

The influence of some of these factors is shown in Table 12.6, which lists the minimum critical mass or size of 93.5% enriched ^{235}U for several different conditions.

TABLE 12.6. MINIMUM CRITICAL MASS OR SIZE OF 93.5% ENRICHED ^{235}U

Condition	Critical mass or size of ^{235}U
Aqueous solution containing not more than 11.94 g ^{235}U per liter	∞
Bare sphere of metallic uranium	48.6 kg
Water reflected sphere of metallic uranium	22.8 kg
Unreflected sphere containing an aqueous solution of 75 g ^{235}U per liter	1.44 kg
Water reflected sphere containing an aqueous solution of 75 g ^{235}U per liter	0.83 kg
Infinitely long unreflected cylinder containing an aqueous solution of 75 g ^{235}U per liter	<8.7 in. diameter
Infinitely long water reflected cylinder containing an aqueous solution of 75 gr ^{235}U per liter	<6.3 in. diameter

Generally, nuclear safety can be assured by limiting at least one of the factors that determine criticality in such a manner that it becomes physically impossible to initiate a sustained chain reaction. The basic control methods for assuring nuclear safety include:

1. *Mass control:* limiting the mass of fissile material to less than the critical mass under any conceivable condition.

2. *Geometry control:* having a geometric configuration that is "always safe", that is, it can never become critical because the surface to volume ratio

is such that excessive neutron leakage makes it impossible to attain a multiplication factor as great as one.

3. *Concentration control:* The size of a critical mass depends strongly on the ratio of fissile atoms to moderating atoms. Thus, if this ratio is kept sufficiently small, that is, if the solution of fissile material is sufficiently dilute, absorption of neutrons by the hydrogen atoms makes a sustained chain reaction impossible. The degree of enrichment of ^{235}U is especially important in concentration control. In this connection it should be pointed out that a homogeneous mixture or a solution of natural uranium in water can never become critical, regardless of the concentration. For an aqueous mixture or solution of uranium, the ^{235}U must be enriched to about 1% before it can be made to attain criticality.

Nuclear safety by one of these three basic methods of control can be maintained by restricting any one of the control parameters to the maximum values listed in Table 12.7. It should be emphasized that the values listed in the table are independent of each other, that is, restriction of any one of the parameters to the value listed in the table allows other restrictions to be relaxed. It is considered good practice, however, to design processes and equipment in such a way that at least two highly unlikely, independent events occur simultaneously before a criticality accident can occur. In this connection, it should be pointed out that it is considered better practice to rely primarily on safety designed into the equipment rather than to rely principally on safety designed into a process. Thus, for example, a process may be designed to meet the concentration or enrichment criteria of Table 12.7. However, it would not be too difficult to inadvertently cause a criticality by using the wrong material. On the other hand, if the equipment were designed according to "always safe" geometry, such as limiting the diameter of a pipe or the capacity of a vessel, then an accidental criticality could not possibly occur under any conceivable operating condition.

When fissile material is being either stored or transported, it is especially important to prevent an interaction between two sub-critical units. This is accomplished by packing the fissile materials (in subcritical quantities) in containers of such design that the fissile materials are properly spaced. Such containers consist of two parts: a unit container in which the fissile material is placed, and a spacing container, whose function is to insure the physical separation of the unit containers. One common type of shipping and storage container, which is called a "bird-cage", is a weldment consisting of a steel pipe with endcaps (the unit container), in the center of a large angle iron frame (the spacing container). A variation of this type of container may be made by welding the unit container coaxially, using appropriate spacers, in a 55-gal. steel drum. The drum, in this case, is the spacing container. The spacing containers should be of such size that the closest distance (surface to surface) between unit containers cannot be less than 8 in. With this much

TABLE 12.7. VALUES OF BASIC NUCLEAR SAFETY PARAMETERS

Isotope	Parameter	Value				Minimum safety factor
		Moderated		Unmoderated		
		Recommended maximum	Minimum critical	Recommended maximum	Minimum critical	
^{235}U	Mass	350 g	820 g	10.0 kg	22.8 kg	2.3
	Diameter of infinite cylinder	5.0 in.	5.4 in.	2.7 in.	3.1 in.	1.1
	Thickness of infinite slab	1.4 in.	1.7 in.	0.5 in.	0.6 in.	1.2
	Volume of solution	4.8 liters	6.3 liters	—	—	1.33
	Concentration (aqueous)	10.8 g/liter	12.1 g/liter	—	—	1.12
	Enrichment of ^{235}U	0.95	1.0	—	—	
^{233}U	Mass	250 g	590 g	3.2 kg	7.5 kg	2.3
	Diameter of infinite cylinder	3.7 in.	4.4 in.	1.7 in.	1.9 in.	1.1
	Thickness of infinite slab	0.8 in.	1.2 in.	0.2 in.	0.3 in.	1.2
	Volume of solution	2.3 liters	3.3 liters	—	—	1.33
	Concentration (aqueous)	10.0 g/liter	11.2 g/liter	—	—	1.12
^{239}Pu	Mass	220 g	510 g	2.6 kg	5.6 kg	2.3
	Diameter of infinite cylinder	4.2 in.	4.9 in.	1.4 in.	1.7 in.	1.1
	Thickness of infinite slab	0.9 in.	1.3 in.	0.18 in.	0.24 in.	1.2
	Volume of solution	3.4 liters	3.5 liters	—	—	1.33
	Concentration (aqueous)	6.9 g/liter	7.8 g/liter	—	—	1.12

From *Nuclear Safety Guide*, U.S. Atomic Energy Commission Report, TID 7016, Rev-1, 1961, U.S.A.E.C.

Notes: 1. Moderation by H_2O is assumed.
2. All values, moderated and unmoderated, assume water reflection.

distance between the unit containers, criticality due to interaction between two or more units is impossible even if the containers are immersed in water. No restrictions on transportation or storage are necessary for:

(a) Uranium enrichment to 0.95% or less as a homogeneous aqueous mixture.
(b) Uranium metal enriched to 5% or less, provided that there is no hydrogenous material within the container.
(c) Aqueous solutions of ^{235}U at concentrations that do not exceed 10.8 g/liter, of ^{233}U at concentrations that do not exceed 10.0 g ^{233}U per liter, or of ^{239}Pu at concentrations that do not exceed 6.9 g/liter.

Problems

1. Cooling water circulates through a water boiler reactor core at a rate of 1 gal/min through a coiled stainless-steel tube $\frac{1}{4}$ in. inside diameter and 7 ft long. The concentration of Na and Cl in the water is 5 atoms each per million molecules H_2O. What is the concentration of induced Na and Cl radioactivity in the cooling water after a single passage through the reactor core, if the mean thermal flux is 10^{11} neutrons per cm^2/sec and the mean temperature in the core is 80°C?

2. If the cooling water in problem 1 circulates through a heat-exchange reservoir containing 100 gal (including the water in the pipes between the core and the reservoir), what will be the concentration of induced activity in the reservoir after 7 days operation of the reactor?

3. If the tank of problem 2 is spherical, what will be the surface dose rate due to the induced radioactivity?

4. A research reactor, after going critical for the first time, operates at a power level of 100 watts for 4 hr. How many curies of fission product activity does the core contain?

5. An accidental criticality occurred in an aqueous solution in a half-filled mixing tank 25 cm diameter × 100 cm high. The energy released during the burst was estimated as 1800 joules. Assuming that, on the average, each disintegration of a fission product is accompanied by a 1-MeV gamma-ray, estimate the gamma-ray dose rate at the surface of the tank (which maintained its integrity during the criticality) and at a distance of 25 ft from the tank at 1 min, 1 hr, 1 day, and 1 week after the criticality accident.

6. A slab of pure natural uranium metal weighing 1 kg is irradiated in a thermal flux of 10^{12} neutrons per cm^2/sec for 24 days at a temperature of 150°F. If the fission yield for ^{131}I is 2.8%, how many millicuries of ^{131}I will be extracted 5 days after the end of the irradiation?

7. What is the uranium concentration of uranyl sulfate UO_2SO_4 aqueous solution that can go critical if the uranium is enriched to (a) 10% and (b) 90%?

8. Calculate η for ^{239}Pu, given that the fission cross section is 664 barns and the non-fission absorption cross section is 361 barns.

9. The blood plasma from a worker who was overexposed during a criticality accident had a ^{24}Na specific activity of 0.001 μCi per ml 15 hr after the accident. The accidental excursion lasted 10 msec. What was the absorbed dose, in rads, due to (a) the ^{14}N (n, p) ^{14}C reaction and (b) the autointegral gamma-ray dose due to the n, α reaction on hydrogen. All the Na in nature is ^{23}Na. The thermal neutron activation cross section at 20° is 0.53 barns.

Suggested References

1. *Criticality Control of Fissile Materials*, I.A.E.A., Vienna, 1966.
2. *Criticality Control*, O.E.C.D., Karlsruhe, 1961.
3. CLARK, H. K.: *Handbook of Nuclear Safety*, Report No. DP-J32, TID-4500, Office Technical Services, Washington, 1961.

4. *Nuclear Safety Guide*, TID-7016, Revision 1, Office of Technical Services, Washington, 1961.
5. HENRY, H. F.: *Guide to Shipment of ^{235}U Enriched Uranium Materials*, TID-7019, Office of Technical Services, Washington, 1959.
6. GLASSTONE, S. and EDLUND, M. C.: *The Elements of Nuclear Reactor Theory*, D. Van Nostrand, Princeton, 1952.
7. MURRAY, R. L.: *Introduction to Nuclear Engineering*, Prentice-Hall, Englewood Cliffs, 1954.
8. McCULLOUGH, C. R.: *Safety Aspects of Nuclear Reactors*, D. Van Nostrand, Princeton, 1957.
9. *Theoretical Possibilities and Consequences of Major Accidents in Large Nuclear Power Plants*, U.S. Atomic Energy Comm., Washington, 1957.
10. *Reactor Safety and Hazard Evaluation Techniques*, I.A.E.A., Vienna, 1962.

CHAPTER 13

EVALUATION OF PROTECTIVE MEASURES

THE effectiveness of protective measures against radiation hazards is evaluated by a surveillance program that includes observations on both man and his environment. Such a surveillance program may employ one or more of a variety of techniques, depending on the nature of the hazard and the consequences of a breakdown in the system of controls. These techniques may include pre-employment physical examinations, periodic physical examinations, estimation of internally deposited radioactivity by bioassay and total body counting, personnel monitoring, radiation and contamination surveys, and continuous environmental monitoring.

Medical Surveillance

The great degree of overexposure required before clinical signs or symptoms of overexposure appear precludes the use of medical surveillance of radiation workers as a routine monitoring device. Nevertheless, medical supervision may play an important role in protecting radiation workers against possible radiation damage. Among the main tasks of medical supervision is the proper placement of radiation workers according to their medical histories and physical condition, and history of previous radiation exposure. Dermatitis, cataracts, and blood dyscrasias, including leukemia are associated with radiation exposure. A pre-employment physical examination, therefore, should be given if the nature of the work, including consideration to possible accidental overexposures, warrants it—in which special attention is paid to physical conditions that may lead to, or be suggestive of, susceptibility to any of these effects. Possible indirect effects from working with radioisotopes also are considered by the examining physician. For example, sensitivity or allergy may contraindicate work that requires the wearing of rubber gloves or that may require washing the hands or body with strong detergents or harsh chemicals in order to decontaminate the skin. In addition to the pre-employment examination, the radiation worker may be routinely examined at periodic intervals to ascertain that he continues to be free of signs that would contraindicate further occupational exposure to radiation. The physician is thus instrumental in preventing damage or injury that could otherwise have arisen, either directly or indirectly, as a consequence of working with radioisotopes or exposure to radiation. Medical supervision of radiation workers may also be necessary to evaluate overexposure, to treat radiation injuries, and to decontaminate personnel. These activities of the physician are, of course, in addition to the routine health services that he provides which are not connected to radiation hazards. It should be pointed out that medical surveillance of workers is not unique to the field of radiation health. All good occupational health programs include pre-employment examinations, consideration of medical findings in job placement, and continuing medical surveillance to help maximize the protection of workers against the harmful effects of toxic substances.

Estimation of Internally Deposited Radioactivity

One of the technics for evaluating a contamination control program is the determination of the body burdens of personnel who are at risk. This determination is done indirectly by bioassay methods, and directly by total body counting in the case of gamma-emitting radionuclides or beta emitters that give rise to suitable bremsstrahlung. The underlying rationale for bioassay is that a quantitative relationship exists among inhalation or ingestion of a radionuclide, the resulting body burden, and the rate at which the radionuclide is eliminated either in the urine or in the feces. From measurements of activity in the urine and feces, therefore, we should be able to infer the body burden. Unfortunately, the kinetics of metabolism of most substances is influenced by a large number of factors, and, as a consequence, the desirable quantitative relationships between body burden and elimination rates are known for relatively few cases. In most instances, therefore, bioassay data give only a very approximate estimate of the degree of internal deposition of radioactivity. Although both urine and feces are available for bioassay measurements in case of an accidental inhalation or ingestion of a large amount of radioactivity, routine bioassay monitoring is usually done only with urine samples, because of the ease of sample collection and also for esthetic reasons.

For purposes of bioassay, we distinguish between readily soluble and difficultly soluble compounds. This distinction is especially important in the case of inhaled particulates that are relatively insoluble, in which the difficultly soluble material is brought up from the lung and swallowed, while at the same time some of the inhaled particulate matter is dissolving and being absorbed into the body fluids. The resulting complexities due to varying pulmonary deposition and clearance rates and the varying urinary to fecal ratios of the radionuclide makes it difficult to quantitatively estimate the lung dose from difficultly soluble air-borne contamination. Nevertheless, an estimate of the minimum amount of difficultly soluble radioactive particulates that was deposited in the upper respiratory tract following a single accidental inhalation could be made from the cumulative fecal activity during the first few days after the inhalation. With this information, and using the ICRP model for lung clearance, a less reliable estimate can then be made of the activity remaining in the deep respiratory tract.

Readily soluble radionuclides may be grouped into three categories according to their distribution and metabolic pathways: (1) those that are uniformly distributed throughout the body, such as ^3H in tritiated water or radiosodium ions, (2) those that concentrate mainly either in specific organs, such as iodine in the thyroid gland, mercury in the kidney, or in the intracellular fluid, such as potassium or cesium, and (3) those which are deposited in the skeleton. Bioassay data are most reliable in the case of the first category, the widely distributed radionuclides. In this case, the radioactivity is excreted exponentially at a rate given by the effective elimination constant, λ_e and the body burden, $A(t)$, at any time t after an intake of $A(0)$ is given by

$$A(t) = A(0)\, e^{-\lambda_e t} \tag{13.1}$$

If a constant fraction of the isotope, f_U, is eliminated in the urine, then the activity in the urine, $U(t)$, at time t after ingestion is given by

$$U(t) = f_U \frac{dA(t)}{dt}, \tag{13.2}$$

$$U(t) = f_U\, A(0)\, \lambda_e\, e^{-\lambda_e t}. \tag{13.3}$$

For tritium (as tritiated water), for example, λ_e is 0.058 per day and f_U is 0.6. Equation (12.3), therefore, for tritium becomes

$$U(t) = 0.035\, A(0)\, e^{-0.058t}. \tag{13.4}$$

The second category, those radionuclides that are concentrated in one or more organs, are not as amenable to quantitative monitoring by bioassay as the widely distributed radionuclides. These radionuclides are absorbed, after ingestion or inhalation, into the body fluids and the blood plasma. From these fluids, they pass into the organs in which they concentrate; a dynamic equilibrium eventually results between the concentration of the nuclide in the organ and in the body fluids. While the isotope is equilibrating between the body fluids and the organ of concentration, it is also being filtered by the kidney into the urine. This leads to a clearance curve, Fig. 13.1, that is the sum of at least two exponential components. The first component, which falls steeply, represents the clearance of the isotope from the body fluids, while the second component represents the clearance of the isotope from the organ of concentration. The slope and magnitude of the first component may be influenced by a number of factors, such as the amount of non-radioactive form of the same element that was inhaled or ingested, the amount of water intake, the physiologic state of the kidney, etc., which makes it extremely difficult to relate the intake of the radionuclide to the urinary excretion data during the first few days after a single intake. The component which represents the clearance from the organ of concentration is much less influenced by these factors. An approximate estimate, therefore, of the radionuclide in the organ of concentration following a single exposure often can

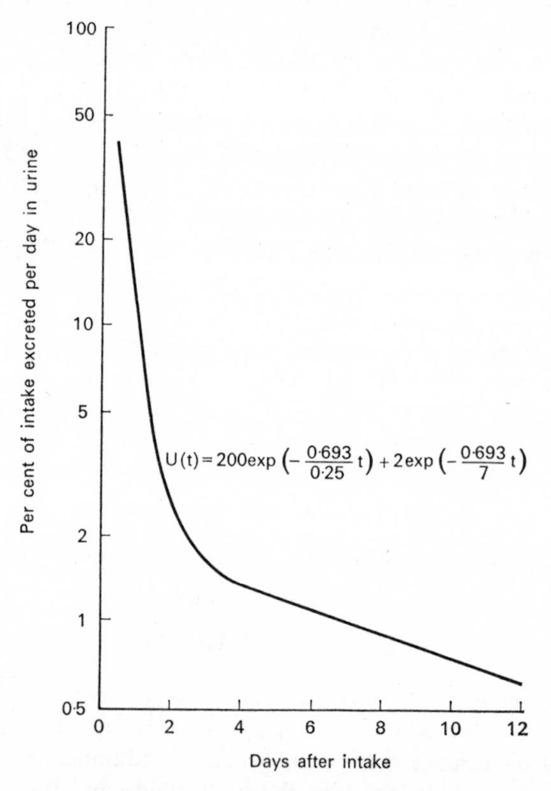

FIG. 13.1. Urinary excretion curve of ^{35}S after a single intake of a soluble, inorganic sulfate. The general shape of the curve is typical for a readily soluble compound that concentrates in a single organ. (From S. Jackson and G. W. Dolphin, Report AHSB (RP) R 51, U.K.A.E.A., 1965.)

be made from the urinary excretion after sufficient data are available to establish the second component of the curve.

The third category, which comprises the elements absorbed into the bone, is a special case of the category of isotopes concentrated in an organ or tissue. Bone seekers differ from other radionuclides mainly in the rate of elimination of the isotope. Whereas clearance half-times for the non-bone seekers are measured in days or weeks, half-clearance times for the bone seekers are measured in years. Furthermore, the rate of clearance from the skeleton is not constant, as is usually the case for the non-bone seekers, but decreases with increasing time. This is due to the fact that the skeleton is not a single "compartment", but rather a number of different "compartments", each of which has its own clearance rate. Over a long period of time ($t \gg 1$), therefore, the sum of all the exponentials representing these different compartments is described mathematically by a power function of the form

$$R(t) = At^{-n}, \tag{13.5}$$

where $R(t)$ = fractional retention t days after intake,
A = normalized fraction of the dose retained at the end of 1 day,
n = a constant.

The fraction of the intake that is eliminated in the urine, per day, if f_U is the fraction of the eliminated isotope that leaves in the urine, is

$$U_f(t) = -f_U \frac{\mathrm{d}R(t)}{\mathrm{d}t}, \tag{13.6}$$

$$U_f(t) = f_U\, Ant^{-(n+1)}. \tag{13.7}$$

For the case of radium, $A = 0.54$, $n = 0.52$, and $f_U = 0.02$. The fraction of the intake that may be found in a 24-hr urine sample, therefore is, from equation (13.7),

$$U_f(t) = 0.0056\, t^{-1.52}. \tag{13.8}$$

The body burden of radium may also be inferred from measurements of the concentration of radon in the breath. Radium decays directly to radon; some of the radon dissolves in the body fluids and in the adipose tissue, and the balance is exhaled. For body burdens on the order of the maximum permissible, i.e. 0.1 μCi, 65% of the radon is exhaled. The exhaled activity, A_e, is related to the body burden q by the equation

$$A_e \frac{\text{pCi}}{\text{min}} = 0.65 \times q\, \mu\text{Ci} \times \lambda\, \text{min}^{-1} \times 10^6 \frac{\text{pCi}}{\mu\text{Ci}}, \tag{13.9}$$

where λ, the decay constant for radium, is 8.1×10^{-10} per min. The concentration of radon in the breath is given by

$$C \frac{\text{pCi}}{\text{liter}} = \frac{A_e\, \text{pCi/min}}{V\, \text{liter/min}}. \tag{13.10}$$

Under resting conditions, the respiration rate is about 20 per min and the tidal volume is about $\frac{1}{2}$ liter; the ventilation rate, V, therefore is about 10 liters/min. Breath radon can be conveniently determined by the method of

Hursch (*Nucleonics* **12**, No. 1, 63, 1954). Radon from a measured volume of exhaled breath is adsorbed on activated charcoal. The radon is then desorbed by heating, and transferred into an ionization chamber for measurement.

Direct determination, by whole-body counting, of body burdens of gamma-emitting nuclides provides a more accurate estimate of the body burden than

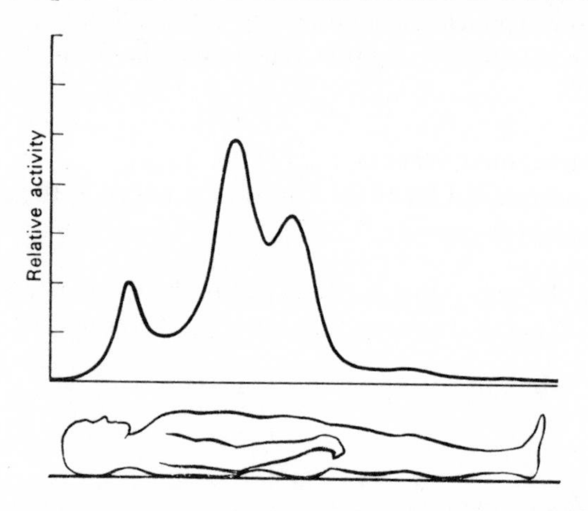

FIG. 13.2. Whole-body scan, with a crystal gamma-ray detector, of a man 2 hr after ingestion of Tc. (From T. M. Beasley, H. E. Palmer and W. B. Nelp, Distribution and excretion of technetium in humans, *Health Physics*, **12**, 1425, 1966.)

does excreta analysis (Fig. 13.2). However, because of the high cost of installation and operation of a total-body counting facility, and also because of its limitation to the determination of gamma emitters or suitable bremsstrahlung generating beta emitters, total-body counters are not generally used for routine monitoring. Their main use, at this time (1968), is in research studies or in the assessment of internal contamination following an accidental exposure.

Personnel Monitoring

Personnel monitoring is the continuous measurement of an individual's exposure dose by means of one or more types of suitable instruments, such as pocket meters, film badges, and thermoluminescent dosimeters (Chapter 8), which are carried by the individual at all times. The choice of personnel monitoring instrument must be compatible with the type and energy of the radiation being measured. For example, a worker who is exposed only to ^{14}C would wear no personnel monitoring instrument, since these isotopes emit only beta-rays of such low energy that they are not recorded by any of the commercially available personnel monitoring devices. Bioassay procedures would be indicated if personnel monitoring were necessary.

The film badge has a number of advantages that makes it widely used for monitoring personnel who may be exposed to X-rays, gamma-rays, high-energy beta-rays, and neutrons. The chief advantage claimed for the film badge is the fact that it provides a permanent exposure record. In addition, it can measure exposure dose over a very large range, from about 10 mrad (X-rays) to several thousand rad, it can be made to record the dose of a wide range of energies on the same film and, conversely, the type and energy of the radiation may be inferred from the film exposure, and it is mechanically rugged. The main disadvantages of film badges are the long delay time between exposure and development and interpretation of the film and the relatively inaccurate exposures that have sometimes been reported by commercial film badge suppliers in several tests in which the films had been exposed to known amounts of radiation (Fig. 13.3). An average accuracy of

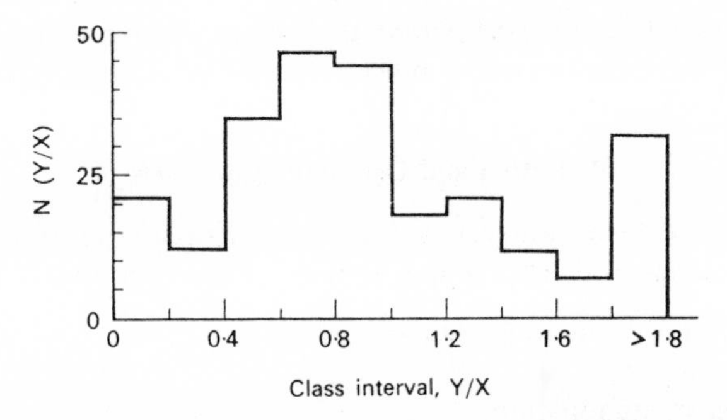

FIG. 13.3. Distribution of errors in reading radiation exposure from film badges. Y/X is defined as the ratio of reported exposure to delivered exposure. (From D. W. Barber, Film badge service performance, *Rad. Health Data and Reports* 7, 623, 1966.)

about $\pm 25\%$ of the true exposure to the film may be expected from a good film badge service. Despite these large inaccuracies reported in tests of routine film, Brodsky *et al.* reported (*Health Physics* **11**, 1071–82, 1965) an accuracy of 10–20% is possible for film badge measurements of annual doses of 100 mR or more.

Many of the alleged disadvantages of the film badge dosimeter (particularly that of its inaccuracy) can be overcome by the use of a thermoluminescent dosimeter (TLD). The accuracy of a LiF TLD in the exposure dose range of 50 mR to 1000 R is about $\pm 9\%$. Furthermore, the TLD is sensitive to exposures as low as several mR, and is reliable for measuring a dose of 10 mR. With a TLD the waiting period before the exposure dose is known is less than that with film badge dosimeter because of the chemical processing required for the film. The exposure information stored in the TLD can be read out in several seconds, thus making it useful for emergency dosimetry.

However, for routine batch processing of large numbers of film, the total processing time per dosimeter of film badges may be about the same as that of TLD. For these reasons, we might expect the TLD (or other solid state dosimeter) to gradually displace the film badge for personnel monitoring purposes. However, it should be pointed out that TLD readout is destructive, while developed films can be stored for possible future reference.

The main purpose of personnel monitoring is to obtain information on the exposure of an individual. In addition to this main purpose, personnel monitoring is also used to observe trends or changes (in time) in the working habits of a single individual or of a department, and thus to measure the effectiveness of a radiation control program. Whereas the distribution of personnel monitoring data might all appear to lie within a normal range when viewed as individual readings, statistical analysis of the grouped data may reveal small but significant differences among different control measures, or different operating procedures or work habits that might otherwise have escaped the attention of the radiation safety officer.

Radiation and Contamination Surveys

A survey is a systematic set of measurements made by a health physics surveyor in order to determine one or more of the following:

(a) an unknown radiation source,
(b) dose rate,
(c) surface contamination,
(d) atmospheric contamination.

In order to make these determinations, the health physics surveyor must choose the appropriate instruments, and must use it properly.

Choosing a Health Physics Instrument

The choice of a surveying instrument for a specific application depends on a number of factors. Some general requirements include portability, mechanical ruggedness, ease of use and reading, ease of servicing, ease of decontamination, and reliability. In addition to these general requirements, health physics survey instruments must be calibrated for the radiation that they are designed to measure, and they must have certain other characteristics:

1. *Ability to respond to the radiation being measured.* This point can be clarified with a practical example: a commonly used side window beta-gamma probe has a window thickness of 30 mg/cm^2. This probe would be worse than useless if one wished to survey for low-energy beta radiation, such as ^{14}C or ^{35}S, or for an alpha contaminant such as ^{210}Po. Each of these radioisotopes emits only radiation whose range is less than 30 mg/cm^2—radiation not sufficiently penetrating to pass through the window of the probe. Incorrect

use of this probe, therefore, may falsely indicate safe conditions when, in fact, there may be severe contamination. Similarly, incorrect inferences may be drawn if a neutron monitor is used to measure gamma radiation or if an instrument designed to measure gamma-rays is used for neutrons. It is essential that radiation survey instruments be used only for the radiations for which they are designed.

2. *Sensitivity.* The instrument must be sufficiently sensitive to measure radiation at the desired level. Thus, an instrument to be used in a search for a lost radium needle should be more sensitive than a survey meter used to measure the radiation levels inside the shielding of an accelerator. In the latter case, where the radiation levels may reach many thousands of mR per hour, an ionization chamber, such as a Juno, whose sensitivity is about 1 mR/hr, is suitable. In searching for the lost radium needle, on the other hand, a sensitivity of 1 mR/hr would greatly limit the area that could be covered in the search; a Geiger counter survey meter that has a sensitivity of about 0.05 mR/hr is much more useful. For example, if a 1-mg radium needle were lost, the distance within which it could be detected with the Juno is about 90 cm, while the Geiger counter will respond to the lost radium at a distance of 412 cm. The Geiger counter can thus cover an area of $53\frac{1}{2}m^3$, while the Juno can cover only about $2\frac{1}{2}$ m^3. Too great a sensitivity, on the other hand, may be equally undesirable. The range of radiation levels over which the instrument is to be used should be matched by the range of radiation levels for which the instrument is designed. Sensitivity is determined mainly by the value of the input resistor across the detector, R in Fig. 9.1 and R_1 in Fig. 9.25. The sensitivity of the detector is directly proportional to the size of the input resistance.

3. *Response time.* The response time of a survey instrument may be defined as the time required for the instrument to attain 63% of its final reading in any radiation field. This time is determined by the product of the input capacity (in farads) of the detector and the shunting resistance (in ohms) across the detector, RC in Fig. 9.1. The time constant is usually expressed in seconds. A low value for the time constant means an instrument that responds to rapid changes in radiation level—such as would be experienced when passing the probe rapidly over a small area of contamination on a bench top or over a small crack in a radiation shield. A fast response time, however, may mean a decrease in sensitivity due to a smaller value of R. Furthermore, a fast response time may result in rapid fluctuations of the meter reading, thus making it difficult to obtain an average level. In practice, the response time of a survey instrument is designed to optimize these divergent factors. Many instruments offer a range of response times, the appropriate one being selected by the surveyor turning the time constant selector switch to the desired value.

4. *Energy dependence.* Most radiation-measuring instruments have a limited span of energy over which the radiation dose is accurately measured.

One of the figures of merit of a radiation dosimeter is the energy range over which the instrument is useful. This information must be known by the health physicist in order to choose a proper instrument for a particular application or if he is to properly interpret his measurements. The energy dependence is usually specified by the manufacturer as "Accurate to ±10% of the true value from 80 keV to 2 MeV", or by means of an energy dependence curve (Fig. 13.4.) The magnitude of the errors that can arise when the energy dependence factor is overlooked is shown in Table 13.1.

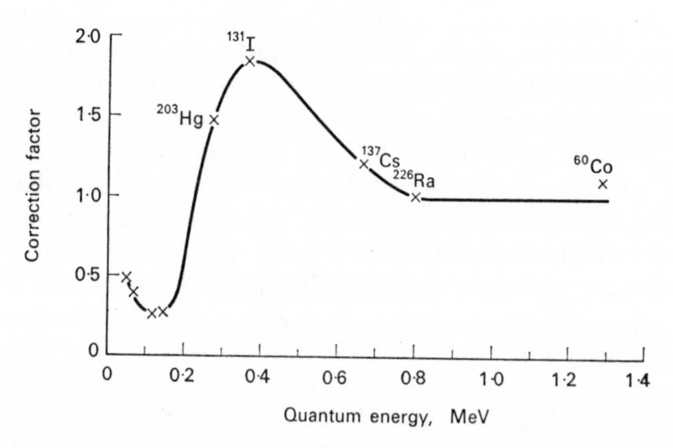

FIG. 13.4. Energy dependence of a Geiger counter survey meter. The meter reading is multiplied by the correction factor appropriate to the quantum energy in order to obtain the true exposure rate. (Courtesy of Electronica Lombarda S.P.A.)

TABLE 13.1. ENERGY DEPENDENCE OF DOSE-RATE RESPONSE OF G.M. AND SCINTILLATION COUNTERS

Meter reading for a true exposure rate of 1 mR/hr

Isotope	Gamma-ray energy, MeV	G.M. counter	Scintillation counter
^{60}Co	1.25	1.15 mR/hr	0.6 mR/hr
^{226}Ra	0.84	1.0	0.96
^{137}Cs	0.661	0.92	1.39
^{198}An	0.411	0.82	2.65
^{203}Hg	0.279	1.29	7.5
^{141}Ce	0.145	2.4	14.1
^{241}Am	0.06	6.0	9.8

From D. H. Peirson, *Physics in Medicine and Biology* 7, 450 (1963).

Surface Contamination

Surface contamination can be located by scanning with a sensitive detector, such as a thin end window Geiger counter. After finding a contaminated spot or area, a dose-measuring instrument may be employed to measure the dose rate at some appropriate distance from the surface. The main hazard from surface contamination is transmission of the contamination from the surface into the body via inhalation or ingestion. To estimate this hazard, a *smear test* is performed to determine whether the surface contamination is fixed or loose, and therefore transmissible. A smear test consists of wiping the suspected area with a piece of filter paper several centimeters in diameter and then measuring the activity in the paper. The area to be smeared varies according to the extent of the suspected contamination and the physical conditions under which the survey is made; a wipe-area of 100 cm^2 is not uncommon. A smear survey, which is a systematic series of smears without first using a scanning instrument to detect the contamination, is often done in a work area that is subject to contamination, and where the background due to radiation sources is high enough to mask the activity due to contamination. It should be emphasized that a smear test is a qualitative, or at best a semiquantitative determination whose chief purpose is to allow an estimate to be made of the degree to which surface contamination is fixed. If significant transmissible contamination is found, and, if in the opinion of the health physicist this contamination may be hazardous, then prompt decontamination procedures are instituted.

Leak Testing Sealed Sources

Sealed gamma-ray, beta-ray, bremsstrahlung and neutron sources are used in a wide variety of applications in medicine and in industry. In all cases, the radioactive material is permanently enclosed either in a capsule or another suitable container. Before being shipped from the supplier, all such sources must pass inspection for freedom from surface contamination and leakage. Either during transport from the supplier or in the course of time, however, the capsule may develop faults through which the radioactive source material may escape into the environment. Because of the serious consequences of such an escape, a sealed source must be tested before being put into use and periodically thereafter for surface contamination and leakage. The testing cycle depends on the nature of the source and on the kind of use to which it is put. However, it is usually recommended that such tests be performed at least once every 6 months. The following technics may be employed to perform these tests: to test for surface contamination, wipe all exposed external surfaces of the source thoroughly with a piece of filter paper or a cotton swab moistened with an appropriate solvent, then measure the activity on the paper or the swab. The source is considered free of surface contamination if less than 0.005 μCi alpha or less than 0.05 μCi beta activity was wiped off. To test for leakage, one of the following tests may be performed:

1. Wipe the source with either a piece of wet filter paper or a cotton swab. Repeat at least 7 days later. If less than 0.005 μCi alpha or less than 0.05 μCi beta activity was wiped off each time, then the source is considered free of leaks.
2. For high activity sources such as those used in teletherapy, where wiping the source might be hazardous, accessible surfaces of the housing port or collimator may be wiped while the source is in the "off" position.
3. Immerse the source in ethanediol, and reduce the pressure on the liquid to 100 mm Hg for a period of 30 sec. A leak is indicated if a stream of fine bubbles issues forth from the source. This method is reliable only for such sources where enough gas would be trapped to produce a stream of fine bubbles.

Air Sampling

Air sampling is considered an important part of a survey where there is a possibility of significant atmospheric contamination. Allowable working levels of contaminated air involve quantities of radioactivity very much less than those which would be considered hazardous if the activity were in a sealed source and if the hazard were limited only to external radiation. Furthermore, even if only sealed sources are used, a program of air sampling is recommended if the nature of the source is such that radioactive gaseous or particulate matter could escape in the event that the source capsule develops a flaw. An air sample, in such a case, might detect the contamination and the leaky source before a significant amount of radioactivity escaped.

An air sampling system consists of three basic elements: (1) a source of suction (a vacuum pump) for drawing the air to be sampled through (2) a collecting device, which usually separates the contaminant from the air, and (3) a metering device for measuring the quantity of air sampled. After collection, the sample of the contaminant is counted to determine the radioactivity content, and then, when this information is combined with the size of the air sample, the concentration of atmospheric radioactivity is calculated. The exact type of the sampling system depends on the nature of the radioactive contaminant—mainly whether the contaminant is gaseous or particulate. Regardless of the nature of the contaminant, however, there exist several problems that are common to all types of contaminants (including non-radioactive contaminants). These common problems include:

1. Obtaining a sample of air *that is representative of the situation under investigation.* In most cases, we are interested in the radioactivity that a person

might inhale. For this purpose, air samples are usually taken in the "breathing zone", that is, at a height of about 6 ft above the ground. If the contaminant is a dust, then the collector should be oriented so that the collection orifice is vertical, in order that it collect respirable dust particles suspended in the air, rather than non-respirable particles that might fall down on the collector. There is a good deal of evidence showing that concentrations of air-borne contamination varies significantly in time and location. Obtaining a "representative sample" of what a person might inhale is thus a fairly difficult task. To simplify this task, a worker may wear a personal air sampler whose collector is as near to his nose as practicable. Measurements made, under actual working conditions, with a personal air sampler and a fixed air sampler have shown that there is little correlation between the activity on the fixed air sampler and that of the personal air samples.

A special problem in obtaining a representative sample arises in sampling dusts that are moving at a high velocity, as in the case of an exhaust duct or a chimney. Air-borne particulates are carried by an airstream, and they tend to follow the streamlines of flow. If the streamlines bend or curve, then the path of an air-borne particle is determined by the ratio of the viscous forces (which tend to keep the particle in the streamlines) to the inertial force (which tends to cause the particle to cut across streamlines). Consider the case where a sampling device is oriented at right angles to the direction of flow in a duct, that is, the gas is blowing directly into the sampling device. If the gas is drawn through the sampler at a velocity less than that of the gas in the duct, then some of the gas must flow around the sampling device. Large particles will, because of their inertia, tend to continue in a straight line, and thus cut across the streamlines and enter into the sampler. This results in an excessive number of particles in the collector, which results in an overestimate of the particulate concentration in the gas. If the gas is drawn through the sampler at a faster velocity than that of the gas, then more of the heavy particles will be undeflected than will the lighter particles, thus leading to a smaller deposition of particles and consequently to an underestimate of the true concentration of particulate matter in the stream of gas. When the velocity of the gas through the sampler is equal to the velocity of gas in the duct, the streamlines of the gas are not disturbed by the sampling orifice, and no sampling error due to the inertia of the suspended particles occurs. This condition is called *isokinetic sampling*, and must be met if the dust sample is to be representative of the dusts in the airstream. The magnitude of the error due to *anisokinetic sampling* conditions increases as the mass of the particles increases and as the difference between the sampling velocity and the gas velocity increases.

2. Obtaining a sample that is large enough to give a reasonably accurate estimate of the mean concentration of dust particles in the air, and also large enough to meet the sensitivity requirements of the radioactivity detector.

Example 13.1

Radioactive $^{90}SrCl_2$ will be produced in an inhalation experiment in which the particles, which are cubic, will have a mean edge length of 1 μ, and a mean activity of 10^{-6} μCi per particle. The atmospheric MPC for ^{90}Sr for 168 hr/week exposure is 10^{-10} μCi/cm^3 or 10^{-7} μCi/liter air. To meet the MPC means, in this case, a mean particle concentration of

$$\frac{10^{-7}\ \mu\text{Ci/liter}}{10^{-6}\ \mu\text{Ci/particle}} = 0.1\ \frac{\text{particles}}{\text{liter}} = 100\ \frac{\text{particles}}{\text{m}^3}.$$

For health physics purposes we wish to detect one-half of this concentration, or 50 particles per cubic meter, within $\pm 20\%$, at the 96% confidence level. The air will be sampled with a membrane filter that is 100% efficient for the collection of these particles.

(i) How large must the air sample be?

(ii) The radiation detector, a windowless proportional counter whose geometry is 50%, requires a net activity of 50 disintegrations per minute in order to make a statistically significant measurement. Will the air sample size calculated from (i) meet the counting requirements?

(a) We wish to determine a mean particle concentration of $50 \pm 20\%$, or 50 ± 10 particles per cubic meter at the 96% confidence level. Since 96% includes two standard deviations, the standard deviation of the mean, σ_c in this case, is $10 \div 2$ or 5 particles per cubic meter.

Since the capture of a particular particle on the filter is a highly unlikely event, Poisson statistics are applicable.

Following the development of equation (9.31), we have, in this example, the mean particle concentration and the standard error of the mean concentration:

$$c \pm \sigma_c = \frac{n}{V} \pm \frac{\sqrt{n}}{V}, \tag{13.11}$$

where n = the number of particles collected in the sample, and V = the volume of air in the sample.

Since

$$\sigma_c = \frac{\sqrt{n}}{V} = \sqrt{\left(\frac{n}{V}\cdot\frac{1}{V}\right)} = \sqrt{\frac{c}{V}},$$

we have

$$c \pm \sigma_c = c \pm \sqrt{\frac{c}{V}}. \tag{13.12}$$

Substituting the appropriate numerical values for σ_c and c into equation (13.12), we find that

$$\sigma_c = \sqrt{\frac{c}{V}},$$

$$5 = \sqrt{\frac{50}{V}},$$

$$V = 2 \text{ cubic meters.}$$

(b) The expected activity of the sample is

$$2 \text{ m}^3 \times 50 \frac{\text{particles}}{\text{m}^3} \times 10^{-6} \frac{\mu\text{Ci}}{\text{particle}} \times 2.22 \times 10^6 \frac{\text{dpm}}{\mu\text{Ci}} = 222 \text{ dpm.}$$

Since there are only two chances in 100 that the sample will be less active than this, and since we require only 50 dpm for counting purposes, this sample is sufficiently large for the radioactivity determination.

3. Choice of collecting device. The main factor in choosing a collecting device is the nature of the contaminant—whether gaseous or particulate. Gas may be collected by a number of different technics including:

(a) Adsorption, in which a monomolecular layer of the gas binds to the surface of certain particulate substances. The capacity of such a substance depends on the specific surface area of the adsorbant (square meters per gram), the partial pressure of the gas, and the temperature. This binding capacity is represented by a curve called an *isotherm* (Fig. 13.5), in which the moles of gas adsorbed per gram of adsorbant is

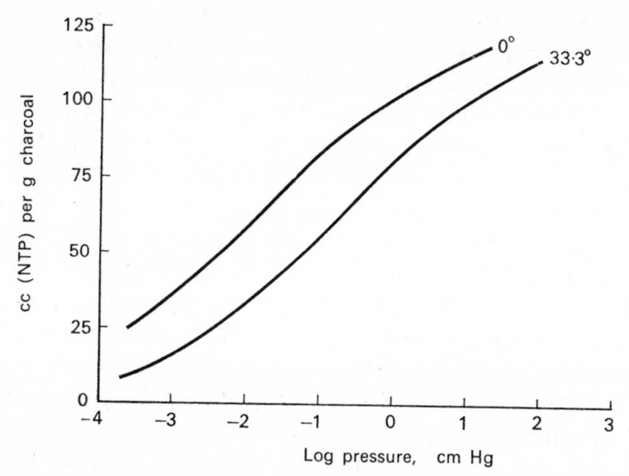

FIG. 13.5. Adsorption isotherm of benzene by charcoal. (From J. W. McBain, *The Sorption of Gases and Vapors by Solids*, George Routledge & Sons, London, 1932.)

plotted against the equilibrium pressure at constant temperature. Commonly used adsorbants are activated carbon, activated alumina, and silica gel. In use, the adsorbant is packed into a suitable container, and the air to be sampled is drawn through the adsorbant. If desired, the adsorbed gas can be driven off in the laboratory by the application of heat.

(b) Absorption, in which the contaminated air is bubbled through a liquid with which the contaminant will interact. For optimum operation, the gas must be brought into intimate contact with the solution, and it must remain in contact long enough to allow the desired chemical reaction to take place. A commonly used absorption collector is the Greenburg–Smith impringer with a fritted glass bubbler. This collector can hold up to 500 ml absorbing solution, and, at a sampling rate of 1–5 liters/min, has a collection efficiency (for the gas appropriate to the absorbing solution) that approaches 100%. A smaller version of this instrument is called a midget impringer (Fig. 13.6).

(c) Grab sample in which an evacuated container is opened in the atmosphere to be sampled, thus permitting contaminated air to enter into the container. The activity of this contaminated air may be determined by transferring the gas to an ionization chamber, and then measuring the ionization current due to the gaseous activity. A vibrating-reed electrometer is useful for this purpose.

Airborne Particulates

Measurement of the concentration of airborne particulates requires the separation of the particles from a known volume of air and their collection in a manner appropriate to the parameter under measurement (mass, radioactivity, particle size, particle number, etc.). Basically, particulates may be sampled by five methods: sedimentation, filtration, electrostatic precipitation, impaction, and thermal precipitation. The first method, sedimentation, is widely used for fallout measurements. A piece of adhesive-bearing paper of known area is exposed for a predetermined length of time. The radioactive fallout per unit area per unit time is then calculated from radiometric measurements of the paper. The last-named method, thermal precipitation, is not widely used in the field by health physicists. It samples the air at a very low rate—on the order of several milliliters per minute—and is thus not very useful for general purpose air sampling. Its main function is to collect dust for accurate determination of size distribution—and this it does extraordinarily efficiently. The thermal precipitator is 100% efficient for particles 0.1 μ and larger. The operating principle in the thermal precipitator is that airborne particles in a stream that passes through a thermal gradiant (which in a thermal precipitator, Fig. 13.7, is maintained between a hot wire and a cold collecting surface made of a glass microscope slide or cover slip) are deflected

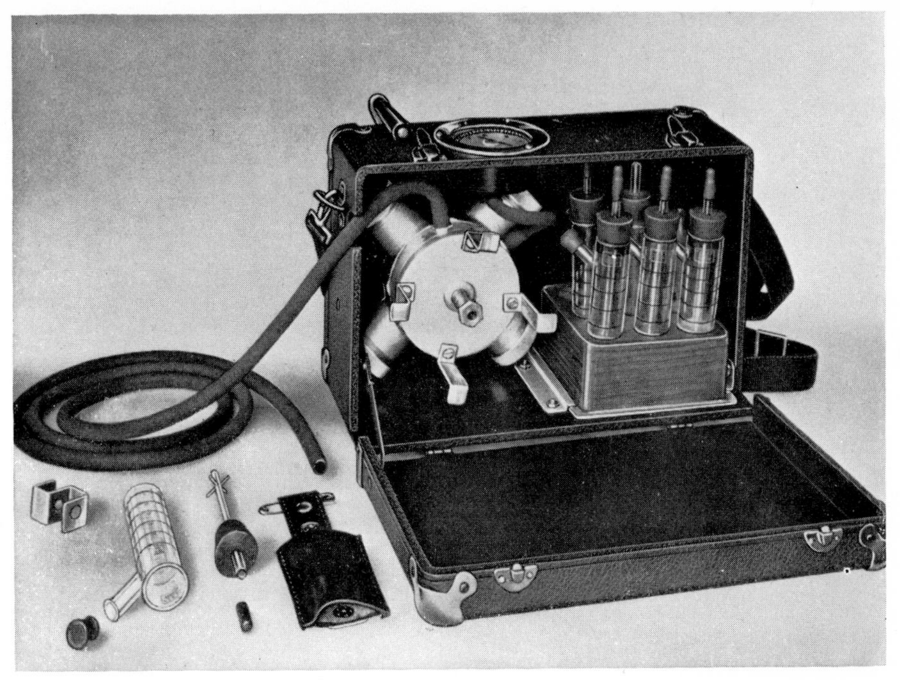

FIG. 13.6. Midget imрringer. A self-contained unit, including a hand-operated pump that samples air at a rate of 0.1 cfm. For sampling gases and vapors, a fritted gas bubbler is used; for sampling dusts, the air is drawn into the collecting liquid at a relatively high velocity through a glass nozzle, causing the dust particles to impinge on the bottom of the flask and to be retained in the liquid. (Courtesy Mine Safety Appliances Co.)

FIG. 13.7. Thermal precipitator. The part on the right has an electrically heated wire across the channel through which the dust-laden air is drawn. The cold collecting plate, a square glass cover slip used by microscopists, is placed into the part on the left. The two parts are then assembled, with the cover slip sandwiched between them. The cover slip is rotated at 1 r.p.m. in order to obtain a uniform size distribution of the collected particles.

by the hot wire with a force that is proportional to the temperature gradient, and are deposited on the cold surface.

Filter

For radioactive particulates, filtration is the most commonly used collection method. Several different types of filter media are available: paper, glass fiber, and membrane (cellulose ester). Each of these has its own flow rate and filtering characteristics. Filters trap particulates mainly by two mechanisms: (1) sieving, which captures particles that are larger than the pore size, and (2) impaction, which captures particles smaller than the pore size. In impaction, the inertia of the particles in the air stream cause the particles to tend to move in straight lines. As the air bends on its tortuous path through the filter's pores, the particles continue in straight lines, and thus strike the filter matrix, and are captured. For this reason, the filtration efficiency of a filter increases as the flow rate through the filter increases. Filters made of glass fiber and paper trap particles within the matrix; membrane filters trap particles on the filter surface. This point is important when sampling for alpha or for low-energy beta particles because of the corrections for self-absorption that must be used. For this reason, as well as its very high retention efficiency for particulates of respirable size, membrane filters are very widely used for sampling of radioactive aerosols. Membrane filters may be made transparent with immersion oil, thus allowing direct microscopic observation of the particles for sizing or for particle concentration measurements. Membrane filters, however, retain a strong electrostatic charge. If the filter is placed into a windowless counter, this electrostatic charge would distort the electric field around the anode, and thus would introduce a counting error. To prevent this, the membrane filter is treated with a mixture of dioxane and petroleum ether, which eliminates the static charges from the filter without interfering with the dust particles on the filter.

The filtration efficiencies, at a sampling velocity of 2 ft/sec, are shown in Table 13.2.

TABLE 13.2. COLLECTING EFFICIENCY OF CERTAIN FILTERS, PER CENT

	<0.4	0.4–0.6	0.6–0.8	0.8–1.0	1–2	>2
Whatman 41	23	28	64	74	80	100
Whatman 4	23	32	38	79	84	100
MSA S	48	47	77	92	94	100
H-70	99.3	99.3				
Glass fiber	99.9	99.9				
Membrane	99.9	99.9				

Electrostatic precipitators

Electrostatic precipitators, which are widely used by industrial hygienists for dust sampling, are also used, though less frequently, by health physicists. The electrostatic precipitator (Fig. 13.8) consists of two coaxial electrodes, a central wire cathode about 1 mm in diameter, and an aluminum anode, several centimeters in diameter, which serves as the dust collector. When a potential difference of about 12,000 volts is applied across these two electrodes (the collecting anode is kept at ground potential) a corona discharge around the central electrode charges the dust particles negatively as they are carried by in the air stream. These negatively charged dust particles are collected by the outer cylinder, which is electrically positive with respect to the charged dust particles. After collection, the collecting cylinder is removed and the particles are washed off the inside of the cylinder for counting. Alternatively, the cylinder may be incorporated into a gas-flow counter (Fig. 13.9), for counting with no further sample preparation. The electrostatic precipitator samples air at a rate of 3 ft^3/min; it is about 99% efficient for particles of 0.2–5 μ. Above 5 μ, the collecting efficiency decreases rapidly.

Cascade impactor

The cascade impactor, as its name implies, collects particles by impaction as a jet of high-velocity air strikes a surface perpendicular to its direction of travel and is thus abruptly deflected. Particulate matter, by virtue of its inertia, tends to continue in the original path, and thus strikes the deflecting surface—which is also a collecting surface. The main advantage of the cascade impactor is that it separates particles according to their size (or their mass). How this is accomplished may be seen in Fig. 13.10. The air is drawn through a series of slits of successively decreasing width, thus resulting in an increasing velocity through the successive slits. Because of their greater inertia, massive particles do not get past the first stage, while lighter particles can be carried by the air stream to the next stage. There the air stream passes through the slit at a higher velocity than through the first slit, thus imparting sufficient momentum to certain particles to cause them to impact on the second stage. In commercially available cascade impactors, this process is repeated four times, and then the air passes through a membrane filter that removes those particles small enough to have escaped impaction on the fourth stage. The various stages are glass slides, coated with a very thin layer of adhesive material. The collected particles can either be examined microscopically, or can be determined either gravimetrically (for non-radioactive particles) or radiometrically. Before use, the cascade impactor stages are calibrated with particles of known density to determine the size distributions collected by each stage (Fig. 13.11). Each stage is thus uniquely associated with a certain mass median diameter (MMD), that is, the particle size that is collected with 50% efficiency by that stage.

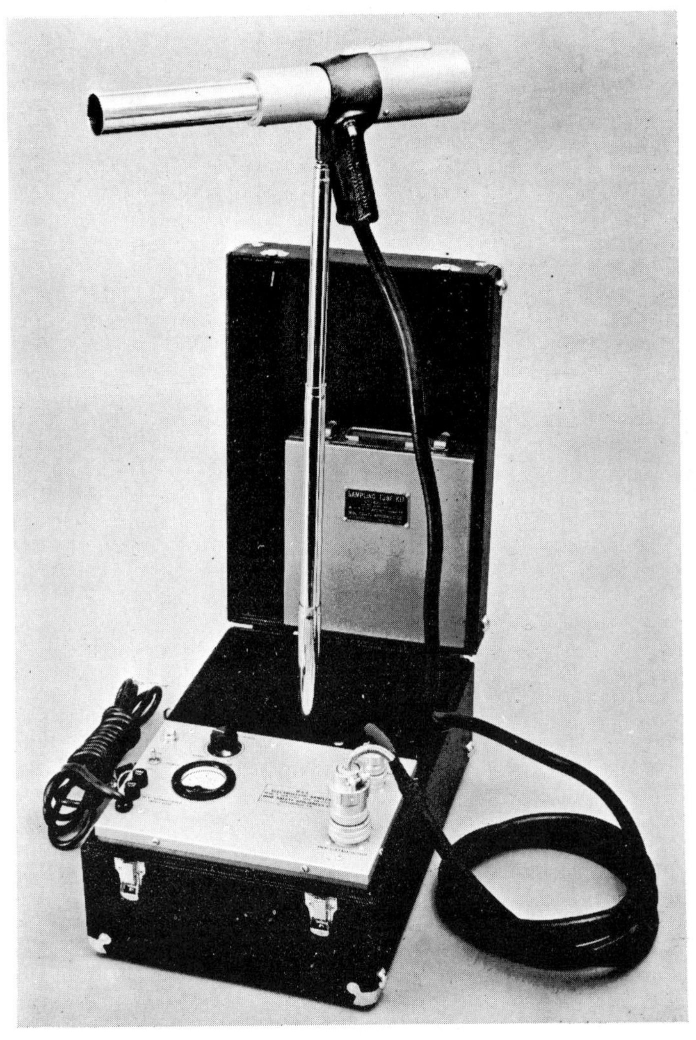

FIG. 13.8. Electrostatic precipitator for collecting samples of atmospheric dust, fumes, and smoke at a rate of 3 ft^3/min. (Courtesy Mine Safety Appliances Co.)

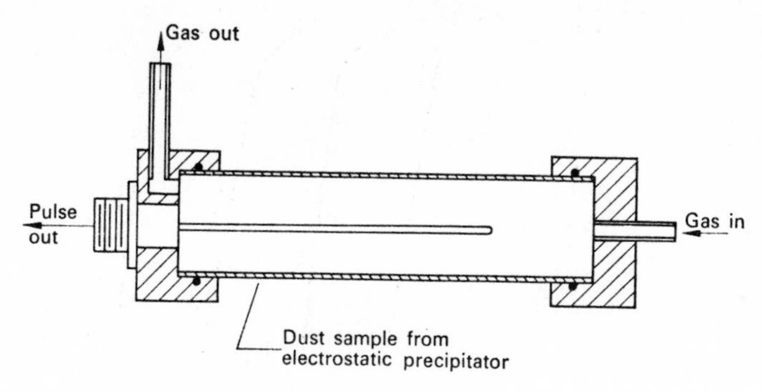

FIG. 13.9. Arrangement for counting the radioactivity in an electrostatic precipitator dust sample.

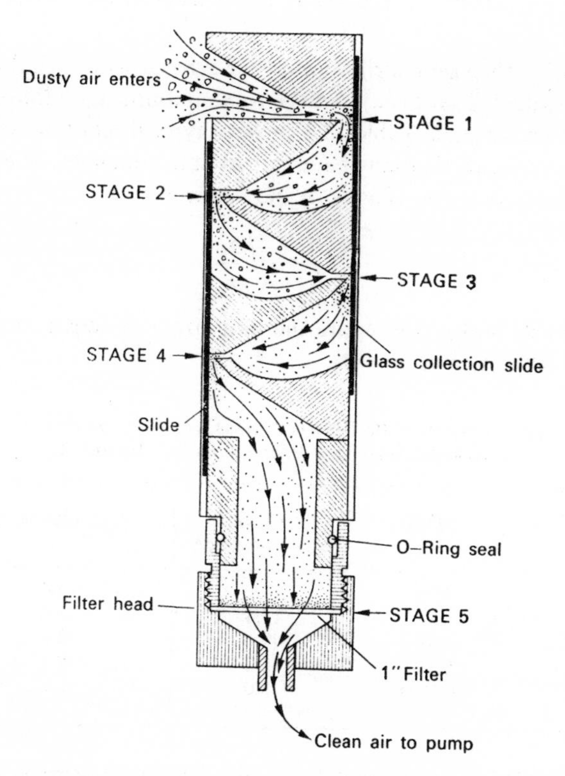

FIG. 13.10. Cascade impactor. The dust particles are collected on two standard 1 × 3 in. glass microscope slides (stages 1, 2, 3, and 4) and on a membrane filter (stage 5). (Courtesy Union Industrial Equipment Co.)

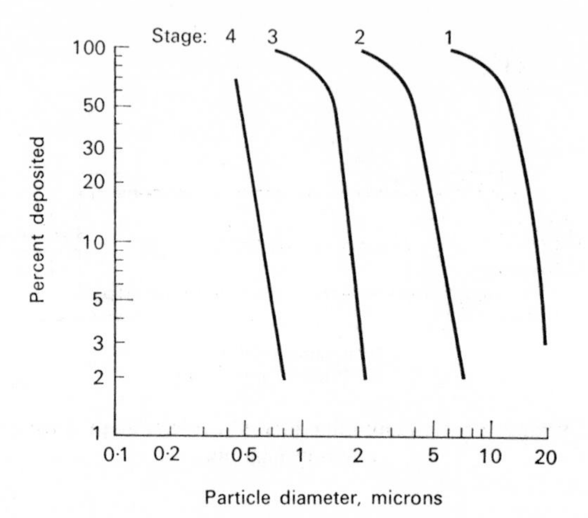

FIG. 13.11. Typical particle size distributions on the first four stages of a cascade impactor for particles of unit density and a sampling rate of 17.5 liters/min.

To determine the size distribution of an aerosol with a calibrated cascade impactor, the dust is sampled, and the total amount of collected material on each stage is determined. Table 13.3 shows a typical set of data for U_3O_8 dust (density $= 8.3$ g/cm^3) that may be obtained in this manner. The median stage diameter is related to the density of the particle by

$$\frac{d_1}{d_2} = \sqrt{\frac{\rho_2}{\rho_1}}, \qquad (13.13)$$

where d_1 and d_2, and ρ_1 and ρ_2 are the unit size and density and sample size and density respectively.

TABLE 13.3. RELATIONSHIP BETWEEN MASS AND SIZE DISTRIBUTION IN AN AEROSOL SAMPLED WITH A CASCADE IMPACTOR

Stage	MMD	% of total mass on stage	Cumulative %
1	4.26	11.2	94.4
2	1.35	12.4	71.0
3	0.52	49.6	40.0
4	0.16	12.2	9.1
5	(filter)	3.0	1.5

The stage MMD's (obtained from the calibration curves and corrected for density by equation (13.13)) are plotted on log probability paper (Fig. 13.12) against the cumulative percentage up to the respective stage—that is

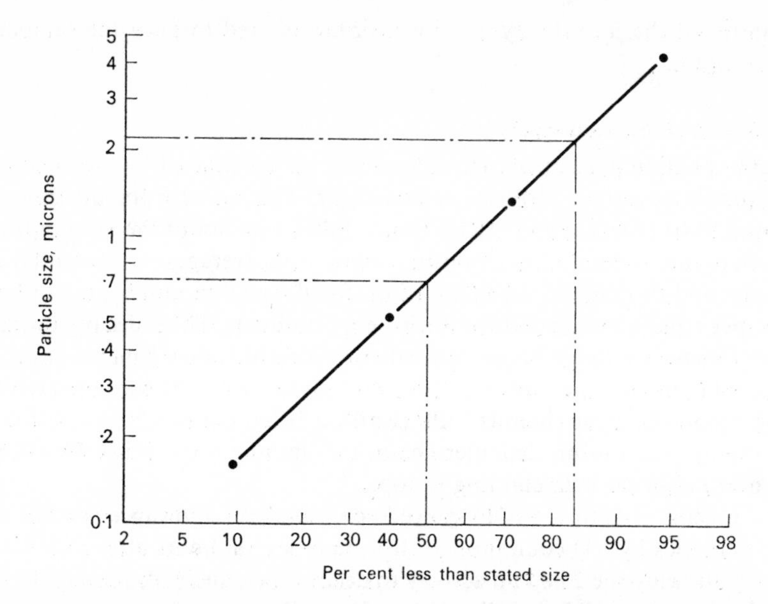

FIG. 13.12. Data of Table 13.3 plotted on log-probability paper. The particle size corresponding to the 50% point is the geometric mean, 0.7 μ, and the ratio of the 84% size to the 50% size gives the geometric standard deviation as 3.

all the weight on the preceding stages plus one-half the weight on the stage in question. When plotting the data, greater emphasis is given to the weights on stages 2, 3, and 4, since the upper limit on stage 1 can be very large and the lower limit on stage 5 (the membrane filter) can be very small. From the curve of Fig. 13.12, the mass median diameter is found to be 0.7 μ, and the geometric standard deviation, σ_g, is 3.

The particle size determined with the cascade impactor is the mass median diameter, that is, the diameter such that 50% of the mass (or volume) of the aerosol is in particles less than this diameter. The linear median diameter, D, that is, the diameter such that 50% of the particles are smaller than this diameter, is related to the mass-median diameter by the equation

$$\log D = \log \text{MMD} - 6.9 \log^2 \sigma_g. \tag{13.14}$$

The standard deviation of the distribution of linear diameters is exactly the same as that of the mass-diameters. For the particles in Table 13.3, the linear median diameter is, from equation (13.14), 0.02 microns, and the geometric standard deviation is 3.

Since the size separation of particles in a cascade impactor depends on the velocity of the air in the jets, the sampling rate which for routine use of a cascade impactor is 17 liters/min, is very important. To avoid errors due to variations in the jet velocity, a critical orifice is used to keep the sampling rate constant.

Natural Airborne Radioactivity

Determination of an airborne contaminant is complicated by the existence of naturally occurring airborne radioactivity. This arises from the gaseous isotopes ^{222}Rn (Radon) and ^{220}Rn (Thoron) that seep out of the ground from the ubiquitous uranium and thorium series (the average concentration of uranium and thorium, in the first 12 in. of soil, is 1 ton and 3 tons respectively per square mile) and from their decay products. These decay products attach themselves to air-borne particulates with which they come in contact, thus contaminating the dust particles with radioactivity. The limiting activity in the radon daughter chain is ^{214}Pb (RaB), a beta emitter whose half-life is 26.8 min. In the thoron daughter chain, the limiting activity is ^{212}Pb (ThB), a 10.6 hr half-lived beta-emitting isotope.

For routine monitoring of long-lived contaminants, allowance is made for these activities by: (1) counting the air sample several hours after collection, thereby allowing the 26.8 min activity to decay away, and (2) counting the air sample about two ThB half-lives later. From these two measurements, the activity of the long-lived contaminant can be calculated. The counting rate, C^1, of the first measurement is due to the natural activity and the long-lived contaminant, C_{LL}:

$$C_1 = C_{n1} + C_{LL}; \tag{13.15}$$

when the second count is made, the natural activity is reduced to

$$C_{n2} = C_{n1}\, e^{-\lambda\Delta t}, \tag{13.16}$$

where λ, the decay constant of ThB, is 0.0655 per hr, and Δt is the time interval, in hours, between the two counts. The second measurement, C_2, includes the following

$$C_2 = C_{n2} + C_{LL}, \tag{13.17}$$

$$C_2 = C_{n1}\, e^{-\lambda\Delta t} + C_{LL}. \tag{13.18}$$

Solving equations (13.17) and (13.18) simultaneously for the long-lived activity gives

$$C_{LL} = \frac{C_2 - C_1\, e^{-\lambda\Delta t}}{1 - e^{-\lambda\Delta t}}. \tag{13.19}$$

Example 13.2

In monitoring for air-borne ^{90}Sr, a 10 m^3 air sample is taken with a membrane filter. The first measurement, taken 4 hr after sampling, gives a net counting rate of 100 counts/min. The second count, 20 hr later, gives a net counting rate of 50 counts/min. If the filter (after treatment to prevent electrostatic charge accumulation) was counted in a windowless 2π gas-flow counter, what was the mean concentration of ^{90}Sr in the air?

Substituting the respective values for C_1 and C_2 into equation (13.19) gives

$$C_{LL} = \frac{50 - 100\, e^{-0.0655 \times 20}}{1 - e^{-0.0655 \times 20}}$$

$$= 31.5 \text{ counts per minute}$$

for the long-lived (^{90}Sr) activity. Since the counter has an efficiency of 50% and the volume of air sampled was 10 m³, the mean concentration of ^{90}Sr is calculated as follows:

$$C = \frac{31.5 \text{ cpm} \times 2 \text{ dpm/cpm}}{2.22 \times 10^6 \text{ dpm}/\mu\text{Ci} \times 10 \text{ m}^3 \times 10^6 \text{ cm}^3/\text{m}^3}$$

$$= 2.84 \times 10^{-12} \frac{\mu\text{Ci}}{\text{cm}^3}.$$

If the contaminant is of such half-life that a significant amount will decay during the waiting period, then due allowance must be made for this fact. In this case, the activity in the sample at the times of the first and second counts is:

$$C_1 = C_{n1} + C_{c1}, \tag{13.20}$$

$$C_2 = C_{n2} + C_{c2}, \tag{13.21}$$

where C_n is the naturally occurring activity and C_c is the contaminant. At the time of the second count the natural and contaminating activities will have decayed to

$$C_{n2} = C_{n1}\, e^{-\lambda_n \Delta t}, \tag{13.22}$$

$$C_{c2} = C_{c1}\, e^{-\lambda_c \Delta t}. \tag{13.23}$$

Solving these equations simultaneously gives

$$C_{c1} = \frac{C_2 - C_1\, e^{-\lambda_n t}}{e^{-\lambda_c t} - e^{-\lambda_n t}}. \tag{13.24}$$

Continuous Environmental Monitoring

Continuous monitoring of the environment is usually done if a breakdown of control measures could lead to a serious hazard. For example, if a threat to life could result from a source inadvertently left unshielded, then a continuous radiation monitor coupled to an alarm would be indicated. Continuous monitoring may also be required in a laboratory where low-level activities are measured, and where precise knowledge of fluctuations in the background is therefore necessary. Another application of continuous monitoring is where the amount of radioactivity discharged into the environment must be known —as in the case of the gaseous and particulate effluent from an incinerator used to burn radioactive waste or the case of a sewage line from a building in which much liquid radioactive waste is disposed of via the sink.

Continuous environmental monitors fall into three classes, those used to measure radiation levels (these are often called area monitors), those used to measure atmospheric radioactivity, and those used to measure liquid radioactivity.

Area monitoring systems usually consist of an appropriate detector, a ratemeter, a recorder, and, if necessary, an alarm that is actuated when a preset radiation level is exceeded. Liquids or gases may be monitored by letting them flow around or through a suitable detector. Basically, a liquid monitoring system is the same as the area monitoring system—except that the read-out is calibrated to read in activity units, such as microcuries per cm^3 rather than units of radiation dose. For air-borne particulates, air is sucked through a filter at a known rate. The filter is placed in close proximity to an appropriate detector which responds to the radioactive dust caught by the filter, and which is shielded against environmental radiation (Fig. 13.13). The pulses from the detector are measured by a rate meter whose output is recorded on a strip chart. Since radioactivity continues to accumulate on the filter as more air is sampled, the rate-meter reading continues to increase. The index of the degree of atmospheric contamination, therefore, is the rate of increase of the count rate rather than the value of the count rate. Monitors for airborne dust are therefore equipped with a derivative alarm which is actuated when the rate of increase of the counting rate exceeds a preset value.

Problems

1. A series of measurements with threshold detectors showed the following spectral distribution of neutrons:

Energy	Per cent neutrons
Thermal	40
1000 eV	20
10,000 eV	10
0.1 MeV	10
1 MeV	10
10 MeV	10

500 mg ^{32}S was irradiated for 2 hr in this field, and when counted in a 2π counter 24 hr after the end of irradiation gave 500 counts/min. What is the dose rate in this neutron field?

1. A film badge worn by a worker in a fast neutron field showed the following distribu
of proton recoil tracks among 100 random microscopic fields of 2×10^{-4} cm² each:

Observed track per field	Frequency
0	40
1	40
2	18
3	2

a) If 2600 tracks per cm² correspond to 100 mrem, what was the fast neutron dose?
b) What is the 95% confidence limit of this measurement?

2. Using ICRP lung model and the physiologic data for the standard man, comput
e to the lungs and to the bone following a single acute exposure of one mCi-se
cubic meter of a respirable (< 5 μ dia) aerosol of (a) strontium titinate, (b) strontiun
oride.

3. The following size distribution was obtained on a sample of an aerosol.

Per cent by number	Class interval, μ
10	0.5–1.0
15	1.0–1.5
15	1.5–2.0
10	2.0–2.5
10	2.5–3.0
10	3.0–3.5
10	3.5–4.5
10	4.5–5.5
10	5.5–6.5

a) Plot the cumulative frequency distributions on linear graph paper, on linear probabil-
paper, and on log probability paper, by number, surface area (assume the particles to be
erical), and by mass (assume the particles to have a density of 2.7 g/cm³).
b) Are the size distributions normally or log-normally distributed?
c) Compute the geometric mean and standard deviations for each of the three types of
tributions.

Suggested References

CADLE, R. D.: *Particle Size*, Reinhold, New York, 1965.
DRINKER, P. and HATCH, T. F.: *Industrial Dust*, McGraw-Hill, New York, 1954.
EISENBUD, M.: *Environmental Radioactivity*, McGraw-Hill, New York, 1963.
GAFAFER, W. M.: *Occupational Diseases*, A Guide to Their Recognition, U.S.P.H.S. Publication 1097, Washington, 1966.
GODBOLD, B. C. and JONES, J. K.: *Radiological Monitoring of the Environment*, Pergamon, Oxford, 1965.
GREEN, H. L. and LANE, W. R.: *Particulate Clouds: Dusts, Smokes, and Mists*, D. Van Nostrand, Princeton, 1964.
HOBSON, W.: *The Theory and Practice of Public Health*, Oxford University Press, London, 1965.
LANZL, L. H., PINGEL, J. H. and RUST, J. H.: *Radiation Accidents and Emergencies in Medicine, Research, and Industry*, Charles Thomas, Springfield, 1965.
SAENGER, E. L.: *Medical Aspects of Radiation Accidents*, U.S.A.E.C., Washington, 1963.
SARTWELL, P. E.: *Preventive Medicine and Public Health*, Appleton-Century-Crofts, New York, 1965.
Methods of Radiochemical Analysis, F.A.O., Rome, 1959.
Safe Handling of Radioisotopes, Medical Addendum, I.A.E.A., Vienna, 1960.
Diagnosis and Treatment of Radioactive Poisoning, I.A.E.A., Vienna, 1963.

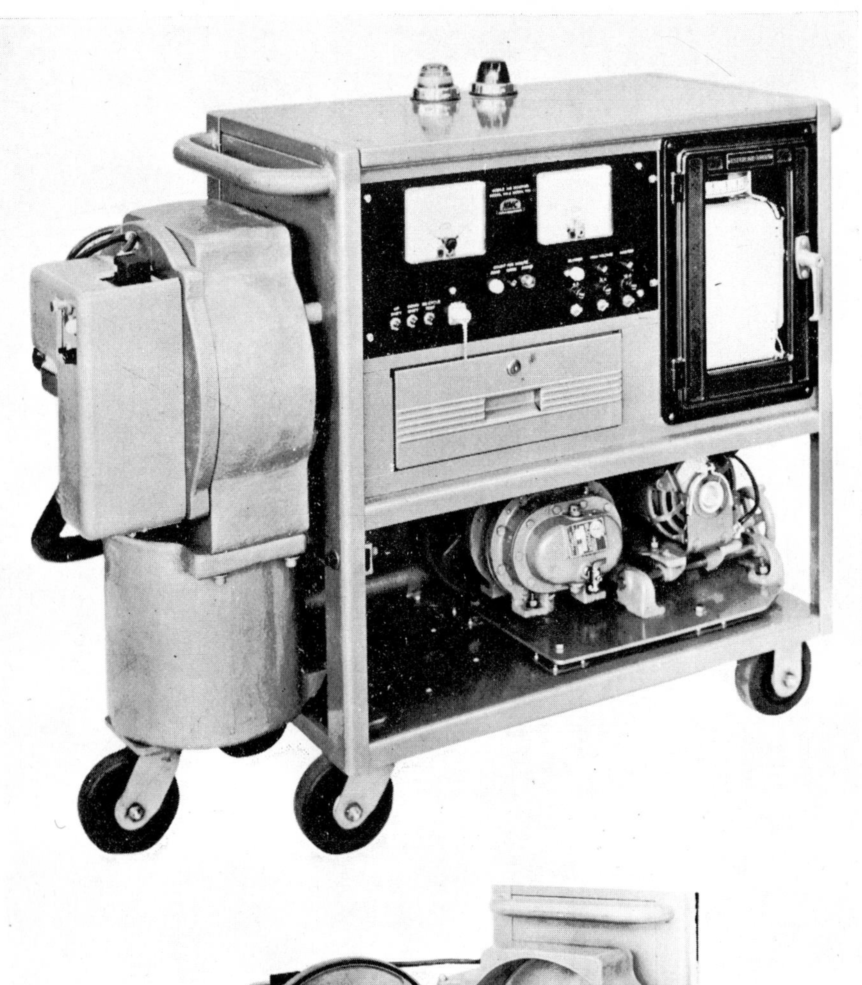

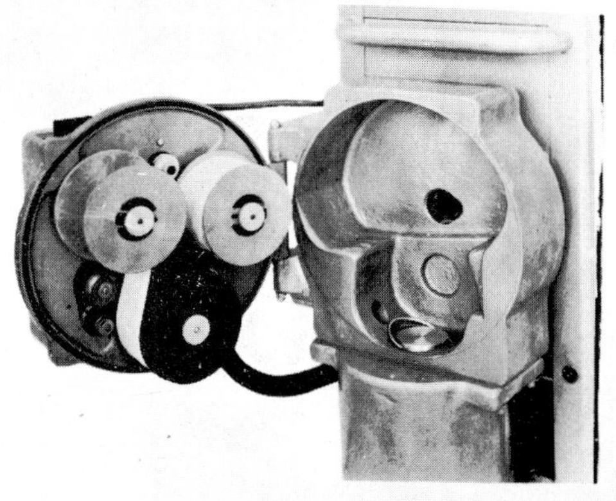

FIG. 13.13. Continuous air particulate monitor for simultaneously detecting alpha, beta, and gamma activity at a rate of 1 ft³ air per min. The sensitivity at this sampling rate is 10^{-12} μCi/cm³ of beta–gamma and of alpha activity. (Courtesy Nuclear Measurements Corp.)

2. A sealed ^{90}Sr source is leak-tested. The wipe, counted in
155 counts in 5 min. The background was 130 counts in 5 min.
is the source contaminated?

3. An air sample on a filter paper was counted in a 2π gas
counts in 5 min. A background count gave 260 counts in 10
deviation of the net counting rate?

4. A worker in the radioisotope lab of a hospital inadverten
"atomic cocktail" that contains 100 mCi tritium. The physical l
and the biological half-life as tritiated water is 12 days.

(a) What is the initial dose rate to the total body (assume
distributed before any elimination occurs)?

(b) What is the total dose during the 13 week period follow

(c) What is the expected urinary output of ^{3}H 30 days after

5. The maximum permissible skeletal burden of ^{90}Sr is 2 μ
disintegrations per minute per 24-hr urine sample that may b
of this skeletal burden if 0.05% per day is eliminated in the urin

6. Using the ICRP two-compartment lung model and the
calculate the ratio of concentration of uranium in the air to ur
air/μCi/ml urine, for the case of continuous inhalation of inso
that all the uranium that is brought up from the lung and is sw
feces.

7. The body burden of ^{137}Cs at time t days following a sir

$$Q(t) = Q(0)(0.1e^{-0.693t} + 0.9e^{-0.011t}$$

If the ratio of urinary to fecal excretion is 9:1, calculate the act
1 day and 10 days after exposure.

8. A chemist accidentally inhaled a ^{14}C-tagged organic solv
from the lungs. The solvent is known to concentrate in the liv
that is eliminated before deposition in the liver leaves in the ur
ducts are eliminated from the liver into the G.I. tract and int
eliminated in the urine, and 75% in the feces. Following the inh
were collected over a 2-week period, and the following data we

Days after inhalation	1	2	3	4	5	6
μCi per sample	2.65	1.55	1.06	0.70	0.55	0.48

(a) How much activity was absorbed into the body?

(b) What was the total dose to the body during the 13 weeks

(c) What was the total dose to the liver during the 13 weeks

9. A health physicist samples waste water to ascertain tha
discharged into the environment. The water analysis is made b
^{90}Sr, allowing the ^{90}Y daughter to accumulate, then extracti
activity. The volume of the sample was 1 liter, the ^{90}Y in growt
^{90}Y activity was determined 15 hr after extraction in an internal
overall efficiency of 50%. The background counting rate, determ
was 35 counts/min. The sample (including background) gave 27
was the ^{90}Sr concentration, at the 90% confidence level?

10. An air sample that was counted 4 hr after collection gave
background was counted for 30 min, and gave a rate of 45 co
counted again 20 hr later, and gave 990 counts in 10 min; a 60-n
2940 counts. If the volume of the air sample was 1.0 m^{3}, and if
50%, calculate the atmospheric concentration of the long-lived c
the 95% confidence limits.

14. *Assessment of Radioactivity in Man*, I.A.E.A., Vienna, 1964.
15. *Basic Requirements for Personnel Monitoring*, I.A.E.A., Vienna, 1965.
16. *Environmental Monitoring in Emergency Situations*, I.A.E.A., Vienna, 1966.
17. *Manual on Environmental Monitoring in Normal Operations*, I.A.E.A., Vienna, 1966.
18. *Assessment of Airborne Radioactivity*, I.A.E.A., Vienna, 1967.
19. *Principles of Environmental Monitoring Related to the Handling of Radio-active Materials*, I.C.R.P. Publication 7, Pergamon, Oxford, 1965.
20. *Radioassay Procedures for Environmental Samples*, U.S.P.H.S. Publication 999-RH-27, Washington, 1967.
21. *Protection of the Public in the Event of Radiation Accidents*, W.H.O., Geneva, 1965.

APPENDIX I

VALUES OF SOME USEFUL CONSTANTS

Constant		Value
Electronic charge	e	4.80294×10^{-10} sC
Electronic mass	m_0	9.1086×10^{-28} g 0.00055 amu
Velocity of light	c	2.997928×10^{10} cm sec^{-1}
Avogadro's number	N	6.0247×10^{23} mole^{-1}
Gas constant	R	0.08205 liter atmos mole^{-1} (°K)$^{-1}$ 8.3144×10 ergs mole^{-1} (°K)$^{-1}$
Boltzman's constant	k	1.38049×10^{-16} erg (°K)$^{-1}$
Planck's constant	h	6.6254×10^{-27} erg sec
Acceleration of gravity (std)	g	980.665 cm sec^{-2}
Gravitational constant	γ	6.670×10^{-8} dyne cm^2 g^{-2}
Faraday's constant	F	96,500 C mole^{-1}
Atomic mass unit	amu	1.649×10^{-24} gm 931.16 MeV
Proton mass		1.67243×10^{-24} g 1.00759 amu
Neutron mass		1.67474×10^{-24} g 1.00899 amu

APPENDIX II

TABLE OF THE ELEMENTS

Name	Symbol	At. No.	Atomic weight
Actinium	Ac	89	227
Aluminum	Al	13	26.98
Americium	Am	95	(243)
Antimony	Sb	51	121.76
Argon	A	18	39.944
Arsenic	As	33	74.92
Astatine	At	85	(210)
Barium	Ba	56	137.36
Berkelium	Bk	97	(249)
Beryllium	Be	4	9.013
Bismuth	Bi	83	208.99
Boron	B	5	10.82
Bromine	Br	35	79.916
Cadmium	Cd	48	112.41
Calcium	Ca	20	40.08
Californium	Cf	98	(251)
Carbon	C	6	12.011
Cerium	Ce	58	140.13
Cesium	Cs	55	132.91
Chlorine	Cl	17	35.457
Chromium	Cr	24	52.01
Cobalt	Co	27	58.94
Columbium, *see* Niobium			
Copper	Cu	29	63.54
Curium	Cm	96	(247)
Dysprosium	Dy	66	162.51
Einsteinium	E	99	(254)
Erbium	Er	68	167.27
Europium	Eu	63	152.0
Fermium	Fm	100	(253)
Fluorine	F	9	19.00
Francium	Fr	87	(223)
Gadolinium	Gd	64	157.26
Gallium	Ga	31	69.72
Germanium	Ge	32	72.60
Gold	Au	79	197.0
Hafnium	Hf	72	178.50
Helium	He	2	4.003
Holmium	Ho	67	164.94
Hydrogen	H	1	1.0080
Indium	In	49	114.82
Iodine	I	53	126.91
Iridium	Ir	77	192.2
Iron	Fe	26	55.85
Krypton	Kr	36	83.80
Lanthanum	La	57	138.92
Lead	Pb	82	207.21
Lithium	Li	3	6.940
Lutetium	Lu	71	174.99
Magnesium	Mg	12	24.32

Name	Symbol	At. No.	Atomic weight
Manganese	Mn	25	54.94
Mendelevium	Mv	101	(256)
Mercury	Hg	80	200.61
Molybdenum	Mo	42	95.95
Neodymium	Nd	60	144.27
Neon	Ne	10	20.183
Neptunium	Np	93	(237)
Nickel	Ni	28	58.71
Niobium (columbium)	Nb	41	92.91
Nitrogen	N	7	14.008
Nobelium	No	102	(254)
Osmium	Os	76	190.2
Oxygen	O	8	16.000
Palladium	Pd	46	106.4
Phosphorus	P	15	30.975
Platinum	Pt	78	195.09
Plutonium	Pu	94	(242)
Polonium	Po	84	210
Potassium	K	19	39.100
Praseodymium	Pr	59	140.92
Promethium	Pm	61	(147)
Protactinium	Pa	91	231
Radium	Ra	88	226
Radon	Rn	86	222
Rhenium	Re	75	186.22
Rhodium	Rh	45	102.91
Rubidium	Rb	37	85.48
Ruthenium	Ru	44	101.1
Samarium	Sm, Sa	62	150.35
Scandium	Sc	21	44.96
Selenium	Se	34	78.96
Silicon	Si	14	28.09
Silver	Ag	47	107.873
Sodium	Na	11	22.991
Strontium	Sr	38	87.63
Sulfur	S	16	32.066
Tantalum	Ta	73	180.95
Technetium	Tc	43	(99)
Tellurium	Te	52	127.61
Terbium	Tb	65	158.93
Thallium	Tl	81	204.39
Thorium	Th	90	232
Thulium	Tm	69	168.94
Tin	Sn	50	118.70
Titanium	Ti	22	47.90
Tungsten (wolfram)	W	74	183.86
Uranium	U	92	238.07
Vanadium	V	23	50.95
Xenon	Xe	54	131.30
Ytterbium	Yb	70	173.04
Yttrium	Y	39	88.91
Zinc	Zn	30	65.38
Zirconium	Zr	40	91.22

APPENDIX III

THE STANDARD MAN[a]

1. Mass and Effective Radius of Organs of the Adult Human Body[b]

	Mass in grams	Per cent of total body	Effective radius in cm
Total body[c]	70,000	100	30
Muscle	30,000	43	30
Skin and subcutaneous tissue[d]	6100	8.7	0.1
Fat	10,000	14	20
Skeleton:			
Without bone marrow	7000	10	5
Red marrow[e]	1500	2.1	—
Yellow marrow	1500	2.1	—
Blood	5400	7.7	—
Gastrointestinal tract[c]	2000	2.9	30
Contents of GI tract:			
Lower large intestine	150	—	5
Stomach	250	—	—
Small intestine	1100	—	—
Upper large intestine	135	—	—
Liver	1700	2.4	10
Brain	1500	2.1	—
Lungs (2)	1000	1.4	30
Lymphoid tissue	700	1.0	—
Kidneys (2)	300	0.43	7
Heart	300	0.43	—
Spleen	150	0.21	7
Urinary bladder	150	0.21	—
Pancreas	70	0.10	—
Salivary glands (6)	50	0.071	—
Testes (2)	40	0.057	—
Spinal cord	30	0.043	—
Eyes (2)	30	0.043	—
Thyroid gland	20	0.029	3
Teeth	20	0.029	—
Prostrate gland	20	0.029	—
Adrenal glands of suprarenal (2)	20	0.029	—
Thymus	10	0.014	—
Miscellaneous (blood vessels, cartilage, nerves, etc.)	390	0.56	—

[a] *Radiological Health Handbook.*

[b] The reports, *Standard Man*, by Hermann Lisco, ANL–4253, Nov. 1948–Feb. 1949, p. 96, and *A Survey Report of the Characteristics of the Standard Man*, Sept. 1948, by M. J. Cook were used as the principal sources of reference in the original selection of values given in this table. (Data taken from report of the International Sub-Committee II on Permissible Dose for External Radiations, K. Z. Morgan, Chairman, ICRP/54/4.)

[c] Does not include contents of gastrointestinal tract.

[d] The mass of the skin alone is taken as 2000 g. The minimum thickness of the epidermis is 0.07 mm.

[e] The average depth of the blood-forming organs is assumed to be at 5 cm.

2. Chemical Composition

Element	Proportion %	Approximate mass in the body in grams
Oxygen	65.0	45,500
Carbon	18.0	12,600
Hydrogen	10.0	7000
Nitrogen	3.0	2100
Calcium	1.5	1050
Phosphorus	1.0	700
Sulfur	0.25	175
Potassium	0.2	140
Sodium	0.15	105
Chlorine	0.15	105
Magnesium	0.05	35
Iron	0.006	4
Manganese	0.00003	0.02
Copper	0.0002	0.1
Iodine	0.00004	0.03

The figures for a given organ may differ considerably from these averages for the whole body. For example, the nitrogen content of the dividing cells of the basal layer of skin is probably nearer 6% than 3%.

3. Applied Physiology

Average data for normal activity in a temperate zone:

1. *Water balance*

Daily water intake

Water of oxidation	0.3 liters
In food	0.7 liters
As fluids	1.5 liters
Total	2.5 liters

Calculations of maximum permissible levels for radioactive isotopes in water have been based on the total intake figure of 2.5 liters a day.

Daily water output

Sweat	0.5 liters
From lungs	0.4 liters
In feces	0.1 liters
Urine	1.5 liters
Total	2.5 liters

(The total water content of the body is 50 liters.)

2. *Respiration*[a]

Area of Respiratory Tract

Respiratory interchange area	50 m^2
Nonrespiratory area (upper tract and trachea to bronchioles)	20 m^2
Total	70 m^2

RESPIRATORY EXCHANGE

Physical activity	Hours per day	Tidal air (liters)	Respiration per min	Volume per 8 hr (m^3)	Volume per day (m^3)
At work	8	1.0	20	10	20
Not at work	16	0.5	20	5	20

Carbon dioxide content (by volume) of air

Inhaled air (dry, at sea level)	0.03%
Alveolar air	5.5%
Exhaled air	4.0%

3. *Retention of particulate matter in the respiratory tract*

Retention of particulate matter in the lungs depends on many factors, such as the size, shape and density of the particles, the chemical form and whether or not the person is a mouth breather; however, when specific data are lacking it is assumed the distribution is as follows:

Distribution	Readily soluble compounds %	Other compounds %
Exhaled	25	25
Deposited in upper respiratory passages and subsequently swallowed	50	50
Deposited in the lungs (lower respiratory passages)	25 (this is taken up into the body)	25[b]

[a] As stated in U.S. Department of Commerce, *Bureau of Standards Handbook* 47, 1950, Appendix I.

[b] Of this, half is eliminated from the lungs and swallowed in the first 24 hr making a total of 62½% swallowed. The remaining 12½% is retained in the lungs with a half-life of 120 days, it being assumed that this portion is taken up into body fluids.

4. Duration of Exposure

1. *Duration of occupational exposure*

The following figures have been adopted in calculations pertaining to occupational exposure:

8 hr per day
40 hr per week
50 weeks per year
50 years continuous work period.

2. *Duration of "lifetime for nonoccupational exposure"*

A conventional figure of 70 years has been adopted.

5. Blood Count

Relative and Absolute Values for Leukocyte Counts in Normal Adults per Cubic Millimeter of Blood[(a)]

Type of cell	%	Absolute number		
		Average	Minimum	Maximum
Total leukocytes	—	7000	5000	10,000
Myelocytes	0	0	0	0
Juvenile neutrophils	3–5	300	150	400
Segmented neutrophils	54–62	4000	3000	5800
Eosinophils	1–3	200	50	250
Basophils	0–0.75	25	15	50
Lymphocytes	25–33	2100	1500	3000
Monocytes	3–7	375	285	500

Normal Range of Values—Human Adults[(a)]

	Male	Female
Red cell count (millions per mm^3)	5.4 ± 0.8	4.8 ± 0.6
Hemoglobin (g per 100 cm^3)	16.0 ± 2.0	14.0 ± 2.0
Hematocrite or vol.-packed R.B.C. (cm^3 per 100 cm^3)	47.0 ± 7.0	42.0 ± 5.0

(a) M. M. Wintrobe, *Clinical Hematology*, Lea & Febiger, Philadelphia, 3rd edition.

INDEX